BROADBAND TELECOMMUNICATIONS HANDBOOK

Broadband Telecommunications Handbook

Regis J. Bates

McGraw-Hill
New York • San Francisco • Washington, D.C.
Auckland • Bogotá • Caracas • Lisbon • London
Madrid • Mexico City • Milan • Montreal • New Delhi
San Juan • Singapore • Sydney • Tokyo • Toronto

Library of Congress Cataloging-in-Publication Data on File

McGraw-Hill

A Division of The **McGraw·Hill** *Companies*

Copyright © 2000 by The McGraw-Hill Companies, Inc. All rights reserved.
Printed in the United States of America. Except as permitted under the United
States Copyright Act of 1976, no part of this publication may be reproduced or
distributed in any form or by any means, or stored in a data base or retrieval
system, without the prior written permission of the publisher.

1 2 3 4 5 6 7 8 9 0   AGM/AGM   9 0 4 3 2 1 0 9

ISBN 0-07-134648-1

This book was set in Century Schoolbook by D&G Limited, LLC.

Printed and bound by Quebecor/Martinsburg.

McGraw-Hill books are available at special quantity discounts to use as premiums
and sales promotions, or for use in corporate training programs. For more information,
please write to Director of Special Sales, McGraw-Hill, 11 West 19th Street, New York,
NY 10011.

 This book is printed on recycled, acid-free paper containing a minimum
of 50 percent recycled de-inked fiber.

Contents

Contents

Contents

Contents

Contents

Contents

Acknowledgments

I would like to take the opportunity to recognize several people who had a considerable influence on my ability to complete this project. One can never plan or predict just what will happen during the course of a year or during the production cycle of a manuscript. The scenarios we experienced with this book really came through and challenged us throughout the entire production.

First I have to readily acknowledge and thank all the folks at McGraw-Hill for their continued support of this author and their exceptional patience. I never had any vision that this book would get as delayed as it was when we contracted to write it. Somewhere it is written that the author of a book is supposed to have unlimited time available and undaunted focus. Wherever this is written, it is not true! Too many challenges and changes crept into our daily lives and postponed the inevitable completion of this project. As TCIC grew and we began introducing new training products on the convergence of the Voice and Data worlds, along with the radical changes in the industry causing the consulting part of the business major upheaval, we had to really struggle to juggle all the schedules to maintain control.

So I put the McGraw-Hill people through the paces, promising to get manuscript to them and missing just about every date. Steven Elliott was my ever forgiving editor-in-charge who had the unenviable job of making excuses to his boss when I failed to deliver on time. I thank Steven for his understanding and his willingness to wait this one out.

During the cycle, **Steven Elliott** was replaced on this project with **Steve Chapman**. What a nightmare for a new editor on a project to find that the author is very late and now the "new kid on the block" had to try to figure out how to light a fire under me and get me moving again. In the middle of this transition, a very large customer ordered several thousand copies of this new book (before it was finished) and published delivery dates based on my feeble attempts to set a schedule. Pressure is one thing, but this was pressure extraordinaire!   I thank Steve Chapman for his patience and his periodic prods to remind me to stick with it. Moreover all the offers for help were appreciated even if I never really said it aloud. I sincerely appreciate the efforts of all the folks I never saw or talked yet who had to ramp up their production schedule to accommodate my schedule rather than work the normal routine production cycles. These unsung heroes of the production department never get their credit, but we all should be grateful to them for their dedication and stick to it attitudes.

Beyond the folks at McGraw-Hill, there is a special person who has held the entire project together seven times now. Her ability to keep track of the dribs and drabs of a book that I wrote, convert them into documents and manipulate the graphics (scribbles on paper) into meaningful inserts, never ceases to amaze me. **Gabriele** has been the anchor in all these books, keeping track of what was done, what is in motion and what needed attention. How many weekends and nights she had to sit around and redo something I decided to change was invisible to me. Yet she never complains, just continues to produce and get things out. Her dedication to the overall success of my books never gets the credit she truly deserves. Gabriele provides the gentle push I need from time to time, keeping me focused and working at it. Even when she knows I am behind, there is no panic—just constant reinforcement and encouragement.

A new person was very helpful to me this time and he deserves special credit. **Tim Kelly,** my contributing author, helped me with a few of the chapters when I got into trouble time-wise. He is a special type of friend and collaborator. How many nights we discussed the book on the phone for hours when he had his own work and commitments to perform. Tim went beyond the norm in helping me out, when he went on vacation to Germany (a special thanks to his understanding wife "E-Beth"), he carted notebook computer along with him to help get some of the materials prepared for five of the chapters. He did the research and writing, graphics and input. He even provided his special tweak on the materials as we discussed (bantered) the material content and intent. Without Tim's help, this book might still be in my computer and not in printed form. I will look forward to bringing his special assists into the next productions and seeing his name on the covers of many books in the future. He certainly has earned the right.

Several vendors and friends were supportive and helpful in garnering information for the development o this manuscript. I thank all of them, too numerous to mention each of them individually. However, they know who they are and can take silent comfort in knowing they got us here.

We all tried to keep focus on this broadband concept from start to finish. In many cases, the ideas of broadband communications are still emerging for some of the areas we discussed herein. However, we hope we were able to capture the spirit and the letter of the concept even before it truly develops. Enjoy this book as you would a version of a 201 series after the Voice and Data Communications Handbook. This book is dedicated to those of you who continue to buy the materials we jointly develop for your understanding. It is you, the reader of this material who should also be praised for the demand for more information. Convergence is the name of our industry today, yet we must continue to seek new ways of providing the information and using the technology. As long as you the reader continue to demand high-speed services, reliability and mobility, I will have a job. That job will be to seek the ways of describing and applying the technologies so that you can use them.

Good luck and Happy Reading!

Introduction to Telecommunications Concepts

Introduction

Welcome to the world of telecommunications again! In this book, we'll attempt to deliver a series of different approaches to the use and application of telecommunications' principles, concepts, and guidelines and offer new approaches to the use of voice and data communications.

Last year, I wrote *The Voice and Data Communications Handbook*, Signature Edition, as a means of introducing several new ways of looking at the telecommunications industry. *The Voice and Data Handbook* was so successful that it begged for a sequel with a more in-depth approach and a more technical view of the use of telecommunications. Therefore, my goal is to delve into the topics of broadband communications. For those who have not read other books on this topic, I will attempt to simplify the concepts discussed. For those who had a chance to read the first book (or others on this topic), I will attempt to pick up where we left off during the first volume. This book is structured by groupings of topics. For example, the first few chapters work with the convergence of voice and data networks as we see the virtual private networks, intelligent networks, and the portability of our systems for today and the future.

After the first grouping of chapters, we step into a discussion of signaling systems that make wonderful things happen in the convergence world—coupled with that discussion is the idea of computer and telephony integration. (What better way to describe convergence!)

After a few ideas have sunk in, we move on to a high-speed data networking strategy, with the use of Frame Relay. After Frame Relay, we discuss the use of ATM for its merits and benefits. Next, we take the convergence a step farther and delve into the Frame and ATM internetworking applications—still a great way to carry our voice and data no matter how we slice and dice it.

Just when we thought it was safe to use these high-speed services across the WAN, we realize that local access is a problem. Entering into the discussion is the high-speed convergence in the local loop arena with the use of CATV and cable modems to access the Internet at LAN speeds. Mix in a little xDSL, and we start the fires burning on the local wires. The use of copper wires or cable TV is the hot issue in data access.

From the discussion of the local loop, we then see the comparisons of a wireless local loop with LMDS and MMDS. These techniques are all based on a form of Microwave, so the comparison of microwave radio techniques is shown.

Wireless portability is another hot area in the marketplace. Therefore, we compare and contrast the use of GSM, cellular, and personal communications' services and capacities. Convergence is only as good as one's ability to place the voice and data on the same links.

Leaving the low-end wireless services behind, we then enter into a discussion of the sky wave and satellite transmission for voice and data. No satellite transmission discussion would be worth anything without paying homage to the TCP and IP protocols on the satellite networks. Yet, the satellite services are now facing direct competition where the Low Earth Orbit satellite strategies are becoming ever popular. The use of Teledesic, Iridium, or Globalstar systems are merely transport systems. These pull the pieces together and will offer voice and data transmission for years to come.

One could not go too far with the wireless-only world, so we back up and begin to contrast the use of the wired world again. This time, we look at T1, T2, and T3 on copper or coax cable, which is a journey down memory lane for some. However, by adding a little fiber to the diet, we provide these digital architectures on SONET or SDH services. SONET makes the T1 and T3 look like fun! Topics include the ability to carry Frame Relay and ATM as the networks are now beginning to meld together. SONET is good, but if we use an older form of multiplexing (wavelength), we can get more yet from the fibers. So, we look at the benefits of dense-wave division multiplexing on the fiber to carry more SONET and more data.

With the infrastructure kicked around, the logical step is to complete this tour of the telecommunications arena with the introduction of the Internet, intranets, and extranets. Wow, this stuff really does come together! Using the Internet or the other two forms of nets, we can then carry our data transparently. What would convergence be without the voice? Therefore, the next step is to look at the use of voice over *Internet Protocols* (IPs). Lastly, we have to come up with a management system to control all the pieces that we have grouped and bonded together. This is in the form of a *simple network management protocol* (SNMP) as the network management tool of choice. If all the converged pieces work, there is no issue. However, with all the variants discussed in this book, we must believe that "Murphy is alive and well!" Thus, all the pieces are formed together by groups, to form a homogenous network of Internets.

Basic Telecommunications Systems

When the FCC began removing regulatory barriers for the long distance and *customer premises equipment* (CPE) markets, its goal was to increase competition through the number of suppliers in these markets. Recently, consumers have begun to enjoy lower prices and new bundled service offerings. The local and long distance markets are examples of the new direction taken by the FCC in the 1980s to eliminate and mitigate the traditional telephone monopoly into a set of competitive markets. Although these two components of the monopoly have been stripped away, barriers still exist at the local access network—the portion of the public network that extends between the IEC network and the end user. The local loop and the basic telecommunications infrastructure are not as readily available as one would like to think.

The growth of private network alternatives improves with facilities-based competition in the transport of communications services. The industry realizes that more than 500 competitive local exchange carriers have grown out of the deregulation of the monopolies. These CLECs include cable television networks, wireless telephone networks, *local area networks* (LANs), and metropolitan area networks. *Incumbent local exchange carriers* (ILECs) indicate that their networks are continually evolving into a multimedia platform capable of delivering a rich variety of text, imaging, and messaging services as a direct response to the competition. Many suggest that their networks are wide open, for all competitors. Imagine an *open network*—a network with well-defined interfaces accessible to all—allowing an unlimited number of entrants a means to offer competitive services limited only by their imagination and the capabilities of the local loop network facilities.

If natural monopolies are still in the local exchange network, open access to these network resources must be fostered to promote a competitive market in spite of the monopolistic nature of the ILECs. The FCC continues to wrestle with how far it has to go and what requirements are necessary to open and equal access to the network.

Network unbundling, the process of breaking the network into separate functional elements, opens the local access to competition. CLECs select unbundled components they need to provide their own service. If the unbundled price is still too expensive, the service provider will provide its own private resources. This is the facilities-based provider. All too often, we hear about new suppliers who offer high-speed services, better than the incumbent. Yet, these suppliers are typically using the Bell System's wires to get to the consumer's door. The only change that occurs is the person to whom we send the bill. Hardly a competitive local networking strategy. As a result, the new providers (CATV, wireless local loop, IEC, and facilities-based CLEC) are now in the mode to provide their own facilities.

Components of the Telecommunications Networks

Telecommunications network components fall into logical or physical elements. A logical element is a *software-defined network* (SDN) or *virtual private network* (VPN) feature or capability. This SDN or VPN feature can be as simple as the number translation performed in a switch to establish a call. Switching systems have evolved into the use of external signaling systems to set up and tear down the call. These external physical and logical components formulate the basis of a network element. Moreover, Intelligent Networks (and Advanced Intelligent Networks) have surpassed the wildest expectations of the service provider. These logical extensions of the network bear higher revenue while opening the network up to a myriad of new services. Number portability can also be categorized into the logical elements because the number switching and logic are no longer bound to a specific system. A physical element is the actual switching element, such as the link or the matrices used internally. A network is made up of a unique sequence of logical elements implemented by physical elements.

Given the local exchange network and local transport markets, open mandates had to be considered because the LEC has the power to stall competition. In many documented cases the LECs have purposefully dragged their feet to stall the competition and to discredit the new provider in the eyes of the customer. This is a matter of survival of the fittest. The ILECs have the edge over the network components because their networks were built over the past 120+ years. This is the basis for the deregulatory efforts in the networks, because the LECs are fighting to survive the onslaught of new providers who are in the cream-skimming mode. If access mandates are necessary, to what degree? These and other issues are driving the technological innovation, competition at the local loop, and the development of higher capacity services in a very competitive manner.

The Local Loop

So much attention has been parlayed on the local loop. Nevertheless, is it a realistic expectation to use the network facilities for future high-speed services? Would the newer providers, such as the CATV companies, have an edge over the ILECs? These issues are the foundation of the network of the new millennium. The new providers will use whatever technology is available to attack the competition, including

■ CATV

■ Fiber-based architectures (FTTC, FTTH, HFC)

■ Wireless microwave systems

■ Wireless third-generation cellular systems

■ Infrared and laser based wireless architectures

■ Satellite and DSS type services

Regardless of the technology used, the demand never seems to be satisfied. Therefore, the field of competitors will continue to metamorphose as the demand dictates and as the revenues continue to attract new business.

The Movement Toward Fiber Optic Networks

A transmission link transports information from one location to another in a usable and understandable format. The three functional attributes of this link are

1. Capacity

2. Condition

3. Quality of Service

The deregulation of the local exchange networks has led to significant improvements in one of the following criteria:

■ Access to network capacity

■ Access to intermediate points along the transmission path

The transmission path may include pieces of the existing copper or newer fiber-based network architectures. The current copper-based loop limits opportunities.

■ The transmission distances associated with the subscriber loop limit the amount of bandwidth available over twisted wire pair roughly to the DS1 rate of 1.5 Mbps. As broadband services become increasingly popular, the copper network severely constrains the broadband services.

■ The current switched-star architecture runs at least one dedicated twisted pair from the central office to each customer's door without any intermediate locations available to unbundle the transport segment. This precludes a lot of the innovation desired by the end user.

Although the current copper-based network is unattractive to unbundle the physical transmission components, fiber-based networks offer many more opportunities. The local access network can be improved by telephone companies by deploying fiber in the future. The central office, nodes at remote sites and the curbside pedestal can all be improved with fiber-based architectures. These nodes serve as flexibility points where signals can be switched or multiplexed to the appropriate destination. A small percentage of lines are served by *digital loop carrier* (DLC) systems that incorporate a second flexibility point into the architecture at the remote node. The third flexibility point at the pedestal has been proposed for fiber-to-the-curb systems in the future.

The bandwidth limitations of a fiber system are not due to the intrinsic properties of the fiber, but the limitations of the switching, multiplexing, and transmission equipment connected to the fiber. This opens the world up for a myriad of new service offerings when fiber makes it to the consumer's door. Third parties like Qwest and Level 3 are becoming the carrier's carrier. They will install the fiber to the pedestal, the door, or to the backbone and sell the capacity to the ILEC or CLEC. This produces many attractive alternatives to the broadband networks for the future. No longer will bandwidth be the constraining factor; the application or the computer will be the bottleneck.

Because of the tremendous bandwidth available with fiber optic cable and the technological improvements in SONET and Dense Wave Division Multiplexing, virtually unlimited bandwidth will be available. This statement of course is contingent on the following caveats:

■ The abundance of bandwidth is not likely to appear for some time.

■ This bandwidth is available only over the fiber links. Yet, installation of new technology is a slow process. Fiber will be deployed in hybrid network architectures, which continue to utilize existing portions of the copper network.

Consequently, until fiber is deployed all the way to the customer premises, portions of the network will continue to present the same speed and throughput limitations.

Digital Transfer Systems

The switching and multiplexing techniques characteristic of the transmission systems within the network are all digital. Currently, the network employs a *synchronous transfer mode* (STM) technique for switching and multiplexing these digital signals. The broadband networks of the future will continue to utilize a synchronous transmission hierarchy using the SONET standards defined by the ITU. SONET describes a family of broadband digital transport signals operating in 50 Mbps increments. As a result, wherever SONET equipment is used, the standard interfaces at the central office, remote nodes, or subscriber premises will be multiples of these rates.

Above the physical layer, however, changes are now underway that move away from the synchronous communications modes. The *asynchronous transfer mode* (ATM) is the preferred method of transporting at the data link layer. ATM uses the best of packet switching and routing techniques to carry information signals, regardless of the desired bandwidth, over one high-speed switching fabric. Using fixed-length cells, the information is processed at higher speeds, reducing some of the original latency in the network. These cells then combine with the cells of other signals across a single high-speed channel like a SONET OC−48. In *time division multiplexing* (TDM), timing is crucial. In ATM, timing is *statistically multiplexed* (STDM) so the timing is less crucial at the data link layer. The cells fit into the payload of the SONET frame structure for transmission where the timing is again used by the physical layer devices. ATM will use a combined switching and multiplexing service at the cell level. Continued use of SONET multiplexers will combine and separate SONET signals carrying ATM cells.

What distinguishes ATM from a synchronous approach is that subscribers have the ability to customize their use of the bandwidth without being constrained to the channel data rates.

When the intelligent networks are fully implemented, the logical network components will be separated from the physical switching element—where the physical component of a current digital switch consists of 64 Kbps (DS0) access to the network switch.

ATM should improve the capability to separate the physical switching elements of the network. The attributes of the ATM switch, which could facilitate more modularity, is the bandwidth flexibility. Because each information signal is segmented into cells, switching is performed in much smaller increments. Current digital switching elements switch a DS0 signal whether the full bandwidth is needed or not. With ATM, the switching element resources can be much more efficiently matched to the bandwidth requirements of the user. Access to the ATM switch will be specified according to the maximum data rate forecasted for the particular access arrangement, instead of specifying the number of DS0 circuits required, as is the case today with digital switches.

The Intelligent Networks of Tomorrow

The ILECs have been developing the AIN to provide new services or to customize current services based on the user demand. The Central Office switches contain the necessary software to facilitate these enhanced features. The manufacturers of the systems have fully embodied their application software with the operating systems software within the switch to create a simple interface for the carriers. When new features are added, the integrated software must be fully tested by the switch manufacturer.

The limitations of a centralized architecture caused the vendors and manufacturers concern. Now, as intelligent services are deployed, the movement is to a distributed architecture and Intelligent Peripheral devices on the network. The LECs use a network architecture, which enables efficient and rapid network deployment.

The single most important feature of AIN is its flexibility to configure the network according to the characteristics of the service. The modular architecture allows the addition of adjunct processors, such as voice processing equipment, data communication gateways, video services, and directory look-up features to the network without major modifications. These peripheral devices (servers) provide local customer database information and act like the intelligent centralized architectures of old.

The basic architecture of the AIN takes these application functions and breaks them into a collection of functionally specific components. Ultimately, AIN allows modifications to application software without having to alter the operating system of the switch.

SUMMARY

The telecommunications systems include the variations of the local loop and the changes taking place within that first (or last) mile. As the migration moves away from the local copper-based cable plant (a slow evolution for sure), the movement will be to other forms of communications subsystems to include the use of

- Fiber optics
- Coax cable
- Radio-based systems
- Light-based systems
- Hybrids of the preceding

These changes will take users and carriers alike into the new millennium. Using the CATV modem technologies on coax, the fiber-based SONET architectures in the backbone (and ultimately in the local loop), and copper wires in the xDSL technologies all combine to bring higher speed access.

After access is accomplished, the use of the SONET-based protocols and multiplexing systems creates an environment for the orchestration of newer services and features that will be bandwidth intensive. The SONET systems will be used to step up to the challenges of the 2000s. ATM will add a new dimension to the access methods and the transport of the broadband information through the use of STDM and cell-based transmission. No longer will the network suppliers have to commit specific fixed bandwidth to an application that only rarely uses the service. Instead, the services will merely use the cells as necessary to perform the functionality needed.

Wireless local loop services are relatively new in the broadband arena but will play a significant role in the future. The untethered ability to access the network no matter where you are will be attractive to a large new population of users. Access to low-speed voice and data services are achievable today. However, the demand for real-time voice, data, video, and multimedia applications from a portable device is what the new generation of networks must accommodate. The broadband convergence will set the stage for all future development.

Today speeds are set up in the kilobits to megabits per second range. The broadband networks of the future will have to deal with demands for multimegabit speeds up to the gigabit per second speeds. Through each interface, the carriers must be able to preserve as much of their infrastructure as possible so that forklift technological changes are not forced upon them. The business case for the evolution of the broadband convergence is one that mimics a classical business model. Using a 7–15 year return on investment model, the carriers must see the benefit of profitability before they install the architectural changes demanded today.

CHAPTER 2

Telecommunictions Systems

Introduction

Before going into the overall technologies of this book, now is a good time to review the goal of the book. First, we plan to discuss technologies that are based on the current world of voice and data convergence. This convergence is one that has been sliding along for two decades, yet seems to have caught everyone by surprise. Second, we will be talking about applications and some cost issues throughout the book. Regardless of which discipline you come from, you cannot escape the ultimate strategy management expects: increase productivity yet hold the line on costs. Lose either one of these in the equation and you will be sitting there trying to figure out why management never buys into any of your great ideas. The answer comes to us in the form of packaging. No matter how great your ideas are, if you cannot sell them, you cannot implement them.

So as we proceed through this material, try not to get frustrated with the constant mix of services, technological discussions, and costing issues. From time to time, we may also introduce some extra technical notes that are for the more technically astute but can be ignored by the novice trying to progress through the industry. As you read about a topic, do so with a focus on systems, rather than individual technologies. We have tried to make these somewhat stand-alone chapters, yet we have also tied them together in bundles of three or four chapters to formulate a final telecommunications system. Do what you must to understand the information, but do not force it as you read. The pieces will all come together throughout the groupings of topics.

What Constitutes a Telecommunications System

A network is a series of interconnections that form a cohesive and ubiquitous connectivity arrangement when all tied together. That sounds rather vague, so let's look at the components of what constitutes the telecommunications network. The telecommunications network referred to here is the one that was built around voice communications but has been undergoing a metamorphosis for the past two decades. The convergence of voice and data is nothing new; we have been trying to run data over a voice network since the 1970s. However, to run data over the voice network, we had to make the data look like voice. This caused significant problems for the data because the voice network was noisy and error-prone. Reliability was a dream and integrity was unattainable, no matter what the price.

Generally speaking, a network is a series of interconnection points. The telephone companies over the years have been developing the connections throughout the world so that a level of cost-effective services can be achieved and their *return on investment* (ROI) can be met. As a matter of due course, whenever a customer wants a particular form of service, the traditional carriers offer two answers:

1. It cannot be done technically.
2. The tariff will not allow us to do that!

Regardless what the question happened to be, the telephone carriers were constantly the delay and the limiting factor in meeting the needs and demands for data and voice communications.

In order to facilitate our interconnections, the telephone companies installed wires to the customer's door. The wiring was selected as the most economical to satisfy the need and the ROI equation. Consequently, the telephone companies installed the least expensive wiring possible. Because they were primarily satisfying the demand for voice communications, they installed a thin wire (26-gauge) to most customers whose locations were within a mile or two from the central office. At the demarcation point, they installed the least expensive termination device (RJ-11), satisfying the standard two-wire unshielded twisted pair communications infrastructure. The position of the demarcation point depended on the legal issues involved. In the early days of the telephone network, the telephone companies owned everything, so they ran the wires to an interface point and then connected their telephone equipment to the wires at the customer's end. The point here is that the telephone sets were essentially commodity-priced items requiring little special effect or treatment.

When the data communications industry began during the late 1950s, the telephone companies began to charge an inordinate amount of money to accommodate this different service. Functionally, they were in the voice business and not the data business. As a matter of fact, to this day, most telephone companies do not know how to spell the word *data*! They profess that they understand this technology, but when faced with tough decisions or generic questions, few of their people can even talk about the services. How sad, they will be left behind if they do not change quickly.

New regulations in the U.S., in effect since the divestiture agreement, changed this demarcation point to the entrance of the customer's building. From there, the customer hooked up whatever equipment was desired. Few people remember that in early 1980, a 2400-bps modem cost $10,000. The items that customers purchase from a myriad of other sources include all the pieces we see during the convergence process.

In the rest of the world today, where full divestiture or privatization has not yet taken place, the telephone companies (or PTTs) still own the equipment. Other areas of the world have a hybrid system under which customers might or might not own their equipment. The combinations of this arrangement are almost limitless, depending on the degree of privatization and deregulation. However, the one characteristic that is common in most of the world to date is that the local provider owns the wires from the outside world to the entrance of the customer's building. This local loop is now under constant attack from the wireless providers offering satellite service, *local multipoint distribution services* (LMDS), and *multi-channel multi-point distribution services* (MMDS). Moreover, the CATV companies have installed coaxial cable or fiber, if new wiring has been installed, and they offer the interconnection to business and residential consumers alike.

The CLECs have also emerged as formidable foes to the local providers. They are installing fiber to many corporate clients (or buildings) with less expense and long-term write-off issues. The CLECs are literally walking away from the telephone companies and the local loop. Add the xDSL family of products to this equation and the telephone companies are running out of options. The telephone companies today have approximately 15,000 xDSL connections across the U.S., whereas the CLECs have over 200,000. Moreover, the cable TV companies have close to 800,000 cable modems installed for high-speed Internet access. This is where the CATV companies see the convergence taking place. CLECs are also seeing the convergence in the local loop, and with xDSL in their potpourri of offerings, they are actually nudging the local telephone companies aside.

A Topology of Connections Is Used

In the local loop, the topological layout of the wires has traditionally been a single-wire pair or multiple pairs of wires strung to the customer's location. Just how many pairs of wires are needed for the connection of a single line set to a telecommunications system and network? The answer (one pair) is obvious. However, other types of services, such as digital circuits and connections, require two pairs. The use of a single or dual pair of wires has been the norm. More recently, the local providers have been installing a four-pair (eight wires) connection to the customer location. The end user is now using separate voice lines, separate fax lines, and separate data communications hookups. Each of these requires a two-wire interface from the LEC. However, if a CATV provider has the technology installed, they can get a single coax to satisfy the voice, fax, data, and high-speed Internet access on a single interface, proving the convergence is rapidly occurring at the local loop.

It is far less expensive to install a coax running all services (TV, voice, and data) than multiple pairs of wire, so the topology is a dedicated local connection of one or more pairs from the telephone provider to the customer location or a shared coax from the CATV supplier. This is called a *star* and/or *shared star-bus configuration*. The telephone company connection to the customer originates from a centralized point called a *central office* (CO). The provider at this point might be using a different topology. Either a star configuration to a hierarchy of other locations in the network layout or a ring can be used. The ring is becoming a far more prevalent method of connection for the local Telcos. Although we might also show the ring as a triangle, it is still a functional and logical ring. These star/ring or star/bus combinations constitute the bulk of the networking topologies today.

Remember one fundamental fact: the telephone network was designed to carry analog electrical signals across a pair of wires to recreate a voice conversation at both ends. This network has been built to carry voice and does a reasonable job of doing so. Only recently have we been transmitting other forms of communication, such as facsimile, data, and video.

The telephone switch (such as DMS-100 or #5ESS) makes routing decisions based on some parameter, such as the digits dialed by the customer. These decisions are made very quickly and a cross-connection is made in logic. This means that the switch sets up a logical connection to another set of wires. Throughout this network, more or fewer connections are installed, depending on the anticipated calling patterns of the user population. Sometimes there are many connections among many offices. At other times, it can be simple with single connections.

The telephone companies have begun to see a shift in their traffic over the past few years. More data traffic is being generated across the networks than ever before. As a matter of fact, 1996 marked the first year that as much data was carried on the network as voice. Since that time, data has continued its escalated growth pattern, whereas voice has been stable.

The Local Loop

Our interface to the telephone company network is the single-line telephone line, which has been installed for decades and is written off after 30 or 40 years. Each subscriber or customer is delivered at least one pair of wires per telephone line. There are exceptions to this rule, such as when the telephone company might have multiple users sharing a single pair of wires. If the number of users demanding telephone service exceeds the number of pairs available, a Telco might offer the service on a party-line or shared set of wires.

It is in this outside plant, from the CO to the customer location, that 90 percent of all problems occur. This is not to imply that the Telco is doing a lousy job of delivering service to the customer. In the analog dial-up telephone network, each pair of the local loop is designed to carry a single telephone call to service voice conversations. This is a proven technology that works for the most part and continues to get better as the technologies advance.

What has just been described is the connection at the local portion of the network. From there, the local connectivity must be extended out to other locations in and around a metropolitan area or across the country. The connections to other types of offices are then required.

The Telecommunications Network

Prior to 1984, most of the network was owned by AT&T through its local Bell operating telephone companies. A layered hierarchy of office connections was designed around a five-level architecture. Each of these layers was designed around the concept of call completion. The offices were connected together with wires of various types called *trunks*. These trunks can be twisted pairs of wire, coaxial cables (like the CATV wire), radio (such as microwave), or fiber optics.

As the convergence of voice and data networks continues, we see a revisitation to the older technologies as well as the new ones. Fiber is still the preferred medium from a carrier's perspective. However, microwave radio is making a comeback in our telecommunications systems, linking door-to-door private line services. Carrying voice, data, video, and high-speed Internet access is a natural for a microwave system. Light-based systems, however, are limited in their use by telephone companies. It has been user demand that has brought infrared light and now SONET-based infrared systems in place. Recently, the introduction of an unguided light introduced by Lucent Technologies operates at speeds up to 2.4 Gbps with a promised speed of up to 10 Gbps by end of 2000. This offers the connectivity to almost anyone who can afford the system, because the right of way is no longer an issue.

The Network Hierarchy (Post-1984)

After 1984, ownership of the network took a dramatic turn. AT&T separated itself from the *Bell Operating Companies* (BOCs), opening the door for more competition and new ventures. Equal access became a reality and users were no longer frustrated in their attempts to open their telecommunications networks to competition.

The Public-Switched Network

The U.S. public-switched network is the largest and the best in the world. Over the years, the network has penetrated to even the most remote locations around the country. The primary call-carrying capacity in the U.S. is done through the public-switched network. Because this is the environment AT&T and the BOCs built, we still refer to it as the Bell System. However, as we've already seen, significant changes have taken place to change that environment.

The public network enables access to the end office, connects through the long-distance network, and delivers to the end. This makes the cycle complete. Many companies use the switched network exclusively, while others have created variations depending on need, finances, and size. The network is dynamic enough, however, to pass the call along longer routes through the hierarchy to complete the call in the first attempt wherever possible.

The North American Numbering Plan

The network-numbering plan was designed to enable a quick and discreet connection to any telephone in the country. The *North American Numbering Plan*, as it is called, works on a series of 10 numbers. As progress occurs, the use of *Local Number Portability* (LNP) and *Intelligent Networks* (IN) enables the competitors to break in and offer new services to the consumer. Note that there have been some changes in this numbering plan. When it originally was formulated, the telephone numbers were divided into three sets of sequences. The area codes were set to designate high-volume usage and enabled some number recognition tied to a state boundary. With the convergence in full swing, the numbering plan became a bottleneck.

Now with the use of LNP, the Numbering Plan will completely become obsolete as we know it. No longer will we recognize the number by an area code and correlate it to a specific geographic area. LNP will make the number a fully portable entity. Moreover, 10-digit dialing in the age of convergence becomes the norm because of the multitude of area codes that will reside in a state.

Private Networks

Many companies created or built their own private networks in the past. These networks are usually cost-justified or based on the availability of lines, facilities, and special needs. Often these networks employ a mix of technologies, such as private microwaves, satellite communications, fiber optics, and infrared transmission. The convergence of the networks has further been deployed because of the mix of services that the telephone companies did not service well. Many companies with private networks have been subjected to criticisms because the networks were misunderstood. Often the networks were based on voice savings and could not be justified. Now that the telecommunications networks and systems are merging, the demand for higher-speed and more availability is driving either a private network or a hybrid.

Hybrid Networks

Some companies have to decide whether to use a private- or public-switched network for their voice, data, video, and Internet needs. Therefore, these organizations use a mix of services based on both private and public networks. The high-end usage is connected via private facilities creating a *Virtual Private Network* (VPN), while the lower-volume locations utilize the switched network. Installing private-line facilities comes from the integration of voice, data, video, graphics, and facsimile transmissions. Now VPNs are used on the Internet to guarantee speed, throughput, quality of service, and reliability. This new wave of VPN takes up where the voice VPNs left off. Only by combining these services across a common circuitry will many organizations realize a savings.

Hooking Things Up

The Telco uses a variety of connections to service the customer locations. The typical two-wire interface to the network is terminated in a demarcation point. Normally, Telco terminates in a block; this can be the standard modular block. Another version of connector for digital service is an eight-conductor (four-pair) called the RJ-48X. When a Telco brings in a digital circuit, the four-wire circuit is terminated into a RJ-68 or a smart jack.

Equipment

Equipment in the telephony and telecommunications business is highly varied and complex. The mix of goods and services is as large as the human imagination, yet the standard types are the ones that constitute the ends on the network. The convergence and computerization of our equipment over the years has led to significant variations. The devices that hook up to the network are covered in various other chapters, but here is a summary of certain connections and their functions in the network:

- The *Private branch exchange* (PBX)
- The modem (data communications device)
- The multiplexer (enables more users on a single line)
- *Automatic call distributor* (ACD)
- *Voice mail system* (VMS)
- *Automated attendant* (AA)
- Radio systems
- Cellular telephones

- Facsimile machines
- CATV connections
- Web-enabled call centers
- Integrated voice recognition and response systems

This is a sampling of the types of equipment and services you will encounter in dealing with Telecommunications Systems and convergence in this industry.

Virtual Private Networks

Introduction

The term "Virtual Private Network" can have different meanings, but it usually refers to voice or Internet. In this chapter, we'll learn the meaning of the term in both environments.

History

As corporate communications volumes increased, organizations realized the cost of telephone service was escalating. Originally, all long distance service was charged on a per minute basis. AT&T introduced a volume discount outbound calling plan called *Wide Area Telephone Service*[1] (WATS). For a monthly fixed payment, the organization got 240 hours of service to one of five bands across the country. Each band was priced, based on the distance from the originator's location. A typical company usually had a band 5 line and a band 1 or 2 to cover adjacent state calls. It took some analysis to determine the most cost-effective solution for each company's particular calling pattern.

Foreign Exchange (FX) service provided a fixed rate-calling plan if a company had a large call volume for in-state locations. This is essentially subscribing to telephone service at the "foreign central office" location and leasing an extension cord from the telephone company to the home location. Originally, there were no usage charges on this line so the more you used it, the less expensive it was. Of course, long distance calls made from the "foreign" exchange were billed at the long-distance rate. A FX line is needed to each high volume calling location.

Alternatively, a company could use a leased telephone line between locations. These lines went by several names: *Terminal Interface Equipment* (TIE) line, dedicated line, and a data line, when used for data. These are essentially point-to-point telephone lines that are available in two-wire or four-wire configurations. Because the difference in cost between two and four-wire connections was small (relative to the cost of the line), the four-wire option was preferred unless the company needed many lines.

The next logical step was to use these TIE lines to connect PBXs at the various locations. Here again, there were no usage charges on these dedicated lines. A company with locations in Seattle, Phoenix, Atlanta, and headquarters in Chicago might have a "hub and spoke" arrangement of TIE

---

[1] Some people refer to the term as Wide Area Telecommunications Services.

lines from their headquarters to each regional office. Each location then might have FX lines to adjacent cities; for example, a company based in Seattle might have a FX line to Tacoma, Kent, and Everett (see Figure 3-1).

There were corresponding inbound services where the called party paid. For example, the original Zenith operator provided toll-free calling in the days of manual switchboards. The inbound WATS service, now known as 800 service, was originally structured also in bands. Finally, for local toll service, *Remote Call Forwarding* (RCF) allowed people to sign up for telephone service in a foreign exchange and have them make a long distance call from Tacoma, for example, back to Seattle at your expense. Although this was more expensive (depending on the number of calls) than FX, an advantage of RCF is that you can receive multiple calls at a time.

It soon became apparent to people working in the Phoenix location that they could call their uncle in Kent by first asking the company operator (later by dialing) for the TIE line to Chicago. They would then choose the TIE line to Seattle and finally dial across the FX line to Kent. The PBX, although not smart, did allow a person to dial up the TIE and FX lines.

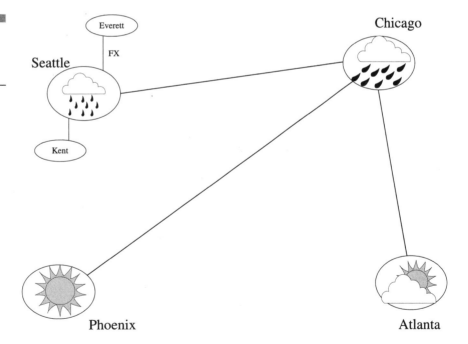

Figure 3-1
Hub and spoke arrangement for TIE lines

The important fly in this otherwise ingenious solution (ointment) to high-cost long distance telephone service is that each TIE or FX line could only handle one call at a time. The challenge for the telecommunications manager was therefore to figure out the optimum number of TIE lines between locations to minimize cost and waiting time for the TIE line, while maximizing savings across the commercial long distance circuits.

About this time, AT&T noticed a small drop in its long distance revenue from such business and a sharp increase in the number of leased lines it was providing. Now, clearly it is much more profitable to rent a telephone channel out at $0.25/min than to lease that capacity to a corporation for $1000/month. Table 3-1 shows somewhat optimistically the amount of revenue that a normal telephone channel could return vs. the lease line.

From Table 3-1, it is clear that the telephone companies much prefer switched service to dedicated service. (This thumbnail sketch focuses only on business hour revenue and ignores after hour revenue and the network providers' cost to provide the service.)

Intelligent PBX Solution

Using these dedicated lines between locations, organizations created a private network. The next step in the evolution of private networks was to devise a corporate-wide numbering plan and have the now intelligent PBX determine the route to the dialed destination via its peers, just like the local telephone office does. After all, other than size, there is little difference between a PBX and a telephone company central office switch!

Table 3-1

Comparison of usage sensitive and leased line costs

Usage	Cost Per Month Based on Usage
Leased line Flat Rate	$1000
8hr/day usage @ .25/min	$2400
4 hr/day usage @ .25/min	$1200

Virtual Private Networks (VPN)

To get corporate America back on the switched network, AT&T devised a marketing strategy. The approach went something like this to the CEO/CFO: "Look, your primary business is banking [building airplanes, trading stocks, selling insurance or whatever], but it is not running a telephone company. Who knows better how to run a telephone system than we do (You can substitute your favorite carrier here. AT&T is chosen here because they were the first to introduce this service.) You think you are saving money by using these dedicated lines. On the surface, it appears that you are. However, who is managing this network? What is it costing you to recover from outages? Do you have back-up facilities for each of your dedicated routes? Your dedicated team of telephony experts is costing you a bundle. Why are you doing this?"

The CFO and CEO look at each other and shrug their shoulders. "Our CIO or CTO[2] sold us on the idea for providing better service at a lower cost," they said in unison.

"Look," said AT&T. "We have the ultimate (outsourcing) deal that will provide all your current capabilities for one low price. We will manage the whole network for you and give you all the service you currently enjoy with your private network with little or no hassle." Our product is called (somewhat obscurely) Software Defined Network™ because you can define the parameters of the network yourself," AT&T said proudly.

Sprint and MCI/Worldcom[3] also offer essentially the same product and call it a *Virtual Private Network* (VPN). We use VPN here because it is both the generally used term, and it is descriptive of the offering.

Here is how the deal works: The company defines the locations that will be part of the virtual private network. The larger the average traffic commitment made between these locations, the lower the price per minute can be. (The catch is that if traffic falls below the average commitment, cost falls into the next higher rate category.)

The big advantage is that organizations no longer have to manage this far-flung network. The carrier will do it. Organizations can now lay off the telecommunications department. (Please note that the staff supporting the

---

[2]CTO is the Chief Telecommunications Officer or Chief Technology Officer depending on the organization.

[3]MCI and WorldCom were different entities at the time of this offering, but for this book are updated to reflect current situations.

Software Defined Network is a registered trademark of AT&T.

PBX in each location is still needed to handle moves, adds and changes. In addition, the staff needed to maintain the dedicated data network is still needed. Even if the organization migrates to a Frame Relay network, some management of the vendor is always required).

All the calls to specifically defined locations (offices) in Chicago, Atlanta, Phoenix, and Seattle are known as on-net calls. These are priced at the reduced rate. Calls to business partners and customers are off-net calls and are charged at a higher rate. If the off-net call volume to these specific locations rises, the organization can still place FX lines into these areas. Again, there is no substitute for knowing the traffic distribution when evaluating any telecommunications plan.

As one can determine from the above description, it takes a sharp pencil to figure out if this is a good deal. It is definitely a good deal for the carrier who gets all those calls and minutes back on the switched network.

The VPN network is more reliable than a dedicated, line-based network because calls are really riding over the *Public Switched Telephone Network* (PSTN), which is rich in multiple paths. One of the features of the private, line-based network was four- or five-digit dialing. This can be preserved intact if we want. Because the switches in the telephone network are computers that have access to a database, they can easily look up how to route a number based on the originating location and number dialed. The VPN then is a special "discount-billing plan," with the carrier managing the network on which we can have a custom dialing plan.

This discussion so far has only considered the case where the corporation owns the PBX and connects it to the VPN. What if a Centrex system is provided by the *Incumbent Local Exchange Carrier* (ILEC) or leased from a reseller? The answer is that one can still implement all the above with a Centrex system at any or all locations.

Because Centrex is essentially a PBX that is physically resident at the local central office, it, too, can have TIE, FX, or RCF trunks. The long distance carrier supplying the VPN will be more than happy to terminate VPN trunks on a Centrex system.

In summary, the important points are as follows

1. Calls are carried over the PSTN.

2. A custom dialing plan is used.

3. Pricing is dependent on the locale.

4. The number of locations

5. The projected or committed traffic volumes

This is all achieved by computer databases in the network.

Internet-Based VPN

One might say that these Internet-based VPNs are the same as voice a VPN but different at the same time. The philosophical point is that a dedicated network will be overbuilt in some areas and underbuilt in others. A shared network offers the hope that we can spread the overall cost out while getting the benefits of a private network. Historically, this accounts for the popularity of shared data networks beginning with X.25, Frame Relay, ATM, and now the Internet.

It is easy to jump to the conclusion that the Internet is free and therefore there are tremendous cost savings to be had from this "free" shared network. Later, we will explore some cost comparisons but as one might guess, the relative cost benefit depends very much on each network's geography and traffic volume.

Goals

The goal of any network is to support users in a flexible, reliable, secure, and inexpensive manner.

- The network manager wants the network to be flexible.
- Users want the network to be reliable and secure.
- Management wants the network to be inexpensive.

A balance of these often-competing goals can be achieved, provided a good dialog is maintained among the participants.

Table 3-2 shows the network goals in terms of applications, users, potential network solutions, and access to the network. It is an exercise left to the reader to select from the list those applications and users who are to be served. The network list indicates that these users and applications could be interconnected by any of these network technologies. As indicated previously, dedicated networks are expensive and rarely fit the need perfectly. Frame Relay and ATM are shared network technologies that can be very cost effective, depending on the geography and traffic volume. Dial-up telephony can be a networking technology for highly mobile, low-volume users. Normally, we would like to have a backbone network with direct access for various users and dial-up access for infrequent users. We will discuss these alternatives in following sections.

Table 3-2

Mix of methods and uses to pick from

Access	Network	Users	Application
Dial-Up	Dial-Up	Road Warriors	E-mail
ISP	Dedicated	Tele-Commuters	db Access
XDSL	X.25	Branch Office	Sales Support
Cable Modem	Frame Relay	Customer	Customer Service
ISDN	Internet	Partners	e-Commerce
Dedicated	ATM		Order Entry

Shared Networks

The advantage of shared networks is that organizations do not have to incur the entire cost of the infrastructure. For that reason, Frame Relay has been extremely popular. Because it (like X.25 before it) is virtual circuit-based, there is little concern about misdirected or intercepted traffic. Still, Frame Relay service is not universally available and access charges to a POP can be expensive. However, compared to the cost of dedicated networks, shared networks offer equivalent performance and a much lower cost.

Internet

The next logical step is to use the Internet as the private network. It is almost universally accessible, minimizing access charges. From our discussion of the Internet in Chapter 29, two things are clear:

1. No one is watching the traffic or performance of the Net as a whole.
2. The path our data takes across the network is quite unpredictable.

This leads to the conclusion that performance will be unpredictable and that our precious corporate data may pass through a router on the campus of "Den-of-Hackers University." (It is not the intent here to malign university students, but only to offer the observation that they are bright, curious, love a challenge, and may have time on their hands and access opportunity to do a little extra curricular research on the vulnerability of data on the Internet.)

There are then two problems: performance and security.

Performance

The performance issue poses the problem of sizing the bandwidth on each link, which becomes a major task as the network grows. Unfortunately, few network managers have a good handle on the amount of traffic flowing between any given pair of locations. Typically, they are typically too busy handling moves and additions to the network, which frequently leads to performance problems. Because the network grew without the benefit of a design plan, invariably it means that portions of the network, including servers, become overloaded.

A dedicated line network is expensive, requires maintenance, and necessitates a backup plan should a line or two fail. Using a shared network does not alleviate the problem of traffic analysis. On the contrary, we now have to worry about the ability of the Internet to provide the bandwidth we need when we need it. Selecting our ISP to provide the performance we need becomes an important issue.

Outsourcing

One solution is to outsource the network to a network provider (the analogy to a voice VPN here is strong). The most popular previous solution was to lease Frame Relay service. The benefit was that the network provider took care of the management of the network and even provided levels of redundancy (for which you paid) within its network. Unfortunately, to make most efficient use of this service, one still needed to have a handle on traffic volumes. For example, a *Committed Information Rate* (CIR) that was too low resulted in lost data and retransmission, while a CIR set too high was a waste of money.

A national or international carrier with its own Internet backbone then becomes a good choice as a VPN provider. One negotiates *Service Level Agreements* (SLA), which include *Quality of Service* (QoS) guarantees. Some ISPs even provide *Virtual IP Routing* (VIPR) in which they permit you to use an internally used, unregistered IP addresses.

If one builds a completely independent, internal (Intranet) network, one could use any set of IP addresses one might choose. This alternative is attractive to large corporations that are constrained to using class C addresses. If these private addresses were to get out onto the Internet, chaos would quickly ensue. VIPR permits the flexibility to continue to use this "unregistered" set of addresses transparently across the Internet. This is strongly analogous to having one's own dialing plan on a voice VPN.

There are many possibilities and choices here. We can outsource the whole network, including the VPN equipment on each site, or outsource pieces.

Standard Outsourcing Issues

A few points are worth making about outsourcing. One must take a realistic look at the task at hand.

1. If the internal staff possess the capability to implement the VPN, do they have the time?

2. If you outsource the whole network, how permanent will the relationship be?

3. To what extent will the internal staff become involved in the design and maintenance of the VPN?

Choose your vendor carefully. Evaluate responsiveness in the areas of presale support, project management, and post sales support. As in any procurement process, writing a system specification and *Request for Proposal* (RFP) is essential. Also, make up the evaluation criteria ahead of time. You may (or may not) choose to publish the evaluation criteria in the RFP. Select the vendor who is most responsive to your requirements. Here is a good opportunity for the vendor to do the traffic analysis so that a traffic baseline for design can be established. Always include growth in the RFP.

Ongoing support will be critical. If the network spans multiple time zones, specify the minimum support requirements. For example, 9 A.M. to 5 P.M. CST is of little use to offices located in Taiwan. What training is offered as part of the package? The more knowledgeable the internal staff can be, the better they will be able to support the VPN—even when they are outsourcing support.

It is important to have a coordinated security plan so that we have an integrated and consistent view across our firewalls, proxy servers, and VPN equipment.

Security

The basic concept of a VPN is to provide a ***secure***, point-to-point connection across the network between communicating entities. A couple of words about security are important to keep our paranoia in check.

The first question is "How much security is enough?"

To answer that question, we must consider the impact on our business if the data we are sending is:

1. Simply Lost. Is there a backup mechanism for sending or recovering the data?

2. Found by another business (not a competitor)

3. Found by a competitor

4. Actively pursued by a competitor

In the last case, we must ask "How much effort is the competitor willing to invest to get our data?"

The answer to these questions will help us decided how much security is enough. Note that in the foregoing example, one can equally substitute the word "hacker" for competitor.

Creating the VPN

There are five ways to create a VPN as follows:

1. Between desktops

2. Between routers

3. Between firewalls

4. Between VPN specific boxes

5. Integrated boxes

Although not normally considered a VPN, one can certainly use desktop PCs to encrypt data and send it across the Internet securely. Additionally, software is available that runs on a desktop capable of creating a VPN to a firewall or stand-alone VPN device. Most VPN equipment vendors offer corresponding software that runs on a laptop or desktop in order to provide a secure path to the home office over the Internet. Most of the discussion then involves creating a VPN between business locations, branch offices, and road warriors.

Encryption

The basic rule is the more secure it is, the less convenient it is to use and the greater impact (negative) it will have on overall system performance. The strength of an encryption mechanism is dependent upon the complexity of the calculation and the length of the key. The most popular mechanism for which hardware is readily available is *Data Encryption Standard* (DES), developed by IBM and now standardized. The basic key is 54 bits long. Triple DES involves simply running the algorithm with a 112-bit key. The question here is as always "How secure do you need to be?"

The more secure, the larger the key used (or the more times the algorithm is run with different keys). This all takes time to encode and to decode. Much has been made lately of the fact that by using thousands of computers, a DES encoded message could be broken in 39 days.

Keep in mind that this is for one key. If we change keys, it would take the crackers and hackers another 39 days. Are they (hackers and competitors) motivated to do this? The method mentioned above used the brute force attack of guessing keys. Changing keys often means that the attackers must start all over again. The other encryption standard (not widely supported) is IDEA, which uses 128-bit keys.

The second basic rule is that encryption performed in hardware is much faster than in software.

Key Handling

A very important part (some say the most important) of an encryption is the mechanism used to disseminate keys. Here again, security is the inverse of convenience. True, keys can be sent in a multiply encrypted file. They can also be sent by "snail mail" or given over the telephone (not very secure). The problem with this private key system is that both communicating parties must have the same key. If all locations are talking to the home office, they all must have the same key, or the central office must keep separate key pairs for each location.

This key management nightmare can be handled in two ways. We could use the X.509 digital certificate system for key management. The other alternative is to use a public key system to encrypt the private key so that they can easily be exchanged.

Public Key Cryptography (RSA)

The layman's version (don't try this at home because it won't work as described here) is that each of us thinks up a couple of prime numbers (the bigger the better). One number we keep for ourselves, and the other number we publish on our Web site along with the product of the two prime numbers as our public key. Anyone wishing to send us something will use the public key to encrypt it with the public key, and only we can decrypt the message with our private key. We can authenticate the source if the sender used his private key to encrypt his signature because only his public key will decrypt his signature.

This system is secure because of the tremendous amount of processing power it takes to factor large prime numbers. (For example, if we could factor the product, we could determine the private key). Unfortunately, performing the encryption and decryption are also processor intensive (read "slow"). But it sure solves the key distribution problem. Therefore, we could use public key cryptography to encrypt and distribute the keys to all our VPN boxes.

Authentication

Authentication is the process of verifying that, "this is the party to whom I am speaking," and that they have authorized access. There are several ways of doing this; however, the most common way is to provide an authentication server that passes out authenticated certificates based on something the user has or knows.

User level authentication   The user has or knows his/her account code (name) and password. User names are public and passwords can be compromised. A more secure system is to use a type of secure ID card. These credit card sized devices are based on an internal clock that generates a different pseudo random code every minute. The authentication server is time synchronized with the card and therefore generates the same number at the same time. When the user calls in, he/she must enter his/her account code and the code from the card as the password.

Packet level authentication   The IPSec standard provides for packet level authentication to prevent man-in-the-middle attacks. (This is where someone intercepts your packets and substitutes his/her own.) An authentication header is created for each packet. The layman's version of this is that a checksum is calculated and encrypted with the data. If the checksum calculated by the recipient doesn't match the one sent by the originator, someone has tampered with the data. The IPSec standard specifies two different algorithms for doing this MD−5 and SHA−1. If your vendor's equipment supports both algorithms, it improves the chances for intervendor compatibility. The other alternative is to simply not use packet level authentication.

In order to guarantee authenticity of the packets, a digital signature is required to authenticate the devices to one another. IPSec has included the X.509 digital certificate standard. Essentially, the X.509 certificate server keeps a list of certificates for each user. When you want to receive data from another device, you first ask for the certificate from the certificate server. The sender stamps all data with that certificate. Since this process is secure, you may be sure that these packets are authentic.

Your vendor then ideally supports both authentication algorithms and X.509. In any case, it is essential that someone in your organization understands in detail how each vendor supports the various levels of security that you intend to use. These authentication and encryption systems all have to work together flawlessly. If the vendors you choose stick to the standards, it improves the chances of, but does not guarantee, an integrated working environment.

Router-Based VPN

Several router vendors offer VPN products based on the ability of the router to perform the requisite security functions. If your VPN is relatively small and the traffic volume not too heavy, then you might consider this option as a cost-effective approach. You need to have compatible routers at each location as shown in Figure 3-2. If there are individuals (for example, laptops or telecommuters that don't have routers), they must have software that is compatible with that provided on the router. Make sure your vendor provides the compatible software that provides the level of security that you require for your VPN. The absence of a firewall in Figure 3-2 may be taken to mean that in this low cost approach we are doing firewall functions on the router. In this case, the network would logically appear as in Figure 3-3.

The general admonition here is that you may be creating a bottleneck in the router. For large networks, let routers route.

Figure 3-2
Compatible routers
are used at each
location for VPN
services

Figure 3-3
Stand-alone firewall

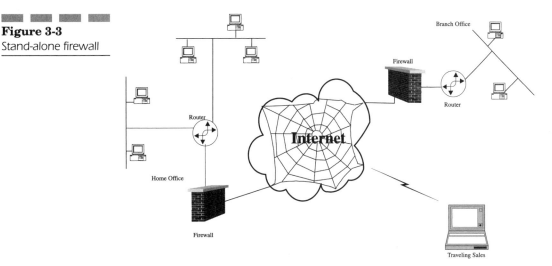

Firewall Based VPN

The very same issues exist here as with routers. One needs to have compatible (preferably the same vendor's) firewalls at each location. Mobile users or telecommuters must have compatible VPN software. Firewalls are always potential bottlenecks, so asking them to perform VPN encryption can adversely affect all other access to your network. Here again, there is no substitute for traffic analysis. We only recommend this solution for small networks where the traffic through the firewall can easily be handled by the firewall hardware.

Figure 3-3 shows a stand-alone firewall hardware that filters all traffic into our network, in addition to VPN functionality.

VPN Specific Boxes

VPN specific boxes are the recommended solution for high volume, large networks. Several vendors offer these solutions in both hardware and software incarnations. The general rule is that hardware boxes will outperform software boxes, and are theoretically more secure because they are based on proprietary technology that is harder to hack than publicly available operating systems. (A hardened Unix-based system is also extremely difficult to hack.) Traffic volume and feature support for remote terminals and industry compatibility will guide your decision here.

These boxes set up secure tunneling by using IPSec encryption and certificates as described previously. They are typically installed in parallel with your firewall. The firewall handles Web (HTTP) requests, while the VPN box handles access to your internal database. Figure 3-4 shows the firewall and VPN box in parallel, reinforcing the division of labor between the two boxes. Because we now have two "holes" into our network, it is imperative that we have the permissions and access rights set up correctly. The firewall should not let users in who would be required to authenticate via the VPN box.

The integrated solution that some vendors are offering is an integrated custom box that does routing firewall and VPN all under one roof. This is an attractive option where traffic volume and performance is not going to be an issue. Again, Figures 3-3 or 3-4 might be used to depict this configuration.

Throughput Comparison

Unfortunately, while there is compatibility testing, there are no consistent performance criteria across the industry. It, therefore, becomes difficult to compare the performance of different vendor offerings. Vendor claims tend

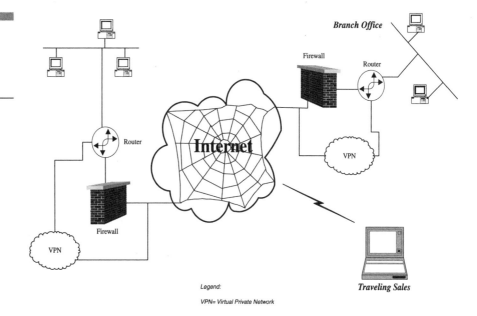

Figure 3-4
Stand alone VPN
equipment divides
access to the
network

Branch Office

Firewall

Router

Internet

VPN

Router

Firewall

VPN

Legend:

VPN= Virtual Private Network

Traveling Sales

to be exaggerated. They will measure their product in the best possible light (for example, maximum-sized packets and data compression turned on, using the simplest encryption algorithm). Our recommendation is to search the periodical literature for tests on the vendors you are considering as a starting point.

Then, in your RFP, specify a test sequence. With encryption and authentication, there is a lot of end-of-packet processing. This causes a significant performance hit when packet sizes are small.

The number of simultaneous sessions also affects performance. Vendors claim thousands of simultaneous sessions, but ask them how many they can set up or tear down at a time, and the number drops to fewer than 100. Notice also that during this peak-processing load of session setup, overall throughput will be affected.

Here again, having knowledge of how your users use the system, when the peak sign-on demand occurs, when the peak traffic occurs, and what kinds of response time you consider to be reasonable all influence your product selection. By the way, being able to set up 100 sessions/second is plenty in a 1,000-user network. (How many of these users are actually using the VPN?) Worst case (which statistically never occurs) means that the last user might have to wait 10 seconds to get a session setup. Most likely, no one except the network manager with the "Sniffer" will ever notice a delay.

Remote Management of VPN Components

If you have only two locations on your VPN, then remote management of policy is probably not an issue. For a large network, visiting each site to install policy rules becomes a burden. For larger networks then, look for the ability to provide remote policy management of not only your VPN devices, but also your firewalls and routers securely.

Cost Considerations

Figures don't lie but liars know how to figure.

Although we're presenting some typical numbers here, you should run the numbers using your own particular configuration. The most beneficial comparisons of a VPN occur when compared to a dedicated, line-based network or one that makes extensive use of long distance dial-up lines. If you are already using a shared network (Frame Relay or ATM), the cost savings are not so striking.

Consider that a VPN box at each location might cost $5,000 including installation; multiplied by seven sites is equal to $35k. Now, how long will it take to save this cost if you substitute your ISP charges for each location and subtract the cost of your existing T1 or frame relay network? If you had six T1s at $5k/mo, you might now have seven T1 access lines from our ISP at $3k or $4k/mo. The $7k/month savings will pay off the $35k investment in 5 months. If your Frame Relay service is costing $1,000/mo per location, the break-even point doesn't happen in any reasonable period.

Using Remote access server and dial-up lines is cheaper to install costing about $6k to $7k for about 20 users to install at the central location. Now comes the big bite, which are the long distance charges from all the remote locations. This could easily grow to $5k/month if each of the users spent two hours online. Each working day at $0.10/min is about $8,000/mo. Plug in your own assumptions as to duration and cost of telephone calls here. (Even at 1hr/day and $0.06/min. that is $2,000/mo for 20 users). A VPN system might cost $14,000 to install, including licenses for PC software at each location. The ISP charges are $20/user/month, plus an ISDN line at the home shop for $100/mo, means that we are saving $1,500 in monthly charges. We can pay off the system in 10 months. Again, do not assume that it will pay off in all cases. But, in all cases, it is worth the effort to perform the calculations.

Your VPN will definitely require more network management than a dial-up system so the cost of perhaps an additional system administrator may have to be added.

Proprietary Protocols

Most VPN products are designed strictly around IP. They will often handle other protocols, such as Appletalk and IPX, by tunneling them inside of IP packets. This introduces both overhead and delay. If the amount of "foreign" protocol traffic is small, then this is not significant. If the bulk of your network is IPX or Apple talk, we recommend you investigate VPN vendors who will support these protocols in native mode.

VoIP VPN

The justification for doing VoIP on a VPN is primarily security, along with the reduced cost of VoIP. Depending on usage, voice generates relatively large amounts of traffic. Be sure to include this additional traffic in your sizing estimates.

Our discussion of VoIP applies to whether we have a VPN or not (see Chapter 30). With a VPN, the delays due to encryption are larger, and therefore we would expect that the performance of voice over the VPN would be worse than voice over IP. If we have chosen a network provider who will offer a SLA with QoS, there is a better chance for success, but the delays due to encryption and basic packet switching will still be there. With the exception of international calling, one must have a very large calling volume to make it worthwhile to put voice over the Internet and suffer the attendant quality reduction.

Summary

VPNs can provide a cost-effective solution to have secure communications across the Internet. Performance can be improved by utilizing a national/international ISP that will offer SLAs and QoS. Choosing hardware-based over software-based VPN equipment will generally provide better performance. Choosing VPN vendors who embrace standards and support multiple standards increases your flexibility to your vendor/equipment choices. Knowing your current and anticipated traffic volumes permits you to make improved cost performance studies.

1. Some people refer to the term as Wide Area Telecommunications Services

2. CTO is the Chief Telecommunications Officer or Chief Technology Officer depending on the organization

3. MCI and WorldCom were different entities at the time of this offering, but for this book things are updated to reflect current situations.

Advanced Intelligent Networks (AIN)

Introduction

The *Intelligent Network* (IN) has been under development since it was first introduced by Bell Communications Research in 1984. The goal of intelligent networking is to integrate the features and benefits on the new generation networks and to allow various types of information to pass through the telephone network without the need for special circuits. Data communications, Internet communications (using Internet Protocols), and voice networking have converged to provide a new and exciting set of services. These services revolve around the backbone of the Intelligent Network, which uses *Signaling System 7* (SS7). Network architects envision one network capable of moving any form of information, regardless of the bandwidth. Data and voice calls will traverse this network the same way—making communication as simple as placing a traditional telephone call.

Intelligent Networks

The Intelligent Networks consist of intelligent nodes (computer peripherals), each capable of processing and communicating with one another over low- to mid-speed data communications links. All nodes in the intelligent SS7 network are called signaling points that work with packet transmissions. A signaling point has the ability to do the following:

1. Read the packet address
2. Determine if the packet is for that node
3. Route the packet to another signaling point

Signaling points provide access to the SS7 network and the various databases on the network. They also act as transfer points.

More information will be explained in the SS7 chapter later. However, the switching network contains *Service Switching Points* (SSP) and provides the basic infrastructure needed to process calls and other related information.

■ The SSP provides the local access because it emerged as the Central Office. The SSP can also be other tandem points on the network or an ISDN interface for the *Signaling Transfer Point* (STP).

- The STP provides packet switching for message-based signaling protocols for use in the *Intelligent Network* (IN) and for the *Service Control Point* (SCP).

- The SCP provides access to the IN database. The SCP is connected to a *Service Management System* (SMS).

- SMS provides a human interface to the database, as well as the capability to update the database when needed.

- Intelligent Peripherals provide resource management of devices such as voice announcers. Intelligent Peripherals are accessed by Signal Control Points when appropriate.

The Intelligent Network enables customers to tailor their specific service requirements within hours instead of days. It is expected that full Intelligent Network implementation will continue to evolve through the new millennium; however, the infrastructure has been laid and work continues. Some of the features available include the following

1. Find-me service
2. Follow-me service
3. Single (personal) number plans
4. Call routing service
5. Computer control service
6. Call pickup service

Business is conducted by using a mix of public services, private networks, the Internet, wireless networks, and specialized carriers. Voice-processing requirements for a telephone switch were quite modest and could be satisfied by the limited announcement capabilities offered by the switch, usually as sequenced announcement of greetings, instructions, and terminating messages. No specialized service creation environment was required to develop and deploy these services. Complex voice-processing capabilities, such as central office-based voice mail were provided, but not as an integrated service of the switch. Complex voice-processing capabilities grew outside of the switching network in the form of interactive voice response systems, voice mail systems, automatic call distributors, and automated attendants. Over time, these systems embraced new technologies, including facsimile, speaker-dependent and speaker-independent voice recognition, text-to-speech, and voice identification.

Advanced Intelligent Networks (AIN)

AIN is a collection of components performing together to deliver complex call switching and handling services. The *Service Switching Point* (SSP) is the central office that provides robust, call-switching capabilities. When switching decisions require complex call processing, the SSP relies on the *Service Control Point* (SCP), a subscriber database, and it executes service logic. The SSP uses SS7 signaling, specifically *Transaction Capabilities Application Part* (TCAP) messages, requesting the SCP to determine the best way to handle the call. The process supports telephony features, including 800 (888/877) and 900 calling, credit/debit card calling, call forwarding, and virtual private networks. AIN has promised an architecture that is amenable to rapid development and deployment of new services. How to maintain the stringent performance requirements of a central office service within this rapidly changing environment is a major challenge in the advancement of AIN.

Intelligent Peripherals (IP) and *Service Nodes* (SN) are elements of the AIN and must be reliable to be deployed and used in the central office. Intelligent Peripherals work in cooperation with *Service Control Points* (SCP) to provide media services in support of call control. Service nodes combine the functions of the SCP and the IP. When viewed as point nodes in the network, these elements (IP and SN) are subject to failure and require redundant components and multiple communication paths. Software and procedures in support of central office reliability are also required.

When switching decisions require complex voice processing services, the SCP cannot always provide all the required services to the SSP. The SCP cannot provide termination of voice circuits and play recorded messages, collect touch-tone input, or perform other voice processing services. The call must be redirected to an intelligent peripheral. The IP provides the voice processing services not available from the SCP.

Advanced Intelligent Networks (AIN) provide more features and functions that are not provided by Intelligent Networks. Advanced Intelligent Network does not specify the features and services, but how they are used by end users. An essential component in AIN is the *Service Creation Element* (SCE). Today, service configuration and changes are handled by Telco personnel at end offices. The SCE specifies the software used to program end office switches. The single, most important change is through a

Graphical User Interface (GUI). Over time, the SCE will reside at the users' organization, allowing customers to tailor their services on an as-needed basis, without telephone company assistance.

Some of these enhanced features available include the following items:

- Calling name delivery
- Call rejection
- Call screening (visual or audio)
- Call trace
- Call trap
- Personal ID numbers (pins)
- Selective call acceptance
- Selective call forwarding
- Spoken caller identification

Some of these features are already available at certain COs, but they are not yet ubiquitous. Limitations to these services are based on the capabilities of end office switching equipment—service offerings and tariffs will not be consistent throughout telephone companies.

Information Network Architecture

Information Network Architecture (INA) is still in development and viewed by many as the successor to AIN. However, there is considerable controversy over this view, and some believe that two architectures will eventually develop. AIN is designed to facilitate the voice network, whereas INA will manage the broadband network. The common belief is that INA will provide better utilities for managing new broadband services offered by telephone companies.

Intelligent networking delivers computer and telephone integration capabilities ***inside*** the network. Two major market forces and architectural frameworks are merging to create the most explosive network services opportunities of the late '90s. Enterprise *Computer Telephony Integration* (CTI) applications and *Advanced Intelligent Network* (AIN) services are being integrated to provide an array of advanced carrier-delivered services. Some of the features that will be available to the executive or professional, especially telecommuters, include the following items:

- Virtual call centers
- Consumer interactive applications
- Centrex productivity enhancements
- Formal and informal call centers,
- Virtual ACDs

Combining AIN and CTI Services

The evolution of AIN and CTI services underpins the marriage of these two architectural frameworks. AIN has its roots in the Local Exchange Carriers' (both ILEC and CLEC) and *Inter Exchange Carriers'* (IEC) desire for vendor and switch independent network architectures.

Improving the speed of service provisioning and delivering advanced network services is crucial to maintain a competitive posture. As early as 1986, Ameritech began proposing a concept called *Feature Node Service Interface* (FNSI). Through successive industry efforts, AIN emerged as a network standard in the early '90s. Figure 4-1 is the basic framework of the AIN architecture.

Figure 4-1
AIN architecture
framework

Legend:
CO= LEC or CLEC Central Office
SSP= Service Switching Point
STP= Signal Transfer Point
IP=Intelligent Peripheral
/IP= Internet Protocol
SMS= Service Management Subsystem

Computer to Telephony Integration is a complement to AIN proposed by the PBX manufacturers. The end user's desire:

- Advanced applications
- Faster feature delivery
- Control and customization of applications all drove the development of the CTI interfaces

Using an external processor, custom applications are configured to enhance the functionality of a PBX, especially in an ACD environment. Figure 4-2 shows the basic connections in a CTI application.

Both CTI and AIN use an external processor to deliver advanced complementary services to the switch. The switch controls call processing access to the external processor via an open interface. Both provide a GUI based *Service Creation Environment* (SCE) for rapid service and application configuration. *Intelligent Peripherals* (IPs) provide additional context information for call treatment. Integrated voice response units are the most widely deployed IPs. Reporting, billing interfaces, and real-time monitoring tools are available in both.

With the introduction of the *Intelligent Network* (IN), these two technologies were brought together by necessity. Services rich in voice processing

Figure 4-2
Basic CTI application

content and requiring an abundance of digitized voice storage could no longer be created solely with the innate capabilities of the switch. The variety of service offerings, their complexity, and, in many cases, the requirement for multilanguage support, quickly outpaced the voice-processing capabilities and capacities of the switch. Voice-processing systems were integrated with switching systems to support these new services.

The Intelligent Peripheral

The Intelligent Peripheral must provide voice circuit support and some mechanism for obtaining information about an incoming call. Several mechanisms are available for capturing call information including the following items:

- ISUP signaling
- ISDN signaling
- In-band DTMF signaling
- DTMF

BellCore developed transaction protocols to handle the call processing, using an 1129-protocol specification. The 1129 transaction is triggered when a call comes in. It is initiated with the delivery of a message from the IP to the SCP. Thereafter, the transaction continues with queries issued by the SCP and synchronous responses to these queries returned by the IP, reporting results of the requested action.

The BellCore recommendations call for multiple intelligent peripherals within the network. Each is capable of working with multiple independent service control points via communication across multiple data paths. Each IP operates as though it is distributed across multiple host platforms interconnected by multiple LANs. Introducing IP into an AIN environment is a major expense, requiring significant investment in hardware, networking, and supporting software.

The BellCore philosophy is to provide redundant components and data paths eliminating single points of failure wherever possible. However, many situations exist whereby an IP or SN provides a service, yet the service does not warrant redundant infrastructure. Therefore, a solution is required for the IP or SN to provide suitable reliability inherently.

Intelligent Peripheral Services

The *Intelligent Peripheral* (IP) must be capable of establishing and maintaining communication with multiple SCPs. Furthermore, it handles the following functions:

- Encoding and decoding complex messages
- Interpreting these messages
- Performing the requested service.

It must also be capable of switching functions including the following:

1. Call setup
2. Transfer
3. Call teardown
4. Detect call presentation and abandonment.
5. Process requests requiring service logic
6. Access databases.

Finally, beyond the capabilities described in the 1129 interface specification, the IP needs support services such as logging, administration, alarm processing, statistics gathering and reporting, and database and network access. The IP falls into two categories of service: application processing and resource processing.

The application processor supports the call processing logic and access to the databases and data networks. It initiates transactions, receives instructions, and reports success or failure of the action. The application processors also require connections to communication networks and hosts. An IP needs multiple data paths because it must be able to contact multiple hosts and must be able to survive the loss of a single communication pathway. It needs to communicate with SCPs to receive instructions and communicate with switching facilities in order to terminate circuits. It also needs to communicate with other network elements, such as the SMS (for provisioning), and to internal and external databases. This requires multiple data paths and alternate routing.

The resource processor manages the switching, voice channel facilities, and media processing resources. The media processing normally requires multiple processors and disk storage to handle the media streams. It may also require switching, such as a time division multiplexer or external switches.

These two elements are very different from each other. They perform different functions; consequently, they require very different hardware and software processing elements.

The architecture for AIN with the SSP, SCP, and IP in place on the stack is shown in Figure 4-3.

Software Architecture: Client, Router, Server

The objective is to build an intelligent peripheral in which application processors drive multiple resource processors. This requires software on the application processor capable of supporting the physical structure. A client-router-server scheme is used.

The client includes the application, media processing logic, controls, manages the resources, and drives a state machine. The intelligent peripheral, driven by the 1129 message set, operates as a state machine. This state machine is driven by messages from the SCP and by trigger events from the network.

The router handles message routing and session management functions. It makes decisions about how to route traffic and load-balance.

Figure 4-3

AIN protocol stack

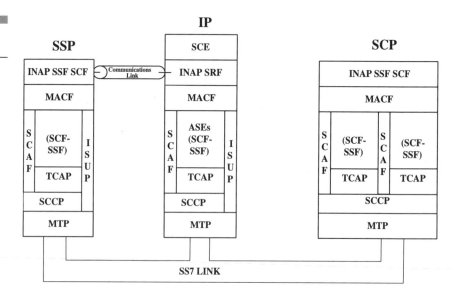

The servers support both local and remote devices. The local devices include local file systems, databases, and locally attached devices such as encryption boxes. The remote devices are networks and hosts. The remote servers must be multiple processor capable.

The Application

The application must be capable of supporting real-time, online updates, which must be done without any disruption of service. The system administrator must be able to introduce new instructions and commands/responses without affecting service. On the fly changes to parameters, such as timers and retry limits, are necessary. The state machine itself must remain operational without call loss while the application logic is being changed.

Results of AIN

It is possible to build intelligent peripherals in support of the AIN. These peripheral devices must be economical and reliable enough to operate in a central office environment. Multiple hardware elements are required. Communication paths must be redundant.

Estimates are that by 2007, telephone industry spending will approximate $27 billion. Products and expenses supporting *Advanced Intelligent Networks* (AIN) will expand exponentially. Approximately $5 billion was spent on AIN products and services in 1998. Telephone industry spending on AIN will include *Signaling Transfer Points* (STP), *Service Control Points* (SCP), *Service Switching Points* (SSP), *Intelligent Peripherals* (IP), *Service Creation Environments* (SCE), and various hybrid products.

Recent studies indicate that the demand for AIN features and functions will be driven by the following factors:

1. Demand for more customer control and services:

   Businesses are placing more strategic reliance on telecommunications. They need features and functions tailored to their specific needs. The number of functions and services required will be dictated by the lifestyle and business competitive environment of the end user. AIN is the only way to support such requirements. Services will have to be user-friendly and somewhat network-centric to meet the demands of the future user.

2. Greater geographical distribution and newer technologies:

Future niche markets will probably emerge. They need to be widely available by the suppliers (LEC and IEC) to be effective and acceptable. AIN services must also become technology stable to support multiple vendor products and service on multiple switching systems. AIN-based services must find a way to work across carrier boundaries.

3. Mobility and mobile applications in a changing world:

Businesses and personal use of communications support from cellular phones, voice mail, pagers, and e-mail are growing exponentially. The need to support a mobile user is now equally important. AIN-based services must cross vendor products and billing mechanisms to be effective. Wireless networks are among the fastest growing applications and services in the industry. The growth rate of 300% per year shows little sign of slowing. *Personal Communications Services* (PCS) networks are growing equally fast and demanding more services. Soon the AIN will have to support the single number for a user, regardless of where that user may reside.

4. Internet, Broadband, and Multimedia:

A CIR[1] report notes that most of the services that AIN is currently concerned with are narrowband and voice-oriented. However, the CIR report projects the belief that many of the services currently identified with AIN will migrate to the Internet. In addition, AIN concepts will increasingly be required for multimedia and broadband services. The market for the next few years around the world is shown in Figure 4-4

Focus

Signaling System 7 (SS7) is an essential technology supporting and developing *Integrated Services Digital Networks* (ISDN), *Intelligent Networks* (IN),

---

[1]Communications Industry Reports is a research firm specializing in market research and analysis of the telecommunications industry.

Figure 4-4
Growth in AIN services around the world

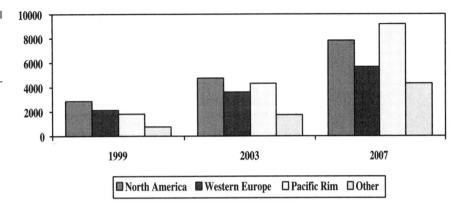

mobile (cellular) telephony, PCS and information systems, personal communications, and many other applications. Currently, this technology is being pursued by the telecommunication industry including carriers, large private networks, switch manufacturers, and an increasing number of software developers. The traditional methods of in-band signaling and other common channel signaling methods in the telecommunication networks have given way to an overlay network using a more capable, layered SS7 protocol.

Although the primary function of SS7 is to handle call control-signaling requirements for voice and data transmission services, it can provide a number of advanced services by use of network databases. These include toll free and alternate billing, rerouting, virtual networks, and other highly sophisticated telecommunications services denoted by Intelligent Network services. Additionally, the transaction capability of the SS7 protocol makes it applicable to a broad range of new services dealing with remote operations. In cellular telephony, SS7 is used for mobility management and handover functions.

Intelligent Network describes architecture designed to facilitate the implementation of highly sophisticated telecommunications services including:

- Features and functions
- Multivendor interworking
- Differing priority services offered by fixed networks, mobile networks, and Personal Communications Systems

The new millennium will hold several benefits and service advantages unavailable in the past. These advantages will all be geared toward satisfying the changing needs and demands of the consuming public. As IN and AIN evolve, newer services will be introduced. The various new providers, such as the CLECs and the CATV companies, will compete to meet the one-stop shopping demands of the customer. AIN will be one of the deciding factors steering consumers to the various providers. The first providers to implement the AIN features will have the edge over taking the customer base.

The edge goes to the Incumbent LECs for now, because the infrastructure is theirs. However, as the new millennium rolls through, changes will occur. This is one of the critical components in being able to capture a niche in the market.

Local Number Portability (LNP)

Introduction

With the Telecommunications Act of 1996 in the United States (and the Telecom Act of 1997 in Canada), a series of competitive changes were required in the network. For years when competitors tried to compete with the Bell System, their efforts were stymied. Part of the reason was the ability to serve a new customer. Whenever the new competitor offered to serve a business or residential customer, customers were required to change their business or residential telephone number to use the new *Local Exchange Carrier* (LEC). *Local Number Portability* (LNP) is therefore an essential issue in the telephony business. Because of the need to change numbers, the incumbent LECs were able to prevent losing their customers to the competition.

Three Flavors of LNP

Local Number Portability gives end users the ability to change Local Service Providers without changing their telephone numbers. Three basic forms of local number portability were introduced to the industry over the past few years: (the only one implemented is service provider portability)

1. Service provider portability allows subscribers to change Local Service Providers without changing their telephone number. This assumes that users can change suppliers and keep their existing telephone number. Still to be completed is the ability to provide LNP in a wireless world.

2. Service portability allows subscribers to change from one type of service to another (for example, analog to digital—ISDN— without changing their telephone numbers) or to be served from a different central office (where the service is available) and not have to take a new telephone number

3. Geographic portability allows subscribers to move from one physical location to another (such as state to state) without changing telephone numbers.

Service provider portability was mandated by the *Federal Communications Commission* (FCC) in July 1996. Service provider LNP involves a circuit-switched network capability, allowing users on one switching system to move or *port* numbers to a different switching system. Congress mandated local number portability, set the regulations governing its implementation, and stated that any network modifications required to comply with these rules were the responsibility of the existing LECs.

In February 1996, President Clinton signed the Telecommunications Act into law. The single most significant characteristic of the new law is that it opens the local exchange market to competition. In an effort to eliminate technical, as well as regulatory market entry barriers, the law requires that all *Local Exchange Carriers* (LECs)—both incumbent LECs (ILECs) and new competitive LECs (CLECs)—provide local number portability. LNP provides "users of telecommunications services with the ability to retain, at the same location, existing telecommunications numbers without impairment of quality, reliability, or convenience, when switching from one telecommunications carrier to another."

The Road to True Local Number Portability

LNP is germane to achieving true local competition. In November 1994, the industry began to seriously investigate methods of providing true Local Number Portability. While the importance of retaining a telephone number was recognized in the early 1960s, it became a significant issue associated with 1–800 (and later 1–888, 1–877 and 1–900) services during the 1980s. The need for portability among telephone companies and providers was never really an issue until the *North American Numbering Plan* (NANP) administration published a proposal for future numbering plans in North America. This NANP document was issued in January 1993. Of course, since that time, the level of interest has increased substantially. As equal access and competition catch up to deregulation, the use of LNP becomes critical.

In late 1994, MCI commissioned a study by the Gallup Organization to assess LNP and determine the following:

1. The potential for businesses and consumers alike to switch local telephone service providers under various market scenarios.

2. The perceived importance of various service factors regarding local telephone services.

Not surprisingly, the results of the MCI study indicated the following for residential customers:

- Nearly 2/3 of all customers were unlikely to switch providers if given the opportunity
- More than 3/4 stated that retaining their telephone number or numbers when switching carriers was very important.
- 80% were unlikely to switch service providers if they would have to change their telephone numbers.

Moreover, business customers showed the following results:

- 57% were unlikely to switch local telephone service providers if given the opportunity.
- 83% said that retaining their telephone number or numbers when switching Local Service Providers was extremely important.
- 90% were unlikely to switch providers if they would have to change their telephone numbers.

With business customers, the cost of changing numbers can be significant. When a business changes telephone numbers, the costs include reprinting stationary, business cards, advertisements, and literature. The statistics are shown in Figure 5-1.

These discussions led to an organization's development of the *Carrier Portability Code* (CPC) model, which was selected by the New York State Public Utility Commission LNP task force as the architecture to be used. MCI with the support of several manufacturers demonstrated the architecture to the FCC via a live test in May of 1995 to prove that true local number portability was indeed technically feasible.

Figure 5-1

Comparison of business and residential user concerns over service provider change

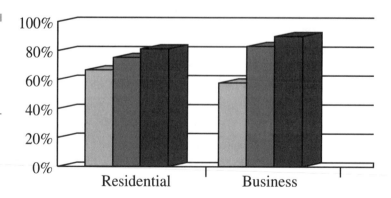

On June 27, 1996, the FCC adopted rules on LNP. The FCC required LNP availability on a permanent basis as of October 1, 1997. This availability must be complete for the top 100 *Metropolitan Statistical Areas* (MSA) by December 1998. After the December 1998 dates, all LECs (ILECs and CLECs) in other areas are required to provide LNP within six months upon request. The FCC actually adopted performance criteria, rather than a specific technology to meet the need. The criteria includes the following information:

- Support of existing services and capabilities
- Efficient use of numbering plans
- No change in numbers for end users
- No requirement to rely on databases of another carrier in order to process (route) calls
- No unreasonable degradation of service or reliability when implemented or when customers switch
- Not resulting in any carriers' proprietary interest
- Ability to accommodate location and service portability in future years
- No adverse impact outside areas where LNP is deployed

States will have flexibility if they meet the criteria listed previously. Wireless carriers have been granted a reprieve from the December dates, originally allowing a June 1999 implementation date. This has since been postponed until June 2000 and may be extended again.

Shortly after the studies were completed, several states began the process of officially selecting the architecture to be used for LNP in their respective states. The Illinois task force requested LNP solutions from a wide array of companies via a *Request For Proposal* (RFP) developed by the carriers that offered service in the state of Illinois at that time. An official voting body, which was comprised of those carriers, was established to select the architecture. After considerable discussion and deliberation, Lucent Technologies' *Location Routing Number* (LRN) architecture was selected.

Basic LNP Networks

The components of LNP are not that much different from the original SS7 networks used for years. The pieces serve different functions, which you can see by looking at the following components:

- The *Switching Service Point* (SSP) is the local CO or tandem switching office.

- The *Signal Transfer Point* (STP) is a packet mode handler that routes data queries through the signaling network.

- The *Signal Control Part* (SCP) is the database for features, routing, and *Global Title Translation* (GTT).

- The LSMS is the *Local Service Management System;* SOA is the *Service Order Administration.*

LSMS and SOA can be provisioned separately, but when combined it is referred to as the *Number Portability Administration Center* (NPAC) connectivity. For wireless providers, the LNP capabilities depend on the MSA served. Wireless providers' schedules keep changing because of the lobbying of the CTIA. See the network overview in Figure 5-2.

Figure 5-2
Basic LNP network

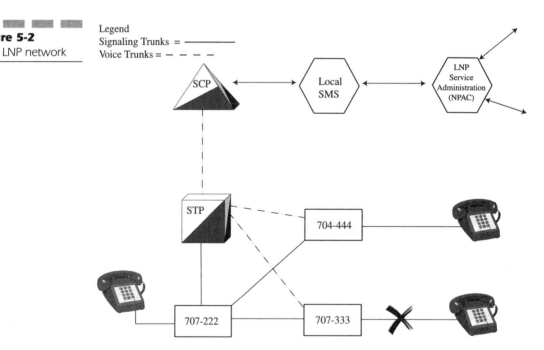

Legend
Signaling Trunks = ⸺⸺
Voice Trunks = ⁻ ⁻ ⁻ ⁻

The Terminology

Several terms are introduced with LNP. Some of the more common ones are as follows:

Portable number   A number that is permitted to move between different service provider switches

Ported number   A *Directory Number* (DN) served by a switch other than the switch that has traditionally served that number

Non-ported number   A Directory Number that is portable, but is currently served by the switch that is identified in the *Local Exchange Routing Guide* (LERG) as serving that number

These three concepts are shown in Figure 5-3 depicting the terminology.

Figure 5-3
LNP terminology

Before Local Number Portability

In the earlier, non-LNP environment, a telephone number performed two functions:

1. It identified the customer.
2. It provided the network with information necessary to route a call to that customer.

Local Number Portability separates these two functions, providing the means for customers to keep their Directory Number when changing Local Service Providers. By separating those two functions, LNP gives customers the flexibility to respond to pricing and service changes offered by rival carriers.

As we have seen, numerous studies conclude that most business and residential customers are reluctant to switch from one service provider to another if they must change their telephone numbers. Without LNP, new entrants (CLECs) would have to price their local exchange service 12–15% lower than the existing local exchange carriers in order to persuade customers to switch carriers. Although the degree to which the lack of LNP hurts competition is arguable, it is clear that LNP is required to provide a level playing field. Interim number portability methods, such as remote call forwarding and direct inward dialing, existed but these methods have several disadvantages:

- Longer call set-up times
- Increased potential for call blocking
- Continued reliance on the incumbent LEC's network
- Use of more directory numbers, which are fast becoming depleted
- Loss of feature functionality
- Substantial on-going costs to the new Local Service Provider.

According to the Telecommunications Act, LNP will promote local exchange competition, which in turn will benefit all customers. As it has done in the long distance market, competition in the local exchange market is expected to do the following:

- Drive down the cost of service
- Encourage technological innovation
- Stimulate demand for telecommunications services
- Boost United States' economic growth

Number Administration and Call Routing in the Network

Telephone numbers in a pre-LNP environment have always been assigned to Local Service Providers' end offices on an area code and exchange code (NPA-NXX) basis. Each NPA-NXX contains 10,000 telephone numbers. Because an NPA-NXX is only served by a single end office in the United States, the telephone number identifies the person, as well as the actual end office that serves that person. In effect, the dialed NPA-NXX is the terminating switch's routing address to the rest of the network. With the implementation of local number portability, which allows any number of Local Service Providers to serve the same NPA-NXX, this routing scheme can no longer be used.

Location Routing Number (LRN)

With LRN, every switch in the network is assigned a unique 10-digit number that is used to identify it to the rest of the network for call routing purposes. An essential advantage of the LRN is that call routing is performed based on today's numbering format. LRN uses the strength of SS7 signaling with *Multi-Frequency* (MF) interworking, and promises to be a longterm solution for LNP. LNP is SS7, *ISDN User Part* (ISUP) oriented, and therefore does not work well with MF interworking.

LRN-LNP on the *Switching Service Points* (SSPs) performs the integral part of the overall network LNP solution. The LRN is shown in Figure 5-2 with a new, ten-digit number assigned to the individual switching points. LRN depends on *Intelligent Network* (IN) or *Advanced Intelligent Network* (AIN) capabilities deployed by the wireline carriers' networks. LRN is a 10-digit number to uniquely identify a switch that has ported numbers. The LRN for a particular switch must be a native NPA-NXX assigned to the service provider for that switch.

LRN assigns a unique ten-digit telephone number to each switch in a defined geographic area. The LRN now serves as the network address. Carriers routing telephone calls to end users who have changed from one carrier to another (and kept their same number) perform a database dip to obtain the LRN corresponding to the dialed telephone number. The database dip is performed for all calls where the NPA-NXX of the called number has been flagged in the switch as a portable number. The carrier then routes the call to the new provider based on the LRN.

Shown in Figure 5-4 is the flow of LRN information, which is the same for both a wireline and a wireless provider (with one modification: the JIP is not used in the wireless network and LRN flow from wireless).

When the caller dials the number (333–3333 in this case), the originating switch (612–222) sends its signaling information (info dialed 333–3333) through the STP to the SCP, which analyzes the route and returns the LRN (612–444–0001). Next an IAM message is forwarded from 612–222 through the STP to the access tandem. The AT translates the LRN and sets up a speech path (trunk) from 222 to 444. Switch 612–444 detects the LRN (612–444–0001) as its address; therefore the called number and the generic address parameter are swapped. From there, the call is connected (terminated) at 333–3333. The donor switch in this scenario has been uninvolved throughout the process.

Similar to the 800-number service, a database is used to store the routing information for subscribers who have moved or ported to another Local Service Provider. The LNP database contains the directory numbers of all ported subscribers and the location routing number of the switch that serves them. The LNP database can be accessed by switches using either *Advanced Intelligent Network* (AIN 0.1) or *Transaction Control Application Part Intelligent Network* (TCAP IN) protocol. Each OLNP-capable switch in

Figure 5-4

The LRN in use

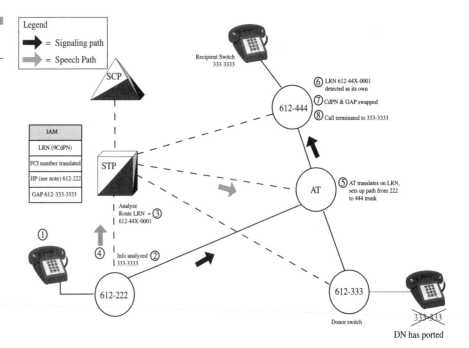

a portability area is assigned a unique ten-digit location routing number. A switch is defined as LNP-capable if it has the capability to launch LNP database queries and route calls based on the returned response. The ten-digit LRN is chosen from an NXX native to that switch (an NXX that was originally owned by that switch prior to Local Number Portability), and is used by other switches to route calls to that switch.

New routing and signaling methods are required to implement this feature because the LRN architecture requires the transport of the network routing address (or LRN of the terminating switch) and the called party number. Only the *Integrated Service User Part* (ISUP) signaling is being modified to carry the additional information needed to support local number portability. Workarounds to use *Multi-Frequency* (MF) trunks, such as sending only the dialed number when a MF route is used, have been included in the requirements.

The LRN feature uses the Location Routing Number returned from the LNP database to route the call and the Called Party Number to complete the call in the terminating switch. From a protocol perspective, the LRN of the end office serving the subscriber is placed into the *Called Party Number* field (CdPN) and the actual Called Party Number is populated in the new *Generic Address Parameter* (GAP) field. As the call traverses the network, all switches will use the LRN to route the call. When the call is delivered to the terminating switch, the terminating switch will compare the LRN received in the called party number field against its own LRN. If these numbers match, then the terminating switch will retrieve the called party number from the GAP parameter and complete the call to the subscriber.

Once a switch performs an LNP database query, that switch must set a new bit called the *Translated Called Number Indicator* (TCNI) in the *forward call indicator parameter* of the *Initial Address Message* (IAM). This will indicate to other switches in the call path that a query has already been performed. This new bit, and the corresponding call-processing logic, ensures that multiple queries will not be performed for the same call.

Using a Database Solution

Many carriers felt that the best way to implement LNP is to establish databases that contain the customer routing information necessary to route telephone calls to the appropriate terminating switches. The LNP database, similar to that already used by the telecommunications industry to provide 800-number portability, will use intelligent network or AIN capabilities.

These capabilities separate the customer-identification information from the call-routing information.

Rather than establishing one national database, similar to the 800-number database, carriers will provide LNP via a system of multiple, regional databases. A national LNP database is not feasible, simply because one database could not store the telephone numbers for the entire United States population or even the subset of the largest metropolitan areas.

Further, a regional database system offers specific advantages for carriers deploying LNP. Regional databases effectively reduce the distance over which carriers must transmit routing information. By minimizing that distance, a regional system reduces the associated routing costs incurred by the carriers. A regional database system also ensures that carriers will have all the number portability routing information they need to route calls between carriers' networks for that regional area.

Because many of the major carriers install their own SCP where the LNP database resides, a single access point must be provided to effectively manage and distribute updates to the common regional LNP database. These updates will provide all carriers with the changes made when end users "port" from one Local Service Provider to another. Most states have decided that a *Number Portability Administration Center* (NPAC), similar to the arrangement used for 800-number portability, should be used. The LNP task force in Illinois issued an RFP to the industry to solicit technical (and financial) proposals for the NPAC. Lockheed-Martin was chosen to develop and administer the NPAC in Illinois. Other companies were selected by different regions, but complications arose and Lockheed-Martin was selected to replace those who could not deliver. Management of the LNP database by a third party will ensure the security of all carriers' customer bases. The full scenario of a call process is shown in Figure 5-5, using the components of the network and the databases.

Triggering Mechanisms

Using local number portability requires *end office* switching systems to determine if a dialed NPA-NXX has been declared open for portability. The switching systems must set triggers on a portable NPA-NXX to cause, or "trigger," a query to the LNP database and retrieve the LRN of the dialed number. Most switching vendors are using AIN 0.1 triggers. AIN 0.1 triggers were defined by industry requirements well before the development of

Figure 5-5
LNP scenario

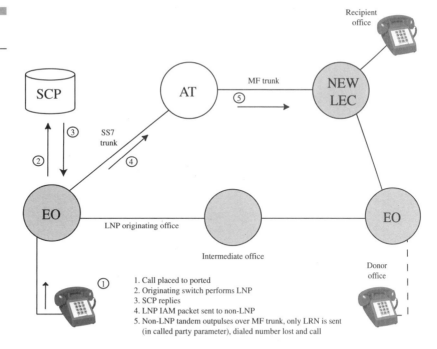

1. Call placed to ported
2. Originating switch performs LNP
3. SCP replies
4. LNP IAM packet sent to non-LNP
5. Non-LNP tandem outpulses over MF trunk, only LRN is sent
   (in called party parameter), dialed number lost and call

LNP. These triggers are administered on the NPA-NXX digit string by using the administration capabilities of the switching systems. Unconditional trigger mechanisms are used during the transition from one service provider to another, such as the following:

1. From a wireline to a wireless provider

2. From an ILEC to a CLEC

3. From a wireless to a wireline provider

These triggers can be assigned at the donor switch (the old provider) and the recipient switch (the new provider). By using an unconditional trigger, the trigger is used at the switch even if the directory number is present (see Figure 5-6).

The industry requirements for the location routing number model state that either a TCAP IN "800-like" or AIN 0.1 query can be used. The protocol defines the information and structure of the query between switching systems and the LNP database. Both of these protocols, defined within industry standards, are deployed in the network today. The AIN model is shown in Figure 5-7.

Figure 5-6
Unconditional trigger

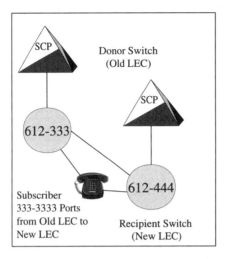

10 Digit Unconditional Trigger

Assigned to a line during the transition from the old to new service provider (LEC).

Can be assigned at both the Donor Switch (Old LEC) and the Recipient Switch (New LEC).

Unconditional trigger: triggers at switch even if DN is present, ESCDN ignored.

Triggers at Donor or Recipient ensuring no subscriber downtime or service outage during service provider transition.

Old LEC		New LEC	
Pre-SCP Update	Post-SCP Update	Pre-SCP Update	Post-SCP Update
Dialed number returned, terminate on this switch (Donor)	LRN returned, route to New LEC (Recipient)	Dialed number returned, route to (Donor) switch	Home LRN returned, terminate on this switch (Recipient)

Figure 5-7
The AIN model

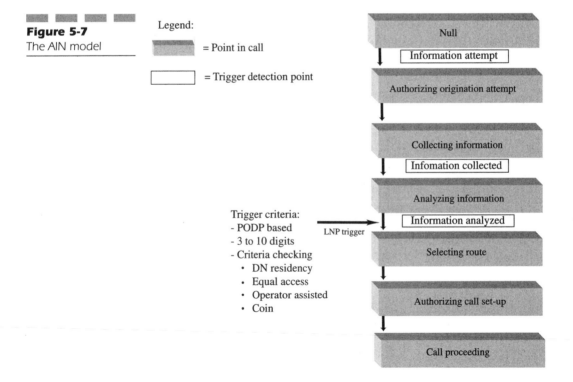

Legend:

= Point in call

= Trigger detection point

Trigger criteria:
- PODP based
- 3 to 10 digits
- Criteria checking
  • DN residency
  • Equal access
  • Operator assisted
  • Coin

LNP trigger

Null

Information attempt

Authorizing origination attempt

Collecting information

Infomation collected

Analyzing information

Information analyzed

Selecting route

Authorizing call set-up

Call proceeding

How Is a Telephone Number Ported?

Many steps are required to move or "port" an end user from one Local Service Provider to another. Database changes will be required in the following locations:

- Old Local Service Provider's central office
- New CLEC central office
- LNP database

Beyond that, the end user's copper pair must be disconnected from the old service provider's central office equipment and connected to the new Local Service Provider's central office equipment or some patch field in a co-located office. All of these activities are controlled by the NPAC.

Several important steps must proceed the actual porting of a subscriber:

1. The NPA-NXX must be opened for porting by the NPAC. This drives the update of the *Local Exchange Routing Guide* (LERG) database, global title translations in the STPs, and end-office translations (to set the LNP trigger for the NPA-NXX) for all Local Service Providers in the portability area.

2. The two service providers must arrange the physical transfer of the subscriber's copper pair. Both providers must have implemented the 10-digit unconditional trigger function in their respective end offices.

3. Once all the pre-move actions have been performed, a subscriber can be ported. The availability of the unconditional trigger in both the new and old Local Service Provider's end offices will preclude the need to synchronize every step of the porting process.

4. The donor switch administers the 10-digit unconditional trigger on the porting subscriber's directory number. This will cause an LNP query under all conditions (even when the subscriber is still served by the old service provider's end office), thereby eliminating the critical timing coordination between the donor and the recipient (new) Local Service Provider. This allows the activities at the donor switch to be performed autonomously with respect to the recipient service provider and route calls to the subscriber based on the LNP query response.

5. The recipient switch administers the translations for the new subscriber, and sets the 10-digit unconditional trigger. This allows

the recipient switch to provision the subscriber prior to the actual physical move of that subscriber's copper pair, without causing calls to be misrouted during that time period. As in the case of the donor switch, the 10-digit unconditional trigger also allows activities at the recipient switch to perform autonomously.

6. After the subscriber's copper wire has been moved, the new Local Service Provider notifies the NPAC of the change. The NPAC then downloads each LNP database (in the portability area) with the LRN of the new Local Service Provider (for the ported subscriber) and records the transaction (for example, date/time). After the new Local Service Provider successfully tests the new subscriber, the unconditional trigger is removed.

Other Issues

Below are but a few of the impacts that local number portability bring to the telecommunications network.

Switching Systems

Substantial changes to call-processing logic and administration software is required in all switching systems in use in today's telecommunications network in order to implement local number portability. The cost of the system upgrades and changes has been accumulated in a pool by the RBOCs and is now being passed on to consumers. The costs typically range from $.27-$.54 per line per month. This pass-through cost will continue for five years (ending in 2004) as the plan calls today. Additional impacts will be evident when a *Line Information Database* (LIDB) requires a *dip* to determine the 0 plus calling card model, as shown in Figure 5-8.

Billing, Administration, and Maintenance Systems

Because local number portability removes the direct association of a subscriber's directory number to a central office, substantial changes will be required in most of the systems in use today in local telephone company networks.

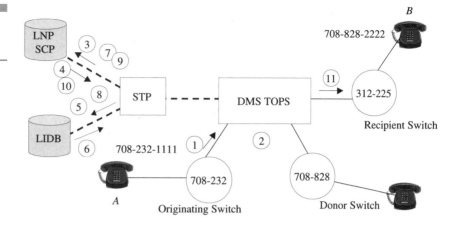

Figure 5-8
LIDB model for 0
plus calling card

Signaling

Local number portability will require an LNP database query for every call
to a ported subscriber that is not served by the originating switch. This will
require capacity increases in the number of *Signaling System #7* (SS7)
links to the signaling transfer points, and deployment of new service control
points to run the LNP database application. These costs have also been
accumulated, and they are now being passed through as part of the fee discussed previously.

Operator Services

Operator calls from subscribers (for example, 0-, 0+, and so on) are routed
directly to the operator services system where an LNP query must be performed to determine the Local Service Provider to which the call must be
routed. Thus, significant modifications are required in operator services
systems (see Figure 5-9).

911 Services

Maintaining the 911 and enhanced 911 databases create another impact for
the processing of calls to emergency response agencies. When we think
about LNP in a wireline environment, the process is straightforward. The
caller places an emergency call from a fixed location, which is easily identi-
fiable. Routing of the E−911 call is to the proper *Public Safety Answering*

Figure 5-9
Operator service
impacts

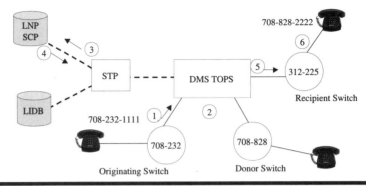

Call flow for 0-Call Model
1. Subscriber 708-232-1111 dials 0- and is connected to TOPS operator.
2. Operator receives request to connect to 708-828-2222 and bill to originator's station.
3. TOPS determines that requested number has been ported and sends an LNP query to LNP SCP.
4. LNP SCP responds with the LRN (312-225-0000) of Recipient Switch.
5. TOPS routes call to Recipients Switch.
6. Recipients Switch terminates call to ported number 708-828-2222.

Point (PSAP), based on a known, prearranged location for each telephone. ANI is mapped one-to-one for all wireline calls by using the appropriate callback number and a location from an *Automatic Location Information* database (ALI). This is shown in Figure 5-10.

Using CAMA trunks (trunks that were originally used for cost accounting and messaging call information), 911 routing is built on limited capability of the trunk. The CAMA trunks will remain in use for some time to come.

Is 911 Important? 80,000 calls per day are made to 911 emergency locations. This places a great burden on the network.

Figure 5-10
E911 calling
with LNP

Computer-to-Telephony Integration (CTI)

Introduction

Throughout the past two decades, business users have sought to find a means of integrating their telephony and computing systems. One can only imagine the great demand for this, assuming the computer and communications budgets are in the same domain. If one were to come up with a totally integrated package, there would be unimaginable savings (or so the story goes). It has been the goal of many organizations to reap the benefit of a single infrastructure.

In the past, the MIS department and the voice communications department were separate entities, yet their operating budgets and staff overlapped in certain areas. The degree of overlap varied by organization, but the overlap itself was the concern. Since the convergence began nearly a decade ago, the fundamental shifts in these operations were somewhat dramatic. These shifts were not explosive in themselves, but more in line with a slowly rolling wave. In many camps within the industry, however, these changes were virtually unnoticeable.

The driving forces behind this convergence combined with voice communications for a commodity-priced service, data communications for an open standard in the TCP/IP world, and the preponderance of LAN technologies overtaking the desktop. From the very beginning of these three disparate technologies, end users looked to the manufacturers to come up with an integrated scheme. It was through the initial offerings of AT&T, IBM, and DEC (albeit all proprietary) that the computer and communications convergence began. Past examples of the manufacturers' attempts to provide an integration scheme included *computer-to-phone integration* (CPI) from both IBM and AT&T and the *computer-integrated telephony* (CIT) from DEC. Unfortunately, the proprietary nature and the blocked architectures of the manufacturers met with disappointing results. Users were not willing to pay for another proprietary solution as they had in the past. Moreover, the nature of the evolving LANs was leaning towards a new set of proprietary solutions from companies like Novell and Banyan, far from the openness the industry was busy praising.

Today two technologies are revolutionizing the way the world communicates: telephones and computers. Telephones are everywhere. For decades, they have been prevalent in offices, public establishments, and homes. And computers are catching up. These two technologies historically have remained separate, however. Often the only thing they share in common has been the desk on which they both sit, but the computer telephony industry has been combining the best of telephone and computer technology to let people exchange information more quickly and easily.

The power driving the computer telephony industry is telephone network access to computer information through almost any convenient, easy-to-use, and available terminal device, including

- Telephones (pulse dial, touch-tone, and wireless)
- Facsimile (fax) machines
- Personal computers (PCs)

These terminal devices access a multi-user computer telephony platform that supports applications that process the information within the call and/or route and monitor the progress of the call. These functions of the computer telephony platform are an integral part of enterprise information systems.

Until recently, our two most widely used business tools, the telephone and the computer, were separated by many factors. For decades, the telephone has been our primary means of business communication. However, the majority of our necessary business data has been accumulated on mainframe, mini-, and personal computers. To make matters even more confusing, we have been bombarded with a multitude of alternative communication methods (such as fax, e-mail, and voice mail), each accessible in different ways from different places.

Computer telephony integration is the merging of the computer and telephone, which will transform the personal computer from being an information-processing device to being a powerful platform for communications. Computer telephony is the "art" of intelligently linking and combining these tools to create systems that enable us to use technology to our advantage. The goal, of course, is to dramatically increase the access we have to the information we need, when we need it. A variety of cost-effective solutions is available to businesses of all sizes.

Linking telecommunications to the processing capabilities and GUI of the computer enables new forms of communications and more robust access to existing types of communication, including

- Voice
- Asynchronous data
- Fax
- Remote access to LANs
- Internet access
- Online services

The communicating PC will redefine how we share ideas and information and will provide a portal to other people, computers, and network services anywhere in the world. Computer telephony integration will also transform the telephone into an information-processing device. By giving the telephone access and control of non-voice media, users will enjoy the convenience and economics of multimedia communications anywhere and at any time.

The Computer World

What had been the bastion of the big iron providers (mainframe and midrange computers) quickly eroded in the early 1980s to a desktop PC-based architecture. Surely, legacy systems, such as the departmental computers, became servers, and the mainframe for the entire organization remained as the mainframe (or became a major server) because of the investments and the nature of the data stored. One cannot just walk in and trash what has been in-place for years. Instead, the integration of the computer and the telephony world requires a slower process that includes a methodology to preserve the legacy data.

Over the years, much time and effort has been spent in trying to develop interfaces that would tie the computer and the telephone network together. As technology has changed, many new interfaces and applications can be created through the integration of these two techniques. Consequently, a whole new industry has emerged. The ability to link computer systems and voice systems together offers some new possibilities on how we approach the office. When one thinks of the *automatic call distribution* (ACD) systems with the capability to link the *automatic number identification* (ANI) and a database to provide screen-popping capabilities, it becomes an exciting opportunity.

Figure 6-1 is a representation of the *computer-to-telephony integration* (CTI) capability using a screen-popping service. In this case, as a call comes into the building, it is initially delivered to an ACD. At this point, the ACD captures the ANI of the calling party. When the called party's number is packaged into a small data packet, it is then sent to the computer as a *structured query language* (SQL) inquiry into a computer database. Assuming the database already exists for the established client, the SQL opens the client file and responds to the query. In this particular case, the telephone number of the calling party is being used to create the query from the database. Once the database entry is recognized based on telephone number, a computer screen will "pop" to the agent's desk. This occurs at the same time the telephone call is being delivered to the agent's desk. Now the agent

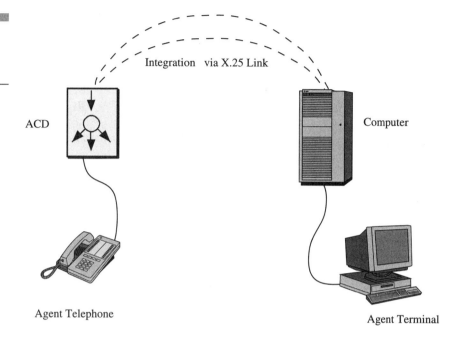

Figure 6-1
The CTI capability using a screen-popping service

Integration via X.25 Link

ACD

Computer

Agent Telephone

Agent Terminal

taking a telephone call does not have to ask the caller for all of the associated and applicable information, such as:

- Name
- Address
- Telephone number
- City and state
- Account number
- Any other pertinent information

By reducing the time spent gathering this information, because it is already available in the database, savings of 20 to 30 seconds of call-processing time can be achieved. However, the 20 to 30 seconds is not the critical part. Instead, the ability to satisfy the customer and provide better service levels is what is important.

When one thinks of other alternatives that might exist here, a couple of thought-provoking ideas come to mind. For example, assume that an agent has received a call and a screen of information from an established customer. In this regard, the established customer has saved time by not having to provide his or her name, telephone number, and so on. The agent gets right to the task at identifying and verifying the customer. Now that the agent has the information, historical files on such things as buying or usage patterns can be developed so that the agent can suggest products and services to the customer based on past buying experiences. Moreover, if there is a problem with a particular customer, such as delinquent payment, the agent can readily obtain this information while talking to the customer.

Let's extend that thought a little further. Assume that the agent encounters a delinquent customer who hasn't paid his or her bill for three months. While talking to this customer or taking an order, which is what the customer called for, the agent sees on the screen that the customer is delinquent and that collection action must be taken. Rather than becoming a collections manager, the agent can immediately suggest to the customer that the call must be transferred to the accounts receivable manager. When the agent transfers the call, not only does the call go to the accounts receivable department, but the screen from the database follows. When the accounts receivable department receives the call, the manager there will have a full screen of information about why that call has been sent to him or her. Answering the call, the accounts receivable manager can then immediately and proactively get to the job at hand, collecting the money. Instead of picking up the telephone and asking for information that has already been determined, totally frustrating the customer, the accounts receivable manager can immediately get to the problem at hand, collecting the money. When the call goes back to the agent, the screen with all of the notes that the accounts receivable manager has entered returns as well.

Again, the agent sees who is on the telephone and what has transpired. This prevents the replication of information gathering and keeps the entire conversation in a proactive mode. It is this positive handling of calls, particularly those of a unpleasant nature, that helps to boost customer service and influence the continued buying relationship.

The above scenario assumes that certain things have taken place. It must be assumed that the customer is calling from a known location and that the database already exists. If, in fact, the database does not exist, the initial data gathering can be accomplished at the receipt of the first call. Thereafter, any time the customer calls from the same telephone number, the database will be activated and brought up with the screen-pop. The first time is the most laborious. Each successive contact between the customer and the organization is much simpler. Using this database and screen-pop arrangement, companies have successfully achieved several things, including

- Improved customer service
- Improved customer satisfaction

- A reduced number of incoming lines
- Reduction in the number of agents necessary to handle the same volume of calls

Through the combined savings of 20 to 30 seconds for each successive attempt, the average hold time (talk time) can be reduced significantly.

Other Possibilities

"Screen popping" in an ACD is one of CTI's primary uses today, but other applications can take advantage of CTI. The newer integrated PBX will become a voice server on the LAN. In Figure 6-2, the integrated PBX is shown as a voice server residing between two separate *local area networks* (LANs).

Now that the various communications vendors have produced an open applications interface, the ability to use a PBX for several components exists. One of its primary applications is to install a third-party bridge or router card inside the PBX. The PBX now serves as the bridging function between the two LANs. Using the PBX also enables an extension of the PBX access to other LAN services. Inside the architecture of the PBX is a high-speed communications infrastructure such as ATM, SONET, or FDDI, depending on the direction chosen by the PBX manufacturer. The PBX

Figure 6-2
The integrated PBX as a voice server between two separate LANs

manufacturers have devised means of running 10 Mbps or 100 Mbps to the desktop. Now they are looking to implement an *Internet Protocol* (IP)-based card, creating a true data application in the telephony architectures. Newer applications will also include the gigabit Ethernet over twisted-pair wires to the desktop (1000 base T). When dealing with the high-speed communications to the desktop, the user will have unlimited access to voice features, LAN features, and multimedia applications running inside the organization. The combined investment in computer and telephony integration offers substantial opportunities for high-speed broadband communications to the desktop. As a true communications server, the PBX equipped with an IP card (router) inside can bring other services such as Internet or intranet access to the PBX.

Still other features can be included with this integration of CTI. Voice messaging, as integral or peripheral to the PBX, can now become available as a feature or a function hooked to the LAN. As an outside caller dials into an organization's PBX, the capability to capture the ANI and hook to the database server on the LAN allows for the delivery of the caller ID to the desktop. This makes it possible for a user who has a screen (PC or other) on his or her desk to be able to see the voice messages in the queue. In reality, what the user sees is not the message, but information regarding the message. This helps to prioritize, or select from a queue, calls that are necessary or important. Moreover, the end user can print out a list of voice messages, a customer database of people who have called, ANI information to be included in a database not yet built, or many other useful and productive types of data.

Beyond the voice messaging capability, connecting to an automated attendant can also be integrated into this CTI application. When an outside caller attempts to reach a particular extension or department, a database can be built regarding who the callers are and what extensions are typically used most frequently. This also extends ANI access to the desktop of the called party so that the database can also link to some form of a screen-pop. By delivering the ANI to a database server on the LAN, the user can immediately run a structured query on the database to pull up the customer file. A simpler way would be to have the database automatically deliver all the information about the calling party to the LAN terminal device. This doesn't work in all applications, but it does offer some unique capabilities and opportunities that never before existed.

Each of the discussed services can all be coupled in the combination of computer and telephony integration. CTI offers the capability to link multiple databases, multiple communications channel capabilities, and the

LAN-to-PBX server functions. The PBX becomes the voice server, whereas the LAN server becomes the PBX's database. Corporate directories, departmental lists, product lines, and other types of usable information can all be made available to the desktop device when a called or calling party needs access. Although the cost is incremental, the overall increases in productivity and satisfaction from both employees and customers may well be worth the effort. This is one of those decisions that a user must make.

In the same manner, you take higher risks by putting all of your eggs in one basket and having a single-server environment. However, the distributed computing architectures that are emerging minimize the risks. Distributed computing architecture can also add to the depth of the PBX by using servers on a LAN that can be voice servers; the functions and the real estate required for a PBX architecture can now be combined and reduced. A large room with special air conditioning is no longer required. The servers can just be deployed wherever they are needed to serve their users.

We can see the excitement and the benefits of CTI when we look beyond the basic functionality of each individual box. Hence, the decision-making process becomes a little more complicated than the purchase of a PBX or a server when thinking about acquiring a combined infrastructure to support the communications needs of an organization.

Why All the Hype?

A lot has already been said about what is going on within the industry and organizations today. Business has reached an intensity and a pace that is extraordinary. The competition and changes in technology are driving the pace exponentially. All of these factors have driven organizations to seek an improved means of dealing with their customers and to answer the needs for increasing productivity, decreasing costs, and enhancing the competitive edge over the existing market segment. Keeping pace with the industry and maintaining the competitive edge has become a crucial element in the survivability of an organization. Therefore, the demands on organizations have escalated dramatically to provide for

- The ability to deal with corporate decision-making processes at a rapid pace
- Readiness to deal with issues, regardless of the time of day, day of week, or availability of executives

- The continued restructuring and reengineering of the organization, which requires fewer people to perform more functions
- The changing workforce, in which talent and skill sets are no longer as available as they were, making technology a necessary solution
- Preparing the administrative assistants who have supplanted the administrative and support staff of old (secretarial support) to deal with issues when the senior staff is not available
- Ensuring the availability of the information that is needed to achieve these goals

Therefore, in order to effectively use the resources that remain within the organization, employees must have information readily available at their fingertips. Using technology of both computers and communications provides the edge to maintain the link between the employee and the customer. It is not necessarily a given that these technologies will always be the solutions, but the right implementation can have significant benefits if done correctly. Several examples in this book show how organizations have tied their computers and communications together through various network technologies to facilitate the timeliness and usability of information. Using a CTI system, employees have the ability to share information on the spot by using updated information on a call-by-call basis. Each of these tools and capabilities is an absolute must if a business is to remain competitive.

The reengineering of an organization must focus on serving the needs of their customers. Examples of this are clearly demonstrated in both the financial community and the airline industry. Customers can no longer be looked on as a nuisance; rather, they must be seen as the most valuable resource and asset that an organization has (excluding its own internal personnel). Providing the necessary treatment and care of the customer becomes one of the primary goals of the organization. It should have been the goal all along, but the focus on driving costs down and increasing productivity began to erode the customer and organizational relationships.

Now that CTI has begun its movement within the industry, it is much more feasible to restructure and reaffirm the relationship between the two parties. Customers expect to be treated as individuals who are special for the size and volume of the business that they do. Competition in any segment of industry is fierce. Therefore, if the necessary care and treatment is not provided, customer loyalty shifts dramatically. Changing market demands due to the rightsizing, downsizing, and capsizing in external organizations places more demand on your organization to meet the customer's needs. In planning our technological innovations, we must therefore look to how we can better satisfy the customer's needs by the following tasks:

- Giving our customers unlimited access to our employees to place orders, check on status, or make inquiries.

- Assuming that the employee/agent will have information readily available and not put the customer on interminable hold.

- Anticipating the demand for products based on sales projections and customer demands so that out-of-stock or back ordering can be curtailed.

- Having technical support available to the agent so that a question can be easily answered by either the technical or the administrative/sales person.

- Considering that the customer's success in business and/or product lines can be matched to your ability to meet the customer's demands for products and services.

- Anticipating customers' needs and satisfying their demands so they will not be exposed by today's diminishing inventories and just-in-time manufacturing processes.

If the above conditions can be satisfactorily met, the customer will feel better about placing orders or checking on the status and availability of products with your organization. This equates to better sales. Telecommunications and computer integration can therefore increase productivity and assist in increasing sales. Using the right technology in the right mix at the right time can enhance a relationship that is ever so fragile in today's competitive marketplace.

Linking Computers and Communications

The architecture of using a three-computer technology is the result of many years of evolution and revolution within the computing platforms. The LAN emerged in the early 1980s as a means of moving away from control by a single entity known as the MIS department. As this evolution began, users were empowered to manipulate their own data, handle their own information, and share that information with others on a selective or exclusive basis. Because of this movement, a lot of pressure was placed on the communications side of the business to provide interconnection using a cabling infrastructure to the various desktop computers and departments that would be interrelated. Management saw an opportunity to produce ad hoc reporting, allow flexible data manipulation, and provide more timely information to managers and departments alike.

As LANs emerged, they were primarily tied to a desktop and an individual server within a department. Later in its evolution, the client/server architecture started to emerge more strongly by allowing multiple departments within a single organization to fulfill their data communications and data access needs by accessing each other's databases or sharing the files they created with their customers. It is this sharing that created additional appliance affiliations within the different groups. The PC, as it attached to the network, enables users to store their applications and files on a single device. By linking these computing systems together with a communications medium, the user can best seek the information wherever it resides within the network. If a new application is needed, either a server can be adapted to fit that need or a new server can be added.

From a communications perspective, devices have emerged that include e-mail within departments, fax servers for transmitting information directly from a desktop to a customer, and voice recognition and response systems that enable customers to call in and literally talk to the server, as opposed to having to talk to a human. Furthermore, the linkage of all of these systems through a communications medium is the glue that binds everything together.

The PC of old, although structured as strictly a desktop device, has changed dramatically. Now, new features include the following

- Extensive processing power
- Sophisticated operating systems
- Integrated voice boards or sound cards
- Integrated video conferencing capabilities
- The client/server software that enables all of the pieces to be pulled together

With the use of CTI, an organization's needs can easily be met using the buildout of the desktop devices connected to the LAN. As already mentioned, the client/server architecture has the capability to bring together the computers that reside on the LAN and telephony services that can be built into either the PCs or the servers that reside on the LAN. This all becomes possible by bonding two technologies that began in the mid-1980s and now have matured in the mid-1990s. The architecture within an organization has dramatically changed. In place of a single provider of information, such as an administrative assistant or a particular sales group, there now exists an integrated homogenous workgroup that can share information, transfer calls among members, or add other departments on calls as needed. This has already been shown in our examples.

While this information process was taking place, organizations began to connect to the outside world and to their customers with their telephony services. In the early 1990s, a new explosion took place within the industry. The emergence of the World Wide Web and the use of the Internet as a commercial resource added to computer telephony development. Attaching an organization to the outside world through the Internet allowed customers to use their telecommunications services to enter the organization and check inventories, catalogs, and other services that would otherwise not normally be available to them. The ability to introduce fax-back servers also enabled customers to obtain catalog information or technical brochures immediately, instead of having to wait for an agent to mail products or brochures through the mail. This increased the availability of information by linking the communications' infrastructure to the computing architecture. Moreover, as organizations started to use *electronic document interchanges* (EDI), they created the ability to place orders, perform functions as dynamic workgroups, and facilitate the ordering and payment process. These services enhanced the true organizational capability.

The mid-1990s has also marked the innovation and growth of this computer and telephony integration with intranets. For now, think of it as the internal Internet to an organization. Using the intranet and communications infrastructures within an organization's client/server architectures, internal organizations can access the same information that might be available to a customer. Furthermore, internal organizations can share files, technical notes, and client notes regarding their customers as well as other valuable information that would not be readily available to the masses. Specific files or bulletin boards can be set up within the organization to achieve that result. Therefore, it is through the integration of our communications systems and the capabilities of the far more powerful and processor-intense desktop devices that the pieces are coming together quickly. No distinction can be made anymore between computing and communications, because the two really draw on each other's resources to facilitate the organization's day-to-day mission: to serve the customer and maximize shareholder wealth. Serving the customer increases sales, which hopefully decreases costs and raises profits. This maximizes the shareholder's wealth, which is the charter for all business organizations.

The Technology Advancement

Through the integration of computer and telephony-type services, the old philosophy of telephony being a necessary evil has passed. Advances in computing technology have enhanced the telephony world. The age of mass pools of telephone operators is gone. Newer technologies enable customers to call and proceed through an organization without ever communicating with an operator or a secretary. The customer can get directly to the person or information that is desired.

In the key system marketplace, the old labor-intensive electromechanical system is now a computer-driven telephone system. However, these computer-driven systems are now rich in features and laden with capabilities undreamed of in the past. The integration of voice messaging, ACDs, and automated attendants in key systems is now commonplace. Key systems are also emerging as voice servers within the organization.

The PBX, serving hundreds if not thousands of employees within an organization, enables an internetwork of services and capabilities to be spread throughout an organization. What was once just a stand-alone telephone system is now the high-end computer that just happens to handle voice. Today's technology provides full digital transmission systems or services and permits linkage to all devices that were exclusively used by the elite. PBXs can now be integrated tightly into your computing systems.

The PBX also acts as the high-end server on a digital trunk capability to the outside world. By linking to a high-speed digital communications trunk from the outside world, users can now access ANI. Using inbound 800/888/877 services and a common channel-signaling arrangement enables the delivery of caller ID directly to the called party. Moreover, enhancements in software (such as call forwarding), overflow arrangements when agents or other individuals are busy, and other services can all be chartered into a single PBX architecture. Another service called *directory number information service* (DNIS) can be used on the 800/888/877 service to direct the call to a specific group, such as technical support or marketing.

The Final Bond

As the client/server architecture and the communications systems were developing throughout the 1980s and into the 1990s, it was the software vendors who became the aggressors. With operating systems that could work on a LAN-based server, the software vendors were aggressive in finding the integration tools necessary for the organization. Microsoft developed a de facto protocol called the *telephony application program interface* (TAPI) in a Windows environment to bring the telephony services right to the desktop LAN-attached device. A TAPI interface on a Windows platform enables users to access information through what is called a *graphical user interface* (GUI, pronounced gooey). Throughout the entire architecture, a server application can handle the distribution of calls to members within workgroups or departments. This includes such services as screening calls, rerouting calls to new agent groups if the primary agent group is busy, or routing calls to the voice messaging and voice response systems as necessary.

Similarly, Novell and AT&T developed what is known as the *telephony services application programmers interface* (TSAPI) as a means of providing the computer with telephony integration capabilities from a LAN. TSAPI works with Novell's NetWare telephony services. The basis for using this particular product is to gain what is known as *third-party call control*. Third-party call control uses the CTI application on behalf of any clients in a workgroup or department. The application is running a shared environment, typically on a server, so there is no direct contact or connection between the user's PC and the telephone interface. What happens, then, is that a logical connection is produced when the PC applications talks to the server, which in turn controls the telephone switch. The server then is the controller and sends the order to the PBX to make calls or connections on behalf of the end user. The shared-server environment can handle individual as well as dynamic workgroup applications, such as directories, individual *personal information managers* (PIMs), and other workgroup functions that occur within a larger organization. Therefore, the server provides the linkage for all calls being handled within a dynamic workgroup. This is a more powerful arrangement regarding call control. A central server can handle the distribution of calls to any member in a workgroup and provide such services as call screening, call answering groups, backing up agents, or routing calls to a supervisory position in the event of overflow. Many of the CTI vendors have seen these capabilities as the benefits and strengths of the CTI applications.

These are the primary applications that have been used within the CTI environment. Thus, developers and manufacturers alike have seen the application working as a telephone set. This is instrumental in designing the capabilities of a call answering/call processing environment, but more can be done through the use of the computer-based technology that is available on the market today. The applications that run in server or computing platforms can use call monitoring features within the PBX and collect information of any type. Using the call-monitoring features, a CTI can watch every keystroke that an agent group enters. This can include such things as

- Dialed digits
- When agents "busy out"
- Answering a call
- How long the call is off hook

Additionally, by monitoring the trunk groups within the PBX, the CTI application can see all the incoming calls and collect the data associated with each call. This application uses ANI and DNIS to see where the call was directed and when it was answered.

The selectivity available in a CTI application therefore enables the supervisor to monitor the activity of each agent within a workgroup and get a clear picture of what transpires during the course of a day. This helps in determining workload effort, staffing requirements to meet a specific demand, or any seasonal adjustments that must be made based on time of year. About 65 to 70 management reports can be generated on an ad hoc basis. Supervisory personnel can monitor the work flow, the productivity of each agent in a group, abandoned calls that were not answered soon enough, or any other anomalies that might occur within a given day. Through these useful tools and reporting structures, the supervisors know whether they have enough, too few, or too many agents on board at any one time. This can aid in workflow scheduling as well as in determining the productivity of the individual agents or of the workgroup as a whole. In the event an agent is not performing satisfactorily, management can then take whatever corrective actions deemed necessary.

If abandoned calls are escalating because the agents are on the phone too long, several other activities might result. These could include such things as new training, analysis of the call type and the information requested, or simply determining the morale and productivity within the workgroup.

Beyond the call-processing and the call-monitoring capabilities, CTI can be implemented to integrate the facilities and capabilities of the PBX as well as the computer systems into one homogenous unit. The typical PBX today has features that are rarely used. There can be as many as 300 to 400 of these features that are designed to either improve productivity or make the job simpler. Although the average user typically activates only three or four of the normal features, many of the functions are available but never used. The CTI applications can customize features and functions for individual users within a workgroup and provide more powerful interfaces using point-and-click GUIs. Through a CTI application, a PC can be used for dialing, activating features, conference calling, call transfers, or any other necessary feature. Simply clicking on a telephone set icon saves the end user from the risk and the unfamiliarity of the PBX features, preventing cutoffs or lost calls and facilitating better utilization of the PBX features.

Taking this one step further, each individual user within a group supported by a CTI application can customize his or her own features and functions in personal folders. These can either be stored on a PC or a server. Therefore, using the CTI application, individuals can select different forms of call screening, call forwarding, or call answering, according to the applications and the individualized services they prefer.

Signaling
System 7 (SS7)

Introduction

The ability of a caller to go off-hook in a telephone world, dial digits, and then miraculously talk to someone anywhere in the world is still a mystique to many. The network's capability to set the call up almost instantly and then tear it down just as fast is what really carries the mystique. How can the network figure out where to send the call, get the connection, and ring the phone on the other end so quickly? All of this happens in under a second and the user is oblivious as to the intricacies of what occurs. What happens behind the scenes constitutes the backbone of the signaling systems. The networks are now dependent on the capability of handling subsecond call set-ups and teardowns.

Several signaling systems have been introduced to the telecommunications networks. The current one in use is called *Signaling System Number 7* (or SS7) in North America. In the rest of the world, this is referred to as *CCITT Common Channel Interoffice Signaling System Number 7* (CCS7 for short). Although the names are different, the functions and the purposes of the two systems are the same. As always, the North Americans do things one way, and the rest of the world does things a different way. This is an age-old problem, but one that we have learned to deal with and adjust to.

The essence of the signaling system boils down to many different factors, but one of the most significant reasons the carriers employ these systems is to save time and money on the network. Following that fact, the carriers are also interested in introducing new features and functions of an intelligent network, as discussed in an earlier chapter. The best signaling systems are designed to facilitate this intelligence in the network nodes that are designated as signaling devices, separate and distinct from the switching systems that carry the conversations.

Pre-Signaling System 7

Prior to the implementation of SS7, *per-trunk signaling* (PTS) was used exclusively in the networks. This method was used for setting up calls between the telephone companies' exchanges. PTS continues to be used in some parts of the world where SS7 has not yet been implemented. Admittedly, the exchanges using the PTS method are declining, as SS7 is gaining in its deployment worldwide. However, the network is always in a state of change and this is no exception. PTS sends tones or *multiple frequencies* (MF), as they are called, to identify the digits of the called party. The trunk also provides all of the intelligence for monitoring and supervision (call seizure, hang up, and answer back) of the call. Telephone systems at the customer's location (PBXs) that are not *Integrated Services Digital Networks* (ISDN) *Primary Rate Interface* (PRI)-compatible use the per-trunk signaling method.

On a long distance call when a call set-up is necessary, each leg of the call repeats the MF call set-up procedure until the last exchange in the loop is reached. In essence, the call is being built by the signaling as the progress is occurring on a link-by-link basis. As each link is added to the connection, the network is building the entire circuit across town or across the country. Each leg of the call set-up takes approximately two to four seconds, with the configuration shown in Figure 7-1, or a total call set-up takes approximately six to 12 seconds (at a minimum) from end to end.

This method works but is an inefficient use of the circuitry in both major and minor networks. Although the call gets to its end destination, several complications could arise, causing extensive delay or incomplete calls. Regardless of the complications, the outcome is the same; the carrier ties up the network and never completes the call. Hence, no revenue is generated for the use of the circuits or the network. This inefficient use of the network costs the carriers a significant amount of money. Therefore, something has to be done to improve this method of call establishment. The call establishment part of the connection could take as much as 24 seconds, then time out, and never get to its end point. However, the carrier ties up parts of the network without getting a completion. This is no big deal when discussing one call, but when a network carries hundreds of millions of calls per day, this accumulated lost time is extensive and expensive.

Figure 7-1

Per-trunk signaling preceded SS7 but was slow

Introduction to SS7

The ITU-TS (once called the CCITT) developed a digital signaling standard in the mid-1960s called *Signaling System Number 6* (SS6) that would revolutionize the industry. Based on a proprietary, high-speed data communications network, SS6 later evolved to SS7. SS7 has now become the signaling standard for the world.

The success of the signaling standards lies in the message structure of the protocol and the network topologies. The protocol uses messages much like the X.25 and other message-based protocols to request services from other entities on the network. The messages travel from one network element to another, independent of the actual voice and data that they pertain to, in an envelope called a *packet*.

The first development of the SS6 in North America was used in the United States on a 2400-bit-per-second data link. Later these links were upgraded to 4800 bits per second. Messages were sent in the form of data packets and could be used to request connections on voice trunks between Central Offices, placing 12 signal units (of 28 bits each) assembled packets into a data block. This is similar to the methods used today in SS7 architectures.

Although SS6 used a fixed-length signal unit, SS7 uses signal units with varying lengths. Additionally, the later version of SS7 uses a 56 Kbps data link. 56 Kbps are used throughout North America, whereas in the rest of the world, SS7 runs at 64 Kbps. The difference in the speeds between 56 and 64 Kbps results in the fact that the Local Exchange Carriers have not yet fully deployed the use of B8ZS on the digital circuits (For a discussion of B8ZS, see Chapter 26, "The T Carrier Systems (T1/T2 and T3")). Consequently, the 56 Kbps is an anomaly in the SS7 networks. Further, SS6 was still being installed by the North American carriers up through the mid-1980s (even though it was invented in the 1960s), while the SS7 deployment began in 1983, leaving two separate signaling systems in use throughout North America.

SS6 networks are slow, whereas SS7 is much faster. 64 Kbps and the possible use of a full DS-1 (1.544 Mbps) is still being considered in the North American marketplace. This is an evolutionary service that is continually being modified.

Purpose of the SS7 Network

The primary purpose of SS7 was to access remote databases to look up and translate information from 800 and 900 number calls (now the addition of 888 and 877 area codes are included). There were several benefits to using this lookup process, such as that carriers do not have to maintain a full database at each switching node but know how to get to the remote database and find the information quickly. The second purpose of the SS7 network and protocols was to marry the various stored program controlled systems throughout the network. This enables the quick and efficient call set-up and teardown across the network in one second. Moreover, this integration provides for better supervision, monitoring, and billing systems integration. Additional benefits of the SS7 network were geared to replacing the SS6 network, which, as of today, is well over 30 years old. Like anything else, the networks have served us well but need upgrading on a regular basis due to technology changes and demands for faster, more reliable services.

SS7 networks enable the introduction of additional features and capabilities into the network. This makes it attractive to the carriers so they can generate new revenues from the added features. SS7 also enables the full use of the channel for the talk path, because the signaling is done out-of-band on its own separate channel. This is more efficient in the call set-up and teardown process.

What Is Out-of-Band Signaling?

Out-of-band signaling is signaling that does not take place in the same path as the conversation. We are used to thinking of signaling as being in-band. We hear dial tone, dial digits, and hear ringing over the same channel on the same pair of wires. When the call connects, we talk over the same path that was used for the signaling. Traditional telephony used to work this way as well. The signals that set up a call between one switch and another always took place over the same trunk that would eventually carry the call.

In early days, the out-of-band signaling was used in the 4 kHz voice grade channel (see Figure 7-2). The telephone companies used band pass filters on their wiring to contain the voice conversation within the 4 kHz channel. The band pass filters were placed at 300 Hz (the low pass) and at 3,300 Hz (the high pass). The range of frequencies above the actual filter was 700 Hz (4000

Figure 7-2
Out-of-band
signaling used the
high frequencies

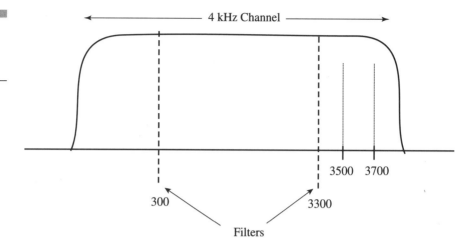

– 3300 = 700). In this additional spectrum, in-band signaling was sent down the wires outside the frequencies used for conversation. Actually, the signals were sent across the 3,500 and 3,700-Hz frequencies. Although these worked and were not in the talk path (out of the band), they were limited to the number of tones that could be sent. The result was also a limit to the states that could be represented by the tones.

Out-of-band signaling has evolved to a separate digital channel for the exchange of signaling information. This channel is called a *signaling link*. Signaling links are used to carry all the necessary signaling messages between nodes. Thus, when a call is placed, the dialed digits, trunk selected, and other pertinent information are sent between switches using signaling links, rather than the trunks that will ultimately carry the conversation.

It is interesting to note that while SS7 is only used for signaling between network elements, the ISDN D channel extends the concept of out-of-band signaling to the interface between the subscriber and the switch. With ISDN service, signaling that must be conveyed between the user station and the local switch is carried on a separate digital channel called the D channel. The voice or data that comprise the call is carried on the B channel. In reality, the out-of-band signaling is "virtual" because the signaling information is actually running on the same path as the B channels. The signaling and the conversational data flows are separated by time slots on the same physical paths. Therefore, it is virtually out-of-band, while it is physically in the same bandwidth.

Why Out-of-Band Signaling?

Out-of-band signaling has several advantages that make it more desirable than traditional in-band signaling:

- It allows for the transport of more data at higher speeds (56 Kbps can carry data much faster than MF out-pulsing).
- It allows for signaling at any time in the entire duration of the call, not only at the beginning.
- It enables signaling to network elements without a direct trunk connection.

The SS7 Network Architecture

If signaling is to be carried on a different path than the voice and data traffic it supports, then what should that path look like?

The simplest design would be to allocate one of the paths between each interconnected pair of switches as the signaling link. Subject to capacity constraints, all signaling traffic between the two switches could traverse this link. This type of signaling is known as *associated signaling*. Instead of using the talk path for signaling information, the new architecture includes the connection from the *Signal Switching Point* (SSP) to a device called the *Signal Transfer Point* (STP). It is then the responsibility of the STP to provide the necessary signaling information through the network to effect the call set-up.

When necessary, the STP sends information to the *Signal Control Point* (SCP) for translation or database information on the routing of the call. The pieces that form the architecture of the SS7 network are described in Table 7-1 and are shown in Figure 7-3 with the connection of the overall components.

The attached drawing shows a typical interconnection of an SS7 network. Several points should be noted:

- Paired STPs perform identical functions and are redundant. Together they are referred to as a mated pair of STPs.
- Each SSP has two links (or sets of links), one to each STP of a mated pair. All SS7 signaling to the rest of the world is sent out over these links. Because the STPs are redundant, messages sent over either link (to either STP) will be treated equivalently.

Table 7-1

Components of the
SS7 Networks

Component	Function
Signal Switching Points (SSP)	SSPs are the telephone switches (end offices and tandems) equipped with SS7-capable software and terminating signaling links. They generally originate, terminate, or switch calls.
Signal Transfer Points (STP)	STPs are the packet switches of the SS7 network. They receive and route incoming signaling messages toward the proper destination. They also perform specialized routing functions.
Signal Control Points (SCP)	SCPs are the databases that provide information necessary for advanced call processing capabilities.

Figure 7-3

SS7 architectural
beginnings

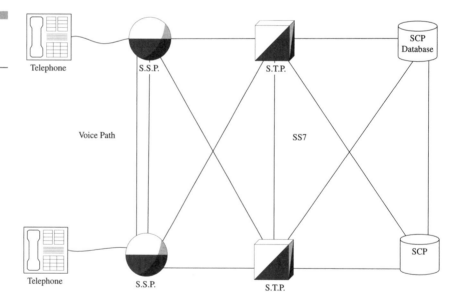

- A link (or set of links) joins the STPs of a mated pair.
- Four links (or sets of links) interconnect two mated pairs of STPs. These links are referred to as a quad.

- SCPs are usually (though not always) deployed in pairs. As with STPs, the SCPs of a pair are intended to function identically. Pairs of SCPs are also referred to as mated pairs of SCPs. Note that a pair of links does not directly join them.

- Signaling architectures such as these that provide indirect signaling paths between network elements are referred to as providing *quasi-associated signaling*.

SS7 Interconnection

The actual linkage enables the local exchange offices to send the necessary information out of band across the signaling links. SS7 therefore uses messages in the form of packets to signal across the network through the STPs. This enables the full use of the talk path for information exchange and the messaging paths for informational dialogue between the switching systems and the transfer points.

The links are used to pass control and billing information, network management information, and other control functions as necessary without interfering with the conversational path.

Basic Functions of the SS7 Network

The basic functions of the SS7 network include some of the following tasks:

- The exchange of circuit-related information between the switching points along the network

- The exchange of non-circuit-related information between the databases and the control points within the network

- The facilitation of features and functions by marrying the stored program control systems together throughout the network into a homogenous network environment

Further, the SS7 network enables these features to be put into place without unduly burdening the actual network call path arrangements. SS7 also accomplishes the following tasks:

- It handles the rerouting of network traffic in the event of circuit failures by using automatic protection-switching services, such as those found in SONET or Alternate Routing information.

- Because it is a packet-switching concept, the SS7 network prevents misrouted calls, the duplication of call requests, and lost packets (requests for service).

- It enables the full use of out-of-band signaling using the ITU Q.931 signaling arrangements for call set-up and teardown.

- It enables for growth so that new features and functions can be introduced to the network without major disruptions.

Signaling Links

SS7 signaling links are characterized according to their use in the signaling network. Virtually all links are identical in that they are 56 Kbps or 64 Kbps bidirectional data links that support the same lower layers of the protocol; what is different is their use within a signaling network. The bidirectional nature of these links enables traffic to pass in both directions between signaling points. Three basic forms of signaling links exist, although they are physically the same. They all use the 56 Kbps DS0A in North America and 64 Kbps DS0C data facilities in nearly every other portion of the world (except Japan where they still use a 4.8-Kbps link). The three forms of signaling links are as follows:

- *Associated signaling links*:   This is the simplest form of signaling link, shown in Figure 7-4. In associated signaling, the link is directly parallel from the end office with the voice path for which it is providing the signaling information. This is not an ideal situation, because it requires a signaling link from the end office to every other end office in the network. Some associated modes of signaling are in use, but they are few and far between.

  The most associated signaling is deployed at the end user location, using a single T1 and common channel signaling. Channel number 24 on a T1 is the associated out-of-band signaling channel for the preceding 23 talk channels.

  In some cases, it may be better to directly connect two SSPs together via a single link. All related SS7 messages to circuits connecting the two exchanges are sent through this link. A connection is still provided to the home STP using other links to support all other SS7 traffic.

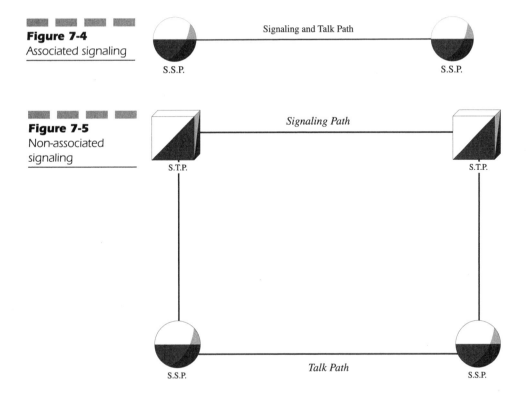

Figure 7-4
Associated signaling

Signaling and Talk Path

S.S.P.   S.S.P.

Figure 7-5
Non-associated
signaling

Signaling Path

S.T.P.   S.T.P.

Talk Path

S.S.P.   S.S.P.

■ *Non-associated signaling links*: In this signaling link arrangement, there is a separate logical path from the actual voice path, as shown in Figure 7-5. Usually, multiple nodes reach the final end destination, while the voice may have a direct path to the final destination. Non-associated signaling is a common occurrence in many SS7 networks.

The primary problem with this form of signaling is the number of signaling nodes that the call must use to progress through the network. The more nodes used, the more processing and delay that can occur. Non-associated signaling involves the use of STPs to reach the remote exchange. To establish a trunk connection between the two exchanges, a signaling message will be sent via SS7 and STPs to the adjacent exchange.

■ *Quasi-associated signaling links*: In quasi-associated signaling, a minimum number of nodes is used to process the call to the final destination, as shown in Figure 7-6. This is the preferred method of setting up and using an SS7 backbone, because each node introduces additional delay in signaling delivery. By eliminating some of the processors on the set-up, the delay can be minimized.

Figure 7–6
Quasi-associated
signaling

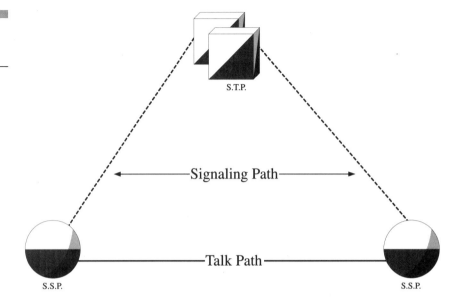

SS7 networks favor the use of quasi-associated signaling. In quasi-associated signaling, both nodes are connected to the same STP. The signaling path is still through the STP to the adjacent SSP.

The Link Architecture

Signaling links are logically organized by link type (A through F), according to their use in the SS7 signaling network. These are shown in Figure 7-7 with the full linkage in place.

- *A Link*:   An *access* (A) link connects a signaling end point (SCP or SSP) to a STP. Only messages originating from or destined to reach the signaling end point are transmitted on an A link.

- *B Link*:   A *bridge* (B) link connects one STP to another STP. Typically, a quad of B links interconnects peer (or primary) STPs (the STPs from one network to the STPs of another network). The distinction between a B link and a D link is rather arbitrary. For this reason, such links may be referred to as B/D links.

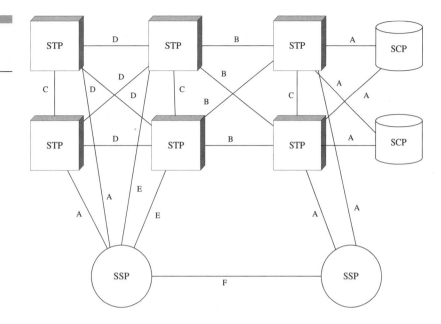

Figure 7-7
The signaling link architecture

- *C Link*: A *cross* (C) link connects STPs performing identical functions into a mated pair. A C link is used only when an STP has no other route available to a destination signaling point due to link failure(s). Note that SCPs can also be deployed in pairs to improve reliability. Unlike STPs, however, signaling links do not interconnect mated SCPs.

- *D Link*: A *diagonal* (D) link connects a secondary (local or regional) STP pair to a primary (inter-network gateway) STP pair in a quad-link configuration. Secondary STPs within the same network are connected via a quad of D links. The distinction between a B link and a D link is rather arbitrary. For this reason, such links may be referred to as B/D links.

- *E Link*: An *extended* (E) link connects an SSP to an alternate STP. E links provide an alternate signaling path if a SSP's "home" STP cannot be reached via an A link. E links are not usually provisioned unless the benefit of a marginally higher degree of reliability justifies the added expense.

- *F Link*: An *fully associated* (F) link connects two signaling end points (SSPs and SCPs). F links are not usually used in networks with STPs. In networks without STPs, F links directly connect signaling points.

Links and Linksets

A *linkset* is a grouping of links joining the same two nodes. A minimum of one link to a maximum of 16 links can make up the linkset. Normally, SSPs have one or two links connecting to their STPs based on normal capacity and traffic requirements. This constitutes a one- or two-link linkset. SCPs have many more links in their linksets to handle the large amount of messaging for 800/888 numbers, 900 numbers, calling cards, and *Advanced Intelligent Network* (AIN) services.

Combined Linksets

A *combined linkset* is a term that defines routing from a SSP or SCP toward the related STP where two linksets share the traffic outward to the STP and beyond. The requirement is not that all linksets are the same size, but the normal practice is to have equally sized groupings of linksets connecting the same end node. Using a linkset arrangement, the normal number of links associated with a linkset is shown below in Table 7-2.

Linksets are a grouping of links between two points on the SS7 network. All links in a linkset must have the same adjacent node in order to be classified as part of a linkset. The switches in the network alternate traffic across the various links to be sure that the links are always available. This load spreading (or balancing) serves many functions and can help you

- Be aware when a link fails.
- Recognize when congestion is occurring in the network.
- Use the links when traffic is not critical to know when a link is down before it becomes critical (Figure 7-8).

Table 7-2

The Configuration of Linksets

Type Link	Number of links
A Links	Maximum of 16 links
B/D Links	Installed in quads up to a maximum of eight links
C Links	Installed individually up to a maximum of eight links

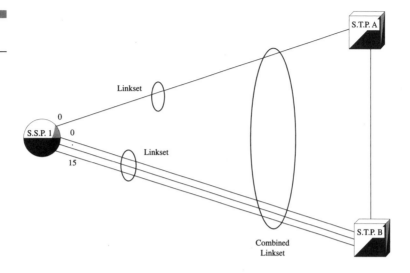

Figure 7-8
Linksets combined

Routes and Routesets

The term *routeset* refers to the routing capability of addressing a node within the SS7 network. Every node within the network has a unique address that is referred to as a *point code*. The addressing scheme or point code is the major routing characteristic of the CCS7 (SS7) network. The terms routeset and point code are somewhat synonymous.

The point code is made up of nine digits broken down into three three-digit sequences. An example of this is 245-100-000. Reading the point code from left to right, we find that

- The first three digits refer to the network identifier (245).
- The next three digits refer to the cluster number (100).
- The final three digits refers to the member number (000).

In any given network, there can be 256 clusters with 256 members. The network number in this case is for Stentor Communications in Canada.

The routing of SS7 messages to a destination point code can take different paths or routes. From the SSP perspective, there are only two ways out from the node, one toward each of its mated STPs. From that point on, the STPs decide which routes are appropriate, based on the time, resources, and status of the network. From the SSP, various originating and terminating (destination) addressing scenarios are defined as follows:

- If the route chosen is a direct path using a directly connected link (SSP1-STPA), then the route is classified as an *associated route*.

■ If the route is not directly connected via links (SSP1-SSP2), the route is classified as a *quasi route*.

All routing is controlled by nodal translations, providing flexible and network specific routing arrangements. This is shown in Figure 7-9.

SS7 Protocol Stack

The SS7 uses a four-layer protocol stack that equates to the seven-layered OSI model (see Figure 7-10). These protocols provide different services, depending on the use of the signaling network. The layers constitute a two-part functionality; the bottom three layers are considered the communications transmission of the messages, whereas the upper portion of the stack performs the data process function.

Figure 7-9
Routes and routesets

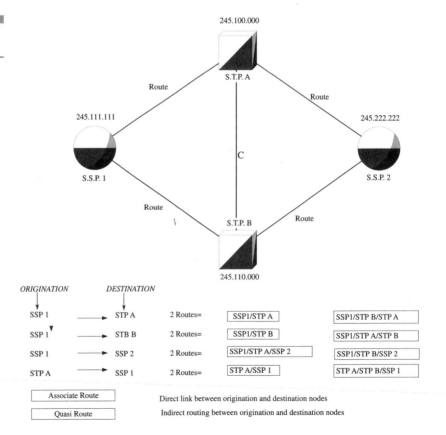

ORIGINATION		DESTINATION				
SSP 1	→	STP A	2 Routes=	SSP1/STP A		SSP1/STP B/STP A
SSP 1	→	STB B	2 Routes=	SSP1/STP B		SSP1/STP A/STP B
SSP 1	→	SSP 2	2 Routes=	SSP1/STP A/SSP 2		SSP1/STP B/SSP 2
STP A	→	SSP 1	2 Routes=	STP A/SSP 1		STP A/STP B/SSP 1

Associate Route	Direct link between origination and destination nodes
Quasi Route	Indirect routing between origination and destination nodes

Figure 7-10
SS7 protocols

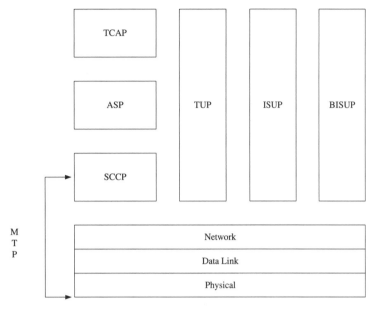

The stack shows that the bottom three layers make up the *Message Transfer Part* (MTP) similar to the X.25 network function. At one time, the SS7 messages were all carried on X.25. Now newer implementations use SS7 protocols, yet in older networks or third-world countries the X.25 may still be the transmission system in use.

The SCCP is used as part of the MTP when necessary to support access into a database and occasionally for the ISDN User Part. This extra link is the equivalent of the transport layer of the OSI model supporting the TCAP.

The SS7 network is an interconnected set of network elements that is used to exchange messages in support of telecommunications functions. The SS7 protocol is designed to both facilitate these functions and to maintain the network over which they are provided. Like most modern protocols, the SS7 protocol is layered. Functionally, the SS7 protocol stack can be compared to the *Open Systems Interconnect Reference Model* (OSI). Although OSI is a seven-layered stack designed to perform several communications and transparent functions, the SS7 protocol stack is similar but different.

Like any other stack, the SS7 protocol stack is specifically designed for the reliable data transfer between different signaling elements on the network. The guaranteed delivery and the prevention of duplication or lost packets are crucial to network operations. To satisfy differing functions, the stack uses various protocols on the upper layers, but consistently use the same lower layers.

Basic Call Set-up with ISUP

The important part of the protocols is the call set-up and teardown. This next example is shown in Figure 7-11 and is described below.

When a call is placed to an out-of-switch number, the originating SSP transmits an ISUP *initial address message* (IAM) to reserve an idle trunk circuit from the originating switch to the destination switch (1a). The IAM includes the originating point code, destination point code, circuit identification code dialed digits, and, optionally, the calling party numbers and name. In the following example, the IAM is routed via the home STP of the originating switch to the destination switch (1b). Note that the same signaling link(s) are used for the duration of the call unless a link failure condition forces a switch to use an alternate signaling link.

The destination switch examines the dialed number, determines that it serves the called party, and that the line is available for ringing. The destination switch transmits an ISUP *address complete message* (ACM) to the originating switch (2a) (via its home STP) to indicate that the remote end of the trunk circuit has been reserved. The destination switch rings the called party line and sends a ringing tone over the trunk to the originating switch. The STP routes the ACM to the originating switch (2b), which con-

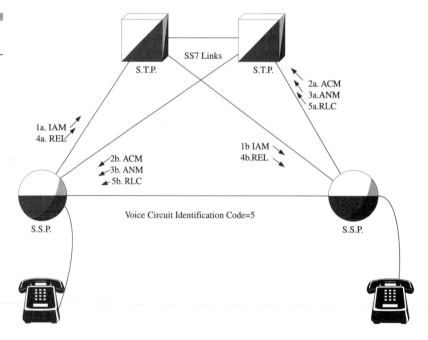

Figure 7-11
Call setup with ISUP

nects the calling party's line to the trunk to complete the voice circuit from the calling party to the called party. The calling party hears the ringing tone on the voice trunk.

In the previous example, the originating and destination switches are directly connected with trunks. If the originating and destination switches are not directly connected with trunks, the originating switch transmits an IAM to reserve a trunk circuit to an intermediate switch. The intermediate switch sends an ACM to acknowledge the circuit reservation request and then transmits an IAM to reserve a trunk circuit to another switch. This process continues until all the trunks that are required to complete the connection from the originating switch to the destination switch are reserved.

When the called party picks up the phone, the destination switch terminates the ringing tone and transmits an ISUP *answer message* (ANM) to the originating switch via its home STP (3a). The STP routes the ANM to the originating switch (3b), which verifies that the calling party's line is connected to the reserved trunk and, if so, initiates billing.

If the calling party hangs up first, the originating switch sends an ISUP *release message* (REL) to release the trunk circuit between the switches (4a). The STP routes the REL to the destination switch (4b). If the called party hangs up first or if the line is busy, the destination switch sends an REL to the originating switch indicating the release cause (such as a normal release or a busy signal).

Upon receiving the REL, the destination switch disconnects the trunk from the called party's line, sets the trunk state to idle, and transmits an ISUP *release complete message* (RLC) to the originating switch (5a) to acknowledge the release of the remote end of the trunk circuit. When the originating switch receives (or generates) the RLC (5b), it terminates the billing cycle and sets the trunk state to idle in preparation for the next call. ISUP messages can also be transmitted during the connection phase of the call, such as between the ISUP ANM and REL messages.

SS7 Applications

At this point, we switch gears and look at some of the applications that are possible because of SS7 implementations. The use of AIN features, ISDN features, and wireless capabilities all are a reality as a result of the functions of SS7 integration. Some of the features are listed here, but remember they are formulated as a result of SS7, even though they may be part of other systems or concepts. These include the following

- 800/888/900 services
- Enhances in 800/888 services within call centers
- 911 enhancements
- Class features
- Calling card toll fraud prevention
- Credit card approvals and authentication
- Software/virtual defined private networks
- Call trace
- Call-blocking features

CTI
Technologies
and
Applications

Introduction

Computer telephony systems ;telephony systems;CTI-basedcan range from simple voice mail to multimedia gateways. The equipment used in these systems includes *Voice Response Units* (VRUs), fax servers, speech recognition and voice recognition hardware, and intelligent peripherals deployed by telephone companies and service bureaus.

Today's businesses need to leverage the power of these diverse, multi-user computer telephony systems to improve productivity, give users more access to information, and provide communications options and services to both customers and employees. Business customers can use the telephone to automatically receive information about a product through a fax machine, while employees can access computer-managed voice, fax, and even data through telephones and computers to connect offsite workers to the office and expand relationships with outside enterprises.

The computer telephony industry offers the power of sophisticated telephone systems to any size businesses the same way the PC industry exploded in the 1980s. In just over 10 years, the computer telephony industry has grown to encompass many diverse applications and technologies. In 1999 alone, analysts estimate the revenue from multi-user computer telephony applications, development toolkits, and services and technologies to be $10 billion worldwide.

Understanding Computer Telephony Technologies

Many manufacturers and VARs are committed to providing technologies and products to customers for achieving success with automated call-processing applications. These products take advantage of technologies and enable users to store, retrieve, and manipulate computer-based information over a telephone network.

Voice Processing

Voice is the fundamental technology at the core of most computer telephony systems. It encompasses both the processing and the manipulation of audio signals in a computer telephony system. Common tasks include filtering, analyzing, recording, digitizing, compressing, storing, expanding, and replaying signals.

Telephone Network Interfaces

Network interfaces enable computer telephony systems to communicate electrically within specific telephone networks. Calls arriving from the public-switched telephone network can be carried on a variety of lines, including the following:

- Analog loop start
- Analog ground start
- Direct Inward Dial (wink start) lines
- T-1/T3
- E-1/E3
- ISDN PRI and BRI lines

Network interfaces interpret signaling, provide data buffering, and include surge protection circuitry.

Tone Processing

Tone processing includes the capability to receive, recognize, and generate specific telephone and network tones. This facilitates an application placing a call and monitoring its progress. Tones that are processed include

- Busy tones
- Special information tones (SIT)
- No answer (RNA)
- Connection
- Ringing
- No ringing
- Dial tones
- Fax tones
- Modem tones

Facsimile (Fax)

Facsimile lets you transmit copies of documents and images over telephone lines to another location. To transmit and receive electronic faxes, PC-based systems use fax boards. This computer-based fax technology can improve productivity by enabling documents to be sent through a broadcast fax to a large number of people in a short period of time. People can also retrieve any number of documents on demand (FaxBack) that reside on a fax server.

Automatic Speech Recognition (ASR)

Automatic speech recognition (ASR) technology (also known as voice recognition) reliably recognizes certain human speech, such as discrete numbers and short commands, or continuous strings of numbers, like a credit card number. ASR can be divided into two groups:

- Speaker-independent ASR, which can recognize a limited group of words (usually numbers and short commands) from any caller.
- Speaker-dependent ASR, which can identify a large vocabulary of commands from a specific speaker. This is popular in password-controlled systems and hands-free work environments.

Text-to-Speech (TTS)

Text-to-speech (TTS) generates synthetic speech from text stored in computer files. TTS provides a spoken interface to frequently updated information and information stored in extensive computer databases. TTS is an economical way of giving customers telephone access to information that would be too expensive or impractical to record using voice technology.

Switching

Switching technology handles the routing, transfer, and connection of more than two parties in a call. Once the domain of PBXs and proprietary switches, switches are now available on boards that can be easily installed in PC-based computer telephony systems.

Understanding Computer Telephony Solutions

Customers use building block components to develop open systems products that are sold in all of the major computer telephony markets. These markets can be grouped into four major sections: Information Access and Processing, Messaging, Connectivity, and *Central Office* (CO)/*Advanced Intelligent Network* (AIN).

Information Access and Processing Applications

Businesses worldwide are expanding their corporate communication systems to automate employee access to information and the capability to process that information. For example, a customer may want to check on the balance of a loan and then get a fax report of the adjusted interest charges for early payment.

Information access and processing applications can improve communication and increase customer service levels. These systems include Audio-Text, *fax-on-demand* (FOD), *interactive voice response* (IVR), *interactive fax response* (IFR), and simultaneous voice and data.

AudioText

AudioText provides prerecorded information to callers. Businesses can offer callers a single message or a choice of messages through touch-tone or ASR.

Voice Recording for Transaction Logging

With the pace of communications escalating and the demand for real-time information increasing, most of us are faced with the challenge to better manage our time and resources by establishing priorities.

One of the toughest elements to manage is inbound telephony, mostly because we do not know who is calling and why. Therefore, we have choices on how to handle the incoming calls:

- Screen the call by a personal assistant
- Route the calls to voice mail
- Answer the calls

Another problem is dealing with our calls while away from the office. Callers are forced to either leave a message or to page us. Unfortunately, we spend a large portion of our valuable time playing "telephone tag." This costs time, opportunities to complete transactions, and money.

Newer systems can handle the phone process for us when they become available. Some of these features include the following:

- **Voice Announce:** Knowing who is calling is the most important criteria in call management. By knowing who is calling, one can decide the priority and nature of the call before taking action. Positive Caller Identification enables us to know who is calling, allowing us to manage and prioritize calls. Callers can be identified by various methods:

  - Matching the Automatic Number Identification with a database or *personal information manager* (PIM).

  - Having the caller speak their name.

  - Having the caller input their telephone or account number prior to transferring the call.

- **Follow-me:** Regardless of where you may be, you can be available to take that important call. Call Management enables you to take calls you want when you want them.

  If an important customer calls your office while you are traveling, they will be identified by the Call Management system and your "Virtual Assistant" will try to locate you by calling previously programmed telephone numbers in the Call Management system. You may have designated your cellular telephone as your primary call-back choice. Your cellular telephone rings, you answer, and your Virtual Assistant says, "You have a call from (caller's voice plays). Would you like to take the call?" At this point, you can take the call, elect to send it to voice mail, or reroute to another associate. You can even choose to play one of many pre-recorded greetings to the caller.

  While using your network PC, you can see the name of the caller on your screen and select the appropriate call handling action. You can connect your PC to the Call Management system remotely via RAS or the Internet and manage your calls remotely as if you were sitting at your desk. The Call Management system can enable you to uniquely handle calls from important business associates and customers. Let's suppose you are expecting a call from your boss and he is expecting you to provide him with an update on a particular project. The Call Management system would enable you to record a message, giving him the desired update, so that when he calls, the update would be played for him.

- *Single Number Availability*   Rather than giving out numbers for your office telephone, fax, cellular, and pager, you (and your customers) can enjoy the simplicity and convenience of a single number. The Call Management system can recognize whether a person or fax is calling and can handle the call accordingly. Your Virtual Assistant can provide your callers with options to contact you, page you, or simply leave you a voice message with a call-back number.

Your Virtual Assistant can also be programmed to locate you and deliver faxes, voice mail, and e-mail messages. You can even have your Virtual Assistant make calls for you when you travel, so you do not have to deal with calling cards. The Virtual Assistant keeps your telephone directory and you can make calls from your directory.

Technology Enhancements

When CTI was first delivered, it was done through a large computing platform, either a mainframe or a high-end midrange computing system. These systems included such things as the IBM mainframe, AS/400, DEC VAX, or HP systems. Although they worked, they were expensive and sophisticated, requiring an extensive investment as well as application programming interfaces that made the service available only to large organizations.

The implementation of server-based or LAN-based platforms, however, has trickled down to the very small organizations. No longer can one determine or assume that a company using CTI applications is very large. As a matter of fact, many of the CTI applications are now rolling out on PC-based platforms for small organizations. Companies with three, five, or as many as 10 call-answering or telemarketing positions have implemented CTI very effectively. These lower-cost solutions have made CTI a reality within organizations around the world.

It was through LAN technology and not PBX technology that CTI actually got a foothold within organizations. Imagine if we had waited for the PBX manufacturers and the telephone companies to roll out CTI integration for us. We would probably still be waiting. In the past few years, however, the computer manufacturers have provided the push and the software developers have contributed the innovations to make CTI a reality.

This relatively new yet rapidly accepted approach to using the server as the instrument to provide CTI has thrust CTI into the forefront of telecommunications technology. As mentioned earlier, the voice server on a LAN is designed to connect directly to the public switch telephone network, handle calls coming into the group, and then process those calls directly to the desktop. Priority customers and special handling arrangements enable specific users to work around the high-end PBX and Centrex platforms and go directly to the individual department or customer service group without proceeding through the corporate platform. This in turn changes the architecture, because of the CTI applications that can work on a server platform.

PBXs, once known for their large investments and proprietary nature, can now remain single-line telephony service providers. When higher-end features and functions are necessary, the end user merely has to buy computer-based software, as opposed to high-end PBX architectural software. Of course, the PBX manufacturers have recognized this shift in the technology implementation and all of the PBX manufacturers are now developing the CTI interfaces or the software with third-party developers to reclaim customers. Because the PBX need only be an uncomplicated telephone system for the masses within an organization, the technology can last significantly longer. In the old days of the PBX, the plan was to keep the system for a period of about 10 years. However, reality dictated that the PBX was changed on a basis of about five to seven years. This involved major investments and changes within architecture, causing a significant amount of corporate stress.

As a quick side note, whenever a new PBX was installed, it usually meant that the Telecom manager within a corporation would be leaving soon. Regardless of how many technological advancements or enhancements were installed, users' expectations were never met. The frustrations of the users and the complaints made to management usually led to the Telecom manager's demise. Now with the features and functions moved to a PC-based platform, the Telecom manager can breathe easier. Without the need to upgrade the PBX or change an entire infrastructure, the Telecom manager can implement which features or functions are necessary and available on a department-by-department basis. On a large scale, all PBX features would be available to all users. By using the CTI implementation on a server, since features are purchased on a department-by-department basis, they are subsequently less expensive. This has been the boon of the 1990s.

Other Technologies

Because of the innovations in the telecommunications environment as well as the server marriage, many other applications and features can be made available by other producers. For example, through the use of the Telecom server on a LAN, the automated attendant, voice messaging, ACD, and IVR functions can all be united in a single server-based platform.

The developers of voice messaging systems' automated attendants recognized this opportunity several years ago. They leapfrogged the market, bypassing the PBX manufacturers, and developed single-card processing systems that could use high-end digital trunking capabilities directly into the servers. Using microprocessor control devices, these companies were able to write the necessary software that would provide the capabilities of all of these features. No longer would an organization have to buy a room-sized voice messaging system; this function can now be performed on a PC-based platform.

When voice messaging was first introduced, the size, heat, and cost of systems were exorbitant. Now, using PC-based systems, just about every vendor offers the capability of allowing several hundred to several thousand active users on a voice messaging system or a call processing system at a much lower cost. Actually, it's becoming much more difficult to tell a PBX performing CTI applications from a CTI server performing PBX capabilities. This convergence is blurring the lines between the various departments. Many of the organizations that now produce systems with these capabilities may have once been niche market providers but are now moving across the border that once separated these two technologies. Just about every feature, function, and capability can now be had using a very low-end server platform at a very reasonable price.

The integration includes features and functions such as

- Voice messaging
- Automated attendant
- IVR
- Text to speech

- Speech to text
- Directory services
- Fax services
- Fax-back services
- Intranet access for catalogs

Taking these one at a time, we'll see how they can all play together and provide unified messaging and integration capabilities. The capabilities of the integrated messaging and unified messaging services enable the desktop user to functionally perform all day-to-day operations at a single interface device, now the desktop PC.

Automated Attendant

With technology moving as quickly as it is, the use of single processing cards in a PC can deliver a combination of voice messaging and automated attendant functions directly to the CTI application. A digital signal-processing capability can literally compress voice calls so that they can be conveniently stored on a hard disk drive. The voice is already in a digital form when it arrives from a digital trunk or digital line card; therefore, storage is a relatively simple technique. The application software used in a voice mail system is basically a file service in which storage and retrieval can be easily accommodated.

Integrated Voice Recognition and Response (IVR)

IVR enables customers to manipulate information in a computer database, such as retrieving an account balance and transferring funds from one account to another. These applications range from AudioText and pay-per-call information systems that deliver a single audio message or a selection of messages to transaction-based systems that enable callers to access accounts and update information on a LAN-based or host-based database. AudioText entertainment lines are popular applications in computer telephony markets.

IVR systems are primarily based on the same type of technology as the auto attendant and voice mail. Using a single digital-processing card, the capability now enables users to arrange for pre-scripted capabilities that will actually walk a caller through a menu. The IVR will play digitally stored messages and solicit a response from the caller at each step, usually in the form of a touch-tone from a telephone set. The response from that tone will then cause the next step of the message to be played in accordance with the script. This is useful when a user is trying to access information from a host-based system, for example, and a played-back message will enable the user to retrieve any form of information.

IVR has been used by several medical providers and insurance providers, enabling a caller to dial in and access information regarding a payment or the processing of claims by merely using a touch-tone telephone. When dialing into the IVR, the user is prompted each step of the way by the system. As the user enters an ID number, a query is sent to a database in a host-computing platform. The appropriate information is then retrieved and played back. Using this CTI application saves an immense amount of time for an organization, because this normally labor-intensive activity can now be achieved through technology.

Fax-Back and Fax Processing

The *digital signal processor* (DSP) card can also be programmed to function as a fax modem that can provide for the sharing of fax services within a single-server environment. A fax image can be downloaded across a LAN, converted by a fax card, and transmitted across the network over a digital trunking facility. In the reverse direction, if an incoming fax is received from the network, it is then converted back to a file format that is easily usable within the PC environment. This can then either be stored in a fax server file for later retrieval by the individual recipient of the message or redirected by a fax operator. Some of these systems and services take more effort to implement and facilitate, but they may well be worth the effort in terms of an organization's needs.

Fax-back capability means that when a user dials into an organization equipped with CTI applications, that user can be directed to a fax server that has a numerical listing of specific documents that the user can retrieve by keying in a telephone number. When the user enters the telephone number, the server retrieves the fax from the file and then automatically transmits it to the designated telephone number the user has just entered. Through this application, catalog information or specific customer information can be retrieved without the use of human intervention. One can see how much time could be saved by using these types of services.

Fax-on-Demand (FOD)

Fax capabilities are indispensable in the business world today. Businesses with dedicated fax-based systems or with fax as an enhancement to their existing communications systems can automatically deliver information on demand to their customers. For example, customers can dial in and listen to a menu telling them which documents are available by fax. They can make a selection by speaking or by pressing a touch-tone digit and then enter the number of their fax machine to receive the document.

Interactive Fax Response (IFR)

Interactive Fax Response (IFR) enables customers to automatically receive a fax in response to a transaction performed through either the telephone or a computer. For example, a customer may receive a printout of an account balance after having transferred funds.

E-mail Reader

An e-mail reader resides on a media server that uses TTS technology. E-mail readers translate the ASCII text of an e-mail message (stripping out unnecessary header information) into voice that can be retrieved by callers through any analog device, such as a telephone.

Text to Speech and Speech to Text

In speech-to-text applications, a pre-stored pattern of words can be used through the CTI application to enable a highly mobile workforce to dial into a server-based platform and literally speak to the machine, as opposed to using touch tones. Speech from the callers, whose voice patterns are already stored in the computer, can then be converted into usable text using a server-based CTI application. This is instrumental when the user cannot access a touch-tone telephone. Without the touch-tone telephone, the user would have to carry a portable touch-tone pad generator, which is very inconvenient. Inevitably, the batteries on these devices die at the very moment the user needs access to information. Consequently, the use of voice patterns or speech patterns that have been pre-recorded with a series of words, such as get, save, retrieve, file, and so on, can be used to facilitate and walk through a computing system.

The TTS applications are comparable in that when a user accesses a particular file, again without a terminal device, for example, the system can convert the text into a speech pattern. What this effectively means is that e-mail and other documents can literally be read back to us no matter where we are. This is exciting because an end user might well dial into the CTI application while traveling on the road and learn that he or she has six voice messages and four e-mail messages waiting. Rather than that user being forced to log on with a different form of terminal device, these e-mail messages can be read right down the telephone line to the end user, facilitating the easy retrieval, storage, or redirection of messages. It is through these types of services that the CTI applications are drawing so much excitement.

Optical Character Recognition (OCR)

Another form of DSP technology can convert scanned images into text. When used with fax machines or fax images, *optical character recognition* (OCR) can change an incoming fax into a document that can easily be edited or incorporated into other types of applications. This would include editing a fax and placing it into a word processing document for easy editing capabilities. The additional storage capabilities could then convert the OCR, which usually would be a file of significant size, into a text-based document, which would be much smaller. Furthermore, by using the OCR to scan a pre-typed document, for example, the application could transform the scanned document into a TTS application, which could be read aloud. One can just imagine the uses and applications of some of these technologies.

Summary

Hopefully, this discussion of CTI will provide you with some appreciation of the capabilities and features that contribute to the merger of computing and PBX architectures. The use of an onscreen interface at a desktop PC enables users to manage and maintain their mailboxes for voice messaging as well as e-mail. Beyond that, a visual display can be received directly to the desktop, outlining the number of faxes, e-mails, or voice messages waiting to be retrieved. With the integration of the voice and text applications, the user can also see who the messages are from and prioritize the receipt of each of these messages based on some preconditioned arrangement. All of this facilitates the integration of the computer and telephony capabilities onto a single, simple platform that empowers the end user to access information more readily.

Moreover, with the implementation of CTI as a front-end processor for the organization's telemarketing or order processing departments, customers have the ability to retrieve information at will. This use of touchtone or voice response systems enables a customer to literally walk through catalogs, check status on orders, check inventories, or even check the process of billing information, all without human involvement. It is not the intent of this discussion to rule out the use of all humans, but rather show how humans can be more productive in performing the functions for which they were initially hired. By taking the repetitive "look-up"-type applications along with the data applications and allowing them to be controlled by the end user (or customer), the organization can save a significant amount of time and money and better utilize the human resources they have. The industry, however, is now facing a severe shortfall of skill sets and talents that could facilitate some of these functions. With the use of the CTI application, this human resources shortfall can easily be supplemented through technology. As things progress even more, additional applications such as video servers may well be added to this architecture and enable callers to view displays on a downloadable file, so that catalog information could be easily retrieved with a video clip that would show exactly what the customer is ordering or buying. Moreover, as the video clips and the fax services and voice messaging capabilities all become integrated into one tightly coupled architecture, customers could see the article, place the order, and literally "construct" the order customized to their needs. One can only imagine some of the possibilities of these features and functions that will be available in the future, but as with anything else, the first steps must be implemented.

In short, an organization must recognize the potential benefits that can be derived from a CTI application. Its capabilities are exciting because of all of the different ways CTI can be used. With a GUI-based system, a point-and-click, mouse-driven application on a desktop enables end users as well as customers to literally walk their way through all catalogs and information.

Integrated Services Digital Network (ISDN)

Introduction

This chapter will describe the concept of the *Integrated Services Digital Network* (ISDN) and will focus specifically on the following topics:

- Its original goals
- How it can be used
- Some of the alternatives

ISDN was conceived in the early 1980s by the world's telephone companies as the next generation network. The existing voice networks didn't deal well with data for the following reasons:

- One had to use modems to transmit data.
- The data rates were around 9600 b/s.
- Connections (worldwide) were unreliable.

Not only would the connection drop without notice but also the error rate was high enough to require a complex protocol to recover from errors. The back end of the network, that is, the interoffice trunks, were practically all digital, with more being installed daily. The switching systems were becoming digital just as quickly. It was expected that the only part of the network that would still be analog was the local loop (our infamous last mile). It was a logical step to provide existing digital capability already in the network directly to the customer.

This created what is always called the local loop problem. The problem is that much of the local loop plant was installed between 40 and 60 years ago and had been designed for normal voice communications only. The local loop problem therefore is how to run high-speed digital data on the local loop. Several solutions to the local loop problem are discussed in this book, and the ISDN solution is a little different from the solutions that will be presented in Chapter 16, "xDSL."

Although the digital network exists and the digital switching systems exist, ISDN has not made large in-roads into the customer premises. Once heralded as the solution for Internet access, it has been overtaken by xDSL (although there is an ISDN like DSL service called IDSL) and cable modems. It has found some success in the business community, specifically for telecommuting and teleconferencing.

Although greeted with enthusiasm by many equipment makers, ISDN has been treated coolly by several of the major North American carriers. The European monopoly carriers implemented ISDN in the major cities, while their rural telephone systems still use electromechanical switching. The committees that defined ISDN concentrated on the interface.

Origins of ISDN

ISDN was a concept developed by the *Consultative Committee on International Telegraph and Telephone* (CCITT). The CCITT has since changed its name to the ITU-T, but it is still the same folks made up primarily of representatives of the world's government-owned monopoly carriers. Recall that in 1980 all telephone companies were monopolies and all were government-owned and run by the post office (thus, the name *Post Telegraph and Telephone* (PTT)). Canada and the U.S. were exceptions because most of the Telcos were private companies that had been granted monopolies.

Although there were about 1,500 Telcos in the U.S., AT&T, GTE, and ITT owned the bulk of the important ones that covered the large population centers. In Canada, the provincial governments owned the telephone company. Exceptions were in British Columbia, Ontario, and Quebec. Canada also has many small independent Telcos in the outlying areas.

Origins of the Standards

The CCITT is a consultative committee to the International Telecommunication Union and have recently changed their name to ITU-T. The ITU-T (CCITT) is a UN treaty organization and, as such, each country is entitled to send representatives to any committee meeting. The representative typically comes from the government-run *post office / telephone monopoly* (PTT). The world's Telcos are becoming privatized and competition is being permitted. This creates an interesting struggle within each country to determine who will represent that country's interest at the ITU-T. Note that when discussing the organization's historical actions and composition, we call it the CCITT. When we discuss its current actions, we call it ITU-T.

The name ITU-T came about due to the privatization trend separating telephone business from the post office and the general elimination of telegraph service. Since its members were no longer PTTs, the organization couldn't be called CCITT. The CCITT was a consulting committee to the

International Telecommunications Union, so the ITU-TSS (ITU-T) is the Telecommunications Standard subsection of the ITU.

The CCITT is comprised of study groups (SG). Each SG has its own area of expertise. Here are some of the better known ones related to ISDN:

- SG VII on public data networks (X.25) X-series standards
- SG VIII terminal equipment for telematic services
- SG XI ISDN and telephone network switching and signaling
- SG XII transmission performance of telephone networks and terminals
- SG XV transmission systems
- SG XVII data transmission over public telephone networks
- SG XVIII digital networks, including ISDN

Although we are calling them standards, technically the CCITT and ITU-T publish recommendations. The philosophy of the ISDN committee (essentially the Telcos) is to specify the customer interface first and then figure out how to support it in the network. The theory being that if we can get all the peripheral or end equipment makers to make equipment based on our specifications, then when we get ready to roll out the service, the store shelves will be stocked with inexpensive ISDN interface equipment.

Interfaces

The customer interface I.45x specifies the *Basic Rate Interface* (BRI). It was intended to become the standard subscriber interface.

The BRI specifies two bearer channels and a data channel. The two bearer channels would bear the customer's information. The initial concept had this as being everything from analog telephone calls (digitized) to tele-conferencing data and these would be switched channels. The only difference between a conventional telephone circuit and a bearer channel is that the bearer channel would be 64 KB/s all the way to the customer. (Note in the current network, your analog telephone circuit is digitized to 64 KB/s at the local Telco office before being switched across the network. It is then turned back into analog at the far end before being delivered to the called party.)

Now with BRI, we have not one, but two such telephone circuits. Since it is digital, we have switched digital 64 KB/s to (theoretically) anywhere in the world.

The problem with the existing Telco network is that the signaling information shares the telephone channel with the user information. With plain voice circuits, a customer doesn't notice or care. With the advent of modems, this represents a loss of channel bandwidth, and with digital transmission, it means a loss of usable bits per second. The customer is therefore stuck with 56 KB/second, instead of the actual channel rate of 64 KB/second.

The BRI interface therefore specifies a multi-functional data channel at 16 KB/s that could handle signaling (its primary function) and network data (X.25) when not needed for signaling. BRI is therefore referred to as 2B + D, two bearer channels and a data channel.

Figure 9-1 shows the BRI graphically and indicates the bandwidth allocation on the ISDN interface. Remember that this is a time-division multiplexed interface where the B, D, and overhead bits are interleaved.

In Figure 9-2, the BRI is created by the *Network Terminal type 1* (NT1). The NT1 creates a four-wire bus called the T interface onto which each ISDN device is connected. We will discuss the interface designations *user* (U), *terminal* (T), *system* (S), and *rate* (R) in the following discussion. Two points should be kept in mind. First, the boxes shown in Figure 9-2 can be combined in any reasonable way. Second, it is not necessary to have an NT2

Figure 9-1

BRI bandwidth
allocation

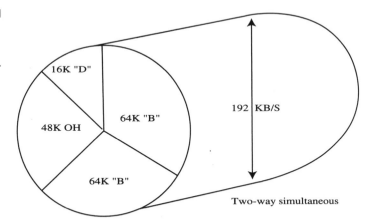

Figure 9-2
NT1 creates the BRI

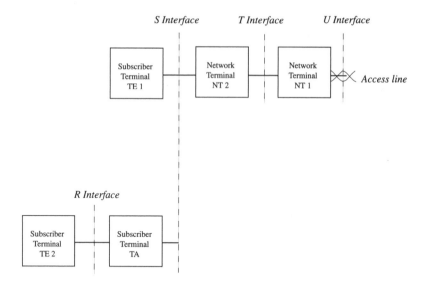

element. This means that the S and T interfaces are logically and physically identical. They have separate identities to allow us to describe the functionality of the NT2 element. The NT2 could create multiple S interfaces and perform the switching to adjudicate access to the B channels on the T interface. Since the two interfaces are the same, they are frequently referred to as the *S/T interface*. As described below, but not shown in Figure 9-2, up to eight devices can be connected to the bus.

It is helpful in the following discussion to have a picture of the ISDN interface architecture handy, and Figure 9-3 serves this purpose. We have outlined the basic access. It provides 2B + D. The two bearer channels shown on the left are simply access to network layer services. Both the link layer protocol and network layer protocol completely depend on the service being accessed. The data channel has a link layer protocol defined as Link access protocol—Data (channel). The link layer (which is essentially HDLC) provides sequenced acknowledged delivery. This reliable link layer service can be used by the network layer services provided across the ISDN interface D channel. The most important of which is the signaling function used to set-up and teardown the B channels.

The packet-switching access is not universally implemented, but it permits access to the worldwide X.25 packet-switching network. The telemetry system access is also generally unimplemented. Several telemetry experiments have been done in Europe and in the U.S., but few resulted in a cost-effective solution. We may have to wait for new lower-cost technology. The concepts, discussed in the following paragraphs, are sound ones.

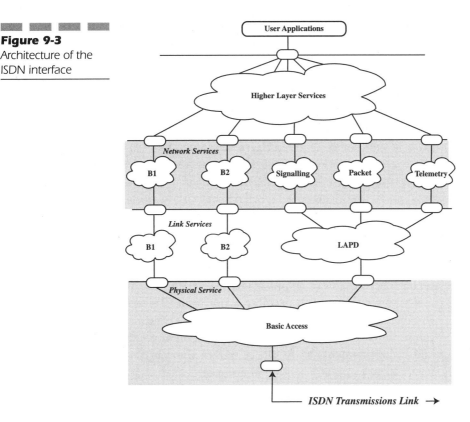

Figure 9-3
Architecture of the
ISDN interface

The higher-layer services could be almost anything. A typical example might be video conferencing. Here the higher-layer services would be the functions of the Codec.

If the ISDN B channels are used for Internet access, then IP would be the network services and TCP, HTTP, and so on would be the higher-layer services. The B and D channels are then just timeslots on the bus that can be grabbed by any of the connected devices. The trick is how the telephone, for example, keeps the computer from grabbing its timeslot while it is off hook. The answer lies in the 48 KB/s of overhead. In this 48 KB/s, the NT1 provides timing to all the devices on the bus. Remember this is a bidirectional channel, so there is 48 KB/s of overhead going back as well. The station that wants to grab the D channel effectively puts its address in the contention slot in the 48 KB/s inbound bus. These bits are echoed by the NT1 as a confirmation of success. Effectively, the highest address will win since the one bits would overlay any zero bits of a device with a lower address. When the terminal sees its address echoed in the outbound overhead channel, it has won the right to send a packet (signaling or otherwise) on the data channel.

If the telephone has signaled the NT1 channel and grabbed a B1 channel, any other device sending a signaling request for that B1 channel will receive a control packet from the NT1 indicating that it is busy. Since one B channel is identical to another B channel, the NT1 will theoretically respond with "you may use the B2 channel." The actual operation depends on the implementation of the NT1 and terminal equipment.

This can get a little sticky when bonding channels. Essentially, bonding (that is, one device using both 64 KB/s channels to get 128 KB/s) must be done at call initiation. One of the channels can be dropped during the call, but it can't be re-added later. This is because the called end (such as ISP) may have already accepted a call on the other channel. The logical question is, "why can't it accept a mated or bonded channel on any other channel?" Although this is a good question, the answer is that when the rebonding takes place, the Telco network treats it like a brand new call. Thus, it can be routed via different offices and the timing relationship between the bonded B channels would not be preserved.

This brings up another interesting issue: Telco implementations vary widely. The number of possibilities is nearly limitless. Each Telco has chosen to implement a subset of ISDN based on what they think they can sell and still implement at a profit. Therefore, there is no such thing as standard ISDN. Every implementation is unique.

Interface Components

The previous interface doesn't allow for switching on premises, which would be necessary if there were multiple telephones.

The European and North American Telephone systems are quite different in operation. In North America, extension telephones simply bridge onto the line, or connect in parallel. Adding a party to a call or transferring from one extension to another is as simple as picking up one telephone and putting down another. In the European system, specific actions (pushing the designated button) are required on the part of the parties to transfer a call. This also implies on premise switching.

NT1

We have mentioned the function of the NT1. It creates the T interface for premise devices (from the U interface). In the original CCITT concept, the

NT1 was provided by the Telco as part of the ISDN service. The U interface was therefore only the concern of the Telcos with open networks (which concerned North America at the time). It is now important to understand on general principles.

NT2

This device would do the switching, permitting more than the standard eight devices to share the T bus by creating perhaps multiple S buses. Therefore, an ISDN (TE) device can't really tell if it is connected to an NT1 or NT2.

TE1

The *terminal equipment type 1* (TE1) is a standard (there is that word again) ISDN terminal that is capable of dealing with the B and D channels. In other words, it can interface with the S/T bus.

TE2

The *terminal equipment type 2* (TE2) is a standard device having an RS-232 or V.35 interface. (In ITU parlance, this is called a V-series interface.) It may be intelligent but doesn't have an ISDN interface capable of handling the D and B channels.

TA

The *terminal adapter* (TA) is the semi-intelligent device that lets a TE2 connect to the S/T ISDN interface. The primary function of the TA is to run the ISDN interface for our TE2. The functionality varies widely due to the manu-facturers. Some are simple and support only one TE2; others support two TE2s and an analog telephone. The TA need not be a stand-alone box; it can come on a PC card and plug into the computer's internal bus. With the proper software to run the card, this instantly creates a TE1 out of your computer.

Thus, the BRI then was designed for *small offices or home offices* (SOHO).

Physical Delivery

One of the more interesting parts of the ISDN service is the solution to the local loop problem. Remember our local loop is ancient and designed for voice. From Figures 9-1 and 9-2, we expect it to support 192 KB/s bidirectionally! This would not be a problem if we only had to go a few hundred feet. Unfortunately, there aren't that many customers within a few hundred feet of the central office.

The problem is approached from the other end actually. If we want to reach 95 percent of the customers, what is the average length of the cable to them and what are the quality and characteristics of that cable?

Some of the answers to these questions you might find surprising. In cities, 95 percent of the customers are within 15,000 cable feet of the central office. When we go to suburbs and the rural areas in particular, then all bets are off.

Figure 9-4 shows a typical local loop layout with emphasis on the fact that local loops, particularly those in older parts of a city, are comprised of different gauges of cable and may have several bridge taps. Although slightly exaggerated, one could imagine that all of the cable taps are on one pair. That is, at one time or another during the 60-year life of this particular cable plant, that pair was used to provide telephone service to each of those different locations. Note also the gauge change from 24- to 19-gauge.

Figure 9-4
Typical local
loop layout

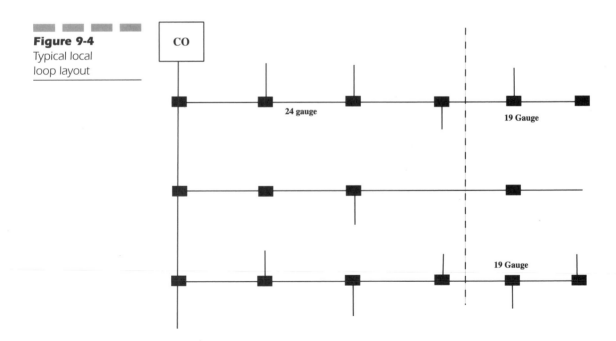

This was obviously not a problem for the analog telephone network, because it worked fine for 60 years on that cable.

High-speed digital transmission presents a whole new set of problems. We will outline just the issues here. (See Chapter 16, "xDSL," for a more in-depth discussion of the transmission problems.) The fundamental problem centers on the fact that a wire is an antenna. This is OK if we are in the radio or TV transmission business, but it is an unfortunate side effect of the telephone transmission business. The problem is worse at higher frequencies that characterize digital signals. If we use our local loop to send digital signals containing high frequencies, the low frequency portion of the signal goes a relatively long way. The high frequency portion of the digital wave is radiated off the wire and is delayed by the characteristics of the wire. The longer the wire, the worse the problem. The result is that the signal dribbling out the end of the wire is weak and distorted because all the components didn't arrive in time or with the right strength. The components of our digital signal that were radiated are picked up by adjacent wires in the cable. We call this *crosstalk* in the telephone business and the unwanted signal picked up is noise. The brute force method can be employed to remedy this. Sending a stronger signal makes more signals available at the destination but also increases the amount of energy radiated.

As the telephone company, we must make sure that any new service added to our outside plant (local loops in aggregate) does not interfere with any existing service. Our goal of wanting to provide more distance must be tempered by the realities of crosstalk.

Interestingly enough, the European Telcos ignored the question of ISDN delivery, preferring to concentrate on defining the S/T interface instead. Their theory was that "we're selling a service and how we deliver it to the customer isn't important—we can always figure that out later." (Also since they were government monopolies, if the cost of delivery was high, they could always raise the tariff.)

The North American Telcos had a different situation. They were not allowed to sell or lease the NT1 interface as part of the ISDN service and had to specify the electrical and logical interface at the Telco demarcation point. Therefore, the signaling on the local loop had to be worked out and standardized before ISDN could be deployed.

The U Interface

The U interface is unique to North America and the "open" telephone network interconnection. Figure 9-5 shows the U interface connecting to the

Figure 9-5
The U interface

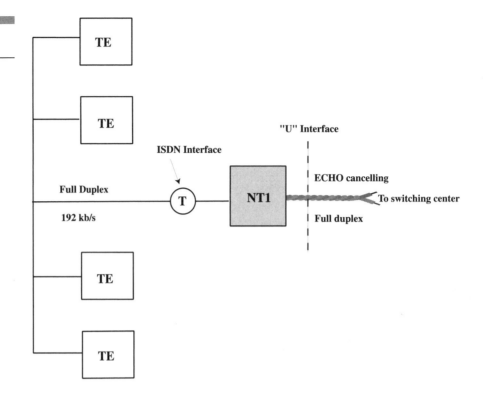

NT1 and the NT1 in turn creating the internal S/T bus to which up to eight ISDN devices can be connected. This U interface can be either a two-wire or a four-wire connection. In the following discussion, we concentrate on the two-wire connection, because it is the more technologically challenging. The four-wire interface requires much less technology and can be delivered over a greater distance. Many early BRI interfaces were installed using the four-wire interface for just these reasons. For wide-scale deployment, however, the two-wire interface has to be perfected, since the Telcos are not about to double the size of their already extensive outside plant.

AT&T (Bell Labs) came to the rescue with a technique known as 2b1Q (At this time, this was all part of AT&T. Since then, Bell Labs was spun off from AT&T as part of Lucent Technologies). Figure 9-6 shows the 2b1Q signal. Although it appears (and is) simple, it solves several problems:

- It is easy to generate (that is, it is a simple wave form).

- It minimizes crosstalk.

- It will work on most local loops.

Figure 9-6
2B1Q technique
for ISDN

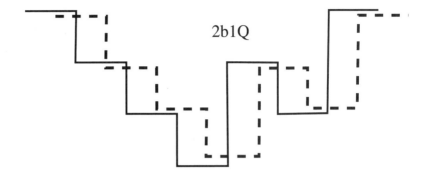

Notice that each of the four levels contains two bits. Therefore, the signaling rate (baud rate) is one-half the bit rate. From our previous discussion of the S/T interface, we might conclude that the bit rate must be 192 KB/s. Fortunately, those 48 KB/s of overhead are only needed on the S/T interface to provide timing and priority (refer back to Figure 9-1).

On the Telco side of the interface, we only need 2B + D and some overhead for timing and control. The total is 160 KB/s. This means in engineering terms that the primary spectral peak is at about 80 kHz. In layman's terms, it means we are only trying to send 80 kHz down the old local loop twisted pair wire. What we have effectively done is halved the bandwidth requirements of the line. This trade-off isn't free though; it means that the signal-to-noise ratio has to be better than if we were to use a simpler encoding system.

The next problem is that the old local loop has different gauge wire and has bridge taps (refer back to Figure 9-5). Unfortunately, when sending pulses down the line, we are going to get reflections from these gauge changes and bridge taps. These reflections show our transmitted signal much lower in amplitude and delayed in time (Figure 9-6 shows this reflection as a dashed line). The reflections will always be

■ Of the same magnitude and

■ At the same relative time from each of the cable plant anomalies

Therefore, when the ISDN NT1 goes "off-hook," it transmits a known pattern. That pattern contains all possible bit combinations (there are only 16 combinations). The receiver at the transmitting end then monitors the resultant complex signal. Since it knows what it sent and can subtract that signal out, it memorizes the resulting reflected energy. The NT1 stops and lets the central office end do the same thing. Each end has learned the reflection characteristics of the local loop.

Now the clever part. Both ends can now simultaneously send data to the party at the other end. However, it is hard to listen to the relatively weak signal from the other end when you are also talking. This is where the learning comes in. Each transmitter can subtract its own transmitted signal *and* the reflections, which it knows to be there. After subtracting our transmitted stuff (and reflections), whatever is left over must be the data from the far end. Yes, it really works!

Unfortunately, we aren't quite home free. This old local loop doesn't handle all frequencies equally, so some of the signal components arrive out of the precise time (or phase) with the other parts of the signal, as mentioned above. This effectively distorts the signal. The U interface hardware therefore uses the old modem technique of "equalizing" the line. While each end trains itself on reflections, the opposite end receives the known signal, recognizes the distortion, and tunes its equalizer to make the signal appear correct. If we were to independently measure the line characteristics (amplitude and phase) and the equalizer, we would (as you might expect) find them equal and opposite. The result is that the equalization process takes the distortion out of the line.

2b1Q works so well that this basic technique is now the primary technology used by the local Telcos to deliver T-carrier service, instead of the old AMI technique. It will go twice as far before repeaters are required. Of course, the name has been changed to confuse us and is now called HDSL.

The Physical Interface

Another clever design feature of both the S/T and U interfaces is that they all use the same RJ-45 type connector. Figure 9-7 shows the standard interface connector.

The S/T and U interfaces carefully select the pin assignments so that accidentally plugging an S/T connector into a U interface and vice versa doesn't hurt anything. However, it won't work either. There was a great deal of discussion about the customer interface concerning how and whether the carrier should provide power. In the current analog system, the carrier provides power to the telephones so that they work, although commercial power is off. Shall this capability be preserved for ISDN? Although the augment lasted several years, three powering mechanisms are provided across the interface:

- The customer provides power to the NT1.

- The carrier provides power to the customer from the NT1.

- The carrier provides a small amount of keep-alive power on the actual bus leads.

Figure 9-7
The standard
interface connector

Here is another opportunity for the implementations of the carriers to diverge. Most carriers don't provide power. The safe bet is for the customer to be able to power his own equipment in case of a power failure.

Applications of the ISDN Interface

The following section describes the areas in which ISDN functions, including multiple channels, telephone services, digital fax, analog fax, computer/video conferencing, signaling, telemetry, and packet switching.

Multiple channels

Figures 9-1 and 9-2 display the logical BRI interface. The plan is to provide access to every possible home device. The original concept was for up to eight devices. After all, you only have two B channel and one D channel to share among eight devices.

Telephone

The obvious starting point is the telephone, which is now a digital telephone. Instead of the telephone conversation being analog from the handset to the central office where it becomes digitized, the conversation can be digitized directly at the source and passed digitally all the way through the network to the other end.

Digital Fax

Fax machines now have to be digital. Therefore, the Group IV fax standard specifies 64 KB/s fax operation.

Analog Fax

Analog fax machines use a modem, so it has to plug into the telephone (or similar device) that would take the analog modem tones and digitize them at 64 KB/s. This would provide compatibility with all existing Group III fax machines.

Computer/Video Conferencing

Our computer or video conferencing equipment can use one of the 64 KB/s or bond both Bearer channels together for a 128 KB/s digital channel across the network.

Signaling

The primary function of the data channel is to provide for signaling, that is, the setting up and tearing down of the switched Bearer channels. At 16 KB/s, the data channel has more bandwidth than is needed for signaling alone. Therefore, when it is not being used for its primary and high priority signaling function, it could be used for other things.

Telemetry

This feature has never been well defined. The concept is that many house-hold devices can be connected to the data channel. This can include an energy management system that would let the power company selectively turn off the refrigerator or air conditioner for an hour or so at peak usage time. The concept also includes connecting the utility meters to permit remote monitoring and billing. Although several proof-of-concept trials of this technology have been conducted, apparently the cost of implementation outweighed the potential savings.

Packet Switching

The 16-KB/s data channel has bandwidth to spare. Therefore, the local carrier can provide a data service on this excess bandwidth. X.25 is just maturing and is the logical packet switching technology to offer. As it turns out, all the data on the data channel, whether it be signaling data, telemetry data, or X.25 data, are always sent in packets anyway. (This packet-handling channel was the logical genesis of Frame Relay. If we can distinguish different kinds of packets, why not frames too?)

These devices are all connected to the same 2B + D interface; therefore, three of them could be in operation at one time. For example, the telephone could be using a B channel, the fax could be using a B channel, and the computer could be doing X.25 packet switching. (Today we run all these services over our Internet connection.)

Primary Rate ISDN

The BRI interface offers 2B + D. The PRI provides 23B + D and, in this case, all are 64 KB/s channels. This should sound familiar. What technology provides 24 64-KB/s channels? Of course, the answer is T1. The physical interface for PRI then is simply a T1. Technically, it is a T1 with *extended superframe* (ESF) framing. The obvious question is, "How is PRI different from an ordinary T1?"

The answer is that at the physical interface it isn't any different. Channel 24 now becomes the signaling channel. One might say that we have added common channel signaling to the T1. PRI then is frequently used for a PBX interface where the full signaling capability of the D channel is needed.

Figure 9-2 shows the S/T interface components for PRI. This same figure is then used to describe the BRI and PRI interfaces, but because it is a T1, we can add familiar labels to the diagram components. The NT1 is now a CSU, and the NT2 is now a PBX. It could also be a router, but typically routers are members of dedicated networks and don't use switched channels. The Telcos charge more for a PRI because it offers many more switched-channel features. If a router needs to be connected to another router, an ordinary T1 will do fine and cost less. Circumstances will occur when we need both our router and another device on the same interface and will periodically adjust the amount of bandwidth to each. For this unique application, PRI is ideal.

If we stick with our PBX example, the NT2 creates multiple S/T interfaces for TE1 or TA devices. TE1s are ISDN BRI telephones. What is interesting about the definition of the PRI interface is that it isn't limited to a single T1. Multiple T1s can be added to the PRI, so that up to 20 T1s can be part of a single interface and all would be controlled by the single 64-KB/s D channel on the first T1. That is, 479 switched B channels could be controlled by a single D channel.

H0 Channels

Equally interesting is that aggregated channels have been defined. These are known as H channels. An H0 channel is 384 KB/s or six B channels treated and switched as a single channel. This means our PRI NT2 could call for an H0 to Atlanta, another to Phoenix, and one to Chicago and still have five B channels to individually switch.

H11 Channels

The H11 channel is a 1.536-Mb/s (T1) switched channel too. If our PRI NT2 had, let's say, five T1s, it could configure any one of them as an H11 channel and another as two H0 channels with 12 B channels. This essentially gives us dial-up T1, 384-KB/s, and 64-KB/s clear channel service.

H12 Channels

The Europeans use an E1 system that has 32 channels each 64 KB/s. One channel is used for timing and alarms and one is used for common channel signaling. In the PRI interface, channel 16 becomes the D channel, so the E1 version gives 30B + D. The H12 channels are only in the E1 system and are 1920 KB/s. Again, theoretically this is switched E1, 384-KB/s, and 64-KB/s service.

Signaling on the D Channel

The signaling packets on the D channel are the same for BRI and PRI. In the early days of defining the interfaces, an argument arose about whether to grant direct access to the signaling system by the customers. As we indicated, SS7 is the mechanism for managing the network and it's logical to simply let the D channel use SS7 packets. Unfortunately, the more paranoid faction won the argument, so the D channel signaling packets require a small amount of conversion to change them from D channel signaling packets to SS7 packets. This is OK, because the D channel link layer frame is standard HDLC (LAP D, Link access protocol for the D channel), which can carry the packets on the D channel whether they are signaling packets or X.25 packets.

Figure 9-8 shows a D channel packet. If it has the look of an X.25 packet, it is not by accident. Since both X.25 and signaling packets are handled (time interleaved) on the D channel, it was thought that a protocol discriminatory byte would be a good idea just in case the packet was received by the wrong entity. (The link layer protocol makes this virtually impossible anyway, but just in case . . .) The protocol discriminator byte pattern never occurs in X.25 and would be immediately recognized as an illegal packet and discarded.

Figure 9-8
D channel packet

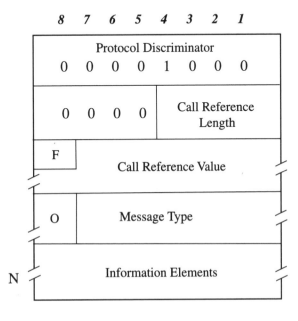

The call reference value is essentially a random number chosen to identify a particular call. Since there could be many (479) calls in progress at once, you might need more than a fixed length value. (In order to prevent confusion about which call is doing what, you don't want to reuse those numbers too quickly.) All signaling packets associated with a given call will have the same call reference value.

The message type indicates the format of the packet. The setup message, for example, would have the called and calling telephone numbers in the information elements field. The call-clearing message would have a single-byte cause code in the information element field.

Of course, all of this wonderful capability isn't of much value if the service provider hasn't implemented the backroom functions to enable these features from the customer's. The carriers have simply implemented that subset that they individually think that they can sell. It should come as no surprise that the offerings vary widely and all features are not generally compatible across carriers.

Installation Problems

It is not our intent here to denigrate the carriers, but rather to point out that installing ISDN for other than plain vanilla applications can be frustrating and time-consuming.

The carriers come from a long tradition of providing voice services. Their entire management system and technical system is designed around providing highly reliable (and profitable) voice services. ISDN is digitally based, the local distribution is digital, and the switching is digital. Historically, the Telcos have strictly partitioned the knowledge of each technical specialty. The installer knows about local drop wires, pedestals, and telephones. The switchman knows about switching. The outside plant folks know about cabling in the streets and on poles. The transmission expert knows about carrier systems. Moreover, no one knows it all!

Therefore, specially trained ISDN technicians are required to deal with the local loop. They must know the parameters for operation and be able to test it out. In the old days, the installer would hook his butt to the line and listen. He could tell a lot about the local loop by the background noise. Sorry, you can't hear digital. Therefore, the technician now has to be trained in handling a special piece of test equipment to make sure the line will support the data rate.

Since PRI is simply T1, the tuning of the local loop is understood; BRI, however, is a different issue. We are now dealing with local loops. T1 is not normally delivered to residential loops, but it could be. T1 is delivered on business loops. There is no real difference between business and residential loops, but the Telcos do tend to run data circuits (T1s) in separate binder groups and they tend to avoid the problems of many bridge taps, shown earlier in Figure 9-4.

Another complicating factor is that each carrier implements its own version of ISDN, making for a frustrating experience getting it installed and working. These different versions often lead to compatibility problems end to end. As a rule, you can count on the 64-KB/s B channels working end to end, and in most cases they can be bonded to 128 KB/s. On the PRI interface, using switched H0 and switched H11 channels is very iffy. The carriers just haven't implemented the switching algorithms to enable this feature.

BRI Application

One of the major uses of ISDN is in video conferencing. This is normally done by installing three BRI lines. The video conferencing equipment has a built-in inverse multiplexer. Figure 9-9 shows a typical configuration. The three BRI interfaces are connected to the inverse multiplexer and the control panel enables the operator to specify how much bandwidth is to be used for the videoconference. Although one could use 64 KB/s for a videoconference, it is quite impractical. A minimal usable videoconference is 128 KB/s, and as long as someone else is paying for the bandwidth, 384 KB/s is acceptable. This is the crux of the problem: how much bandwidth are we willing to afford for the videoconference? The three BRI interfaces give you a choice in increments of 64 KB/s to 384 KB/s.

The inverse mux takes the commands from the control panel and provides the appropriate packets on the D channels to set up the amount of bandwidth required. The inverse mux is needed because it is essentially making up to six independent 64-KB/s calls. These calls are routed independently by the network. Therefore, if we are in Phoenix making conference calls to Atlanta, one of the calls might go through Denver, Minneapolis, and Atlanta. Another might go to Albuquerque, Kansas City, St. Louis, and Atlanta. A third might go through El Paso, Dallas, Birmingham, and Atlanta. The point is that each of these paths has a different length and delay. The inverse mux plays "scatter

Figure 9-9
Typical configuration
of the inverse mux

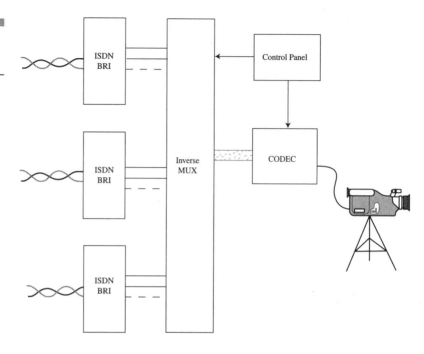

gather" by sending the digitized video in packets (actually frames) alternately over each of the circuits. The peer unit at the other end puts the frames back into order with the proper timing to provide the 384-KB/s channel to the coder/decoder (Codec). If we could afford a PRI interface, we could simply set up an H0 channel to Atlanta.

Broadband ISDN

There has been much confusion over exactly what is broadband ISDN. This confusion stems from at least two sources. First, the definition of the word broadband, and secondly from the fact that even after we are over that hurdle, no one knows what it means.

Definitions

The original meaning of broadband started with high-bandwidth as compared to low-bandwidth telephone lines. The Telcos used the word wideband to describe some of their analog carrier systems that used frequency division multiplexing to put multiple telephone channels on a single transmission facility. So when the Telcos were looking for a new name for their high-data-rate digital service, they had an existing meaning for wideband. They chose the word broadband, despite the fact that this word (outside the Telcos' world) meant a high-bandwidth analog multiplexed system. A cable TV system is a broadband system. In the early days of *local area networks* (LANs), Wang and several others used the analog cable TV technology (using wideband modems) to provide computer connectivity. These were therefore broadband networks. Along comes broadband ISDN and everyone is naturally confused.

One could think of it as a high-bandwidth digital service. Actually, this is exactly how the Telcos would like to sell it. Don't ask what it is or how it works. It is simply a high-data-rate-capable transport service.

Originally then, the Telcos intended to offer this high-rate telephone service. How high a rate was in doubt. What did the customers need? Clearly, we can't sell something that the customers don't need.

There were a couple of stumbles along the way. First, there was a lot of hype about broadband ISDN, yet there was precious little supporting evidence in the Telco infrastructure. Second, it was the intent from the beginning to base the service on a statistical multiplexing type of service, a very high-rate X.25 packet switching technology. It was clear to everyone at the time that such a shared network would offer the customer more bandwidth while requiring less infrastructure bandwidth than the existing circuit-switched, time-division multiplexed network based on T1s, T3s, and T4s.

One of the stumbles grew out of the popularity of LANs. Why not become citywide or even countrywide LAN providers? AT&T invented a thing called SMDS. The LAN committee in the IEEE grabbed the idea and after about 10 years of study accepted the standard 802.6 as the Metropolitan Area Network. This was an ingenious concept that used an access method known as *distributed queue dual bus* (DQDB).

Figure 9-10 is a diagram of the DBDQ system. Although it looks like a ring, it is really two buses (one side of a T1) running in opposite directions. The Telco is the bus controller. It is well known that token passing busses are not efficient, since they effectively poll each station. The time taken in polling when a station has no data is time that a station with data cannot lose. DQDB solves this problem by having a continuous set of packets flowing in both directions, which could be assured by the Telco. These packets are small and are called *cells*. Sending empty cells doesn't hurt the efficiency. Each cell has a header that contains a busy or free bit and a request field. If station D wishes to send something to station B (or to another station) serviced by the Telco on the clockwise (CW) bus, D places a reservation

Figure 9-10
The distributed
queue dual bus
(DQDB) system.

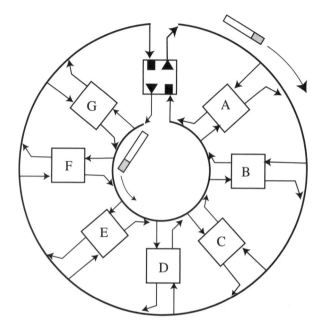

request on the counterclockwise (CCW) bus in the reservation field. (It has been watching the CCW bus for reservations, so it has a queue of prior reservations.) Each free cell arriving on the CW bus causes one to be decremented from the queue. When the queue is empty, the next free cell on the CW bus can be used for transmission. Yes, this system is a little complex. (If you really want to challenge your powers of visualization, try to envision this happening in both directions at the same time.) The point of the discussion is that it is sort of a demand-based system and there are no collisions. If no one else has requested cells, then you can have them all.

The problem is that the rest of the protocol layer built upon the basic DQDB access scheme makes it very inefficient. The service is known as *Switched Multi-Megabit Digital Service* (SMDS). Unfortunately, out of the 1.5 MB/s of a T1, you can only use a little over one MB/s. Very few customers found that the service was cost effective. Another unfortunate aspect of the system is that the cells are not identical to ATM cells. Thus, to use ATM for transport, these cells have to be repackaged into ATM cells. The best part is that the Telco can implement it using existing T1 technology.

SMDS then was part of the broadband ISDN service offering. What about long-haul transport? That was left to ATM, which intended not to be a service per se, but as the implementation technology for broadband ISDN.

You can now see that ATM was intended as the core technology (invisible to the customer) to support the Telco network of the future. Just as narrowband ISDN (PRI and BRI) were intended to provide the network connection of the future, broadband ISDN service was intended to be the high-rate interface. ATM would be the core technology supporting all of the services from circuit-switching to packet-switching.

ATM as the supporting technology was not ready when broadband ISDN was announced, so the carriers couldn't provide the service. All eyes turned to the new kid on the block, ATM. Although not well defined, like a kid, it had lots of promise. Who was to do the heavy lifting until ATM was mature?

Frame Relay was developed to fill the void. The idea for Frame Relay came from both X.25 concepts and from the D channel packet handling of ISDN. Frame Relay filled the need for "broadband" service. Few customers actually needed the nose bleed speeds of ATM, and Frame Relay was a cost-effective replacement for dedicated lines. Being packet-switching, it provided bandwidth on demand. ATM also provided bandwidth on demand at much higher rates.

Conclusion

ISDN therefore was a great technology-driven service that didn't really solve a business (or home) need. It is little wonder that ISDN is not widely implemented or used, but there are, as we have noted, some clear exceptions. The most notable is video conferencing. Internet access is also a possibility, but ISDN can't compete with xDSL technology in performance for the cost. The ISDN primary rate is used extensively in call centers, utilizing computer telephony integration to maximize their efficiency. PRI is also used in PBX applications, where the digital PBX can make use of the network control and status information provided by the PRI interface.

Frame Relay

Introduction

Frame Relay is a fast packet-switching technology introduced in 1992. Its installed base has skyrocketed since its development and introduction in the industry. No one could have ever predicted just how popular this technique would become, but the final outcome is that Frame Relay has become the "bread and butter" service for many carriers. The reason is not as simple as one might believe; therefore, this chapter will explore the details of what has led to the popularity in Frame Relay installations.

Moreover, some carriers have encountered problems with their installations because they did not understand the benefits or the operations of a Frame relay service. Unfortunately, this is the epitaph of the Telecommunications industry: the carriers (and specifically the long-distance carriers) do not understand data communications! Yes, they have some talented people on their staff, but they have always stifled these individuals who knew what was happening. Instead, the carriers let the voice and engineering folks become the spokespersons for the carriers in a data communications arena. How sad, they just never really got the point.

Now in an era when all the voice carriers are trying to become data-literate and the data carriers are trying to become voice-literate, the voice people stand to lose in the overall transition because of the innate ability to mess up the data installation. Whenever we hear about a carrier's capability to handle data traffic, we seem to be mesmerized by the carrier's overall capabilities, instead of understanding specific characteristics for processing and delivering data communications.

Frame Relay Defined

First, a definition of packet switching is in order because Frame Relay falls into the category of a packet-switching family.

Packet switching is a store and forward switching technology for queuing networks where user messages are broken down into smaller pieces called packets. Each packet has its own associated overhead containing the destination address and control information. Packets are sent from source to destination over shared facilities and use a statistical time-division multiplexing concept to share the resources. Typical applications for packet switching include short bursts of data such as electronic funds transfers, credit card approvals, point of sale equipment, short files, and e-mail.

Fast packet switching is a combination of packet switching and faster networking using high-speed communications and low-delay networking. Fast packet is a hold and forward technology designed to reduce delay, reduce overhead and processing, improve speed, and reduce costs. It is designed to run on high-speed circuits with low (or no) error rates. Errors are corrected at the two ends, instead of every step along the route.

Frame relay, as stated, is a fast packet-switching technology used for the packaging and transmission of data communications. Moreover, Frame Relay packages the data into a data link layer frame (LAPF-Core Frame) used to carry the data across the network on a *permanent virtual circuit* (PVC) without all the handling of the X.25 networks. Although X.25 acknowledges every packet traversing the network, Frame Relay does not provide for ACKs or NAKs. Also when an X.25 packet is corrupted, the network node requests a retransmission, which is not so on Frame Relay. Both of the services do, however, use a statistical time-division multiplexing concept. Table 10-1 below is a summary of the comparison of X.25 and Frame Relay services.

By design, Frame Relay is focused on eliminating several of the older networking problems, yet in reality it does more to move the responsibility

Table 10-1

A Summary of the X.25 and Frame Relay Services

Service	X.25	Frame Relay
Statistical time-division multiplexing	Yes	Yes
OSI layer used	Layer 3 (Network)	Layer 2 (Data Link)
ACK and NAK	Yes, extensive	None
Retransmissions	Yes, extensive done at each node on the network	None done by the Frame Relay nodes; retransmissions are requested by higher-level protocols at the end
Packet/Frame Size	Up to 128 bytes average network; up to 512 bytes in some implementations	Up to 1,610 bytes in some networks; up to 4,096 bytes in some vendor products
Speed of transmission	Up to 64 Kbps	Starts at 56 Kbps; up to 50 Mbps, depending on the vendor products

to others than to solve any of the older network problems. What it does, however, is come up with a streamlined type of communications transmission system, eliminating the older overhead. In the original days of data communications, networks were highly unreliable, yet today newer networks are very reliable due to the increased use of fiber in the backbone networks. Consequently, the entire older overhead dealing with error recovery is somewhat superficial. Frame Relay can eliminate this overhead and use the saved capacity to carry more data. Frame Relay also assumes that if an error occurs, the higher-level protocols at the end-user level will be intelligent enough to correct the problem or request a retransmission. This can be a variable, depending on the implementation and what the end user is willing to pay for these services.

What Can Frame Relay Bring to the Table?

Frame Relay in itself is merely a communications protocol designed to eliminate the overhead discussed previously. What one can expect from Frame Relay services is the use of the higher-speed communications, the basis of the newer fiber-based *wide area networks* (WANs). Taking advantage of the capacity improvements, Frame Relay can use the *bandwidth on demand concept* to get faster data across the network. The use of excess capacities in a Frame Relay network also brings newer ideas to the forefront. By allowing the end user to use extra capacity, the bottleneck of the older networks goes away. This excess usage is called the bandwidth on demand concept.

Where People Use Frame Relay

Frame Relay is designed as a WAN technology for data only. When the deployment began, end users and carriers alike all felt that voice data could ride on Frame services. However, that aside, the network and protocols were designed to carry data traffic across the WAN. More specifically, Frame Relay was developed to carry data traffic across the WAN and link *local area networks* (LANs) to other LANs, as shown in Figure 10-1. Transmission of data across the local loop to the local telephone company's central office that is connected to the interexchange carriers' network switching system is handled by a leased T1 or T3 link. In Figure 10-1, a T1 provides the connection.

Figure 10-1

A typical Frame Relay
connection

Note also in this figure, the access device is through a dedicated Frame Relay router on both ends of the connection.

When Frame Relay was first introduced in 1992, the speeds were limited from 56 Kbps up to 1.544 Mbps. However, as time wore on, a small upstart company named Cascade Communications (acquired later by Ascend, who was then acquired by Lucent) decided to set the world on fire by increasing the access speeds and information throughput to approximately 50 Mbps. The industry quickly jumped on this speed and made Cascade the number one supplier in the industry at the time. Figure 10-2 shows the connection at T3 speeds on one end and SONET OC-1 on the other end of the connection. Note this was still used just for data transmission across the network. A higher-end router is installed on each end of the connection to facilitate the data throughput of up to 50 Mbps.

The next step was to use a different device to access the high-speed connection on a Frame Relay network. Devices known as *Frame Relay Access Devices* (FRADs) were introduced to provide the access. This was done through a high-speed CSU/DSU, through a multiplexer, or some form of a switching system. The FRAD enabled the access to be simplified for the end user and the network provider alike. In Figure 10-3, the FRAD is shown. Note

■■ ■■ ■■ ■

Figure 10-2
A higher speed
Frame Relay
connection

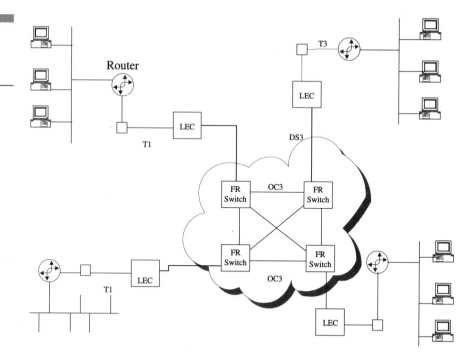

Figure 10-3
A FRAD used on each
end of the circuit at
high speeds

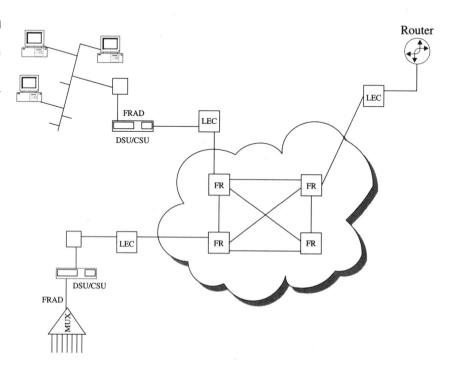

the FRAD used here is through a CSU/DSU on one end and a high-speed T3 multiplexer on the far end. The choices vary for the end user, so the flexibility of the access is one of the strong points for the Frame Relay network.

The Frame

When Frame Relay was developed, the important part of the data-carrying capacity was the use of the frame to carry the traffic and not have the same overhead as an older technology (such as X.25). The frame was filled with data as necessary, but it handled the speed and throughput via the high-speed communications and lower overhead.

In Figure 10-4, the frame is shown. The frame is an HDLC-framed format, as shown in this figure. The beginning of the frame (as with most HDLC formats) starts with an opening flag. Next, a two-byte sequence defines the addressing of the frame. This is called the *Data Link Connection Identifier* (DLCI). By very nature of the title (DLCI), we can assume that Frame Relay works at the data link layer. The DLCI is comprised of several pieces of information, shown later, but is normally a two-byte sequence. Provisions have been made to enable the DLCI to expand to up to four bytes, but very few implementations occur using more than the two-byte address. Following the DLCI is the information field. This is a variable length field. The initial standard allowed for a variable amount of data is up to 1,610 bytes. This is sufficient for most installations, but change always occurs when things are stable. The reason for the 1,610-byte field is to handle a frame from a LAN using a full frame of Ethernet traffic.

The Ethernet frame can be as large as 1,518 bytes (with overhead) and some subnet access protocol overhead (snap); the full frame should therefore accommodate the 1,610 bytes. The Ethernet frame is shown in Figure 10-5. This frame is the same size for an 802.3 IEEE frame or for a DIX Ethernet frame. Therefore, the variable data frame is sufficient to carry the traffic loads necessary.

Following the data field in the Frame Relay frame is a CRC used only to check for corruption. The CRC determines if the frame or the address

Figure 10-4
A Frame Relay frame

Figure 10-5

Ethernet and IEEE 802.3 frames fit into the Frame Relay frame

Figure 10-6

Modified frame size of the Frame Relay information field

information is corrupt. If so, the frame is discarded; if not, the frame is forwarded. There is no ACK or NAK in the Frame Relay transmission along the route. Lastly, there is a closing flag on the frame, indicating that the transmission of the frame is ended and the switching system can then process the entire frame. In many cases when a variable data field is used, the switches must allocate enough buffer space to hold a full frame, regardless of how full each frame is. This is somewhat wasteful across the WAN but does provide the necessary flexibility to handle the traffic.

Shortly after Frame relay was introduced with the 1610-byte information field, a new issue cropped up. What about the clients who use an IBM token ring? The token ring LANs can carry a variable amount of information up to 4,068 bytes. This means that a frame in the Frame Relay world is not large enough to carry a full token and the data must be truncated into three tokens to accommodate the token ring. To solve this problem, the frame was expanded to accommodate up to 4,096 bytes in the information field. Not all suppliers supported this change, but the two major suppliers of Frame Relay products (Cascade and Nortel) both adjusted their systems to accommodate the larger frame. This frame is shown in Figure 10-6 with a variable frame size of up to 4,096 bytes.

The OSI Protocol Stack and Frame Relay

When we discuss the use of the data link protocols, one always compares the OSI to whatever other protocol is being discussed. This book is no different, because one needs to understand where Frame Relay falls on the OSI stack and what Frame Relay's purpose is. The development of any new set of standards is usually done to improve network performance. Frame Relay works at the data link layer to reduce the overhead associated with the movement of data across the wide area. Because we refer to Frame Relay as a wide area networking technology, it is natural that the protocols will work with the improvements made in the network over the past decades.

In the older days, data was shipped across the layer three protocols (such as X.25) to assure the reliability and integrity of the data. This is because the networks back in the 1970s were unreliable, so the protocols were put in place to accommodate this network flaw. The X.25 protocol worked at layer three, as shown in Figure 10-7. The overhead associated with the transmission and reception of the data on the X.25 networks was inordinate. To facilitate better data throughput and eliminate some of the overhead, Frame Relay was developed. The comparison of Frame and X.25 to the OSI model is shown in the figure. Two things were in place, however, to allow the use of Frame Relay instead of X.25:

■ The networks were now improved through the mass deployment of fiber-based networking technologies and the use of SONET protocols.

■ The networking strategies of many end users were based on router technologies and LAN to WAN communications, instead of the older terminal to host intercommunications.

These two changes actually revamped the way we communicate. No longer did we have to use a timing relationship, as in the older data networks. Any form of data transmission could be accommodated across the newer improved techniques and protocols.

With this comparison in mind, one will note that the Layer 2 protocol (in this case, Frame Relay) eliminates some of the overhead associated with

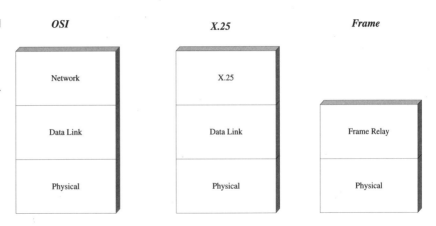

Figure 10-7
OSI compared to Frame and X.25 stacks

the transmission of data. The need for network addressing using Layer 3 is reduced, because many of the link architectures are based on point-to-point circuits or private networking techniques. Moreover, where the network address was required to send the data to its end destination, Frame Relay uses the DLCI as the PVC connection. Therefore, by using PVCs, the routing of the data traffic is predetermined to occur across a highly reliable direct connection to the far end. Switching and routing decisions are not required once the connection is established, because all the traffic for this connection between two end nodes follows the same path. This mapping is done through the use of the DLCI address to pre-subscribe the connection in a virtual circuit connection, as shown in Figure 10-8. The only time the traffic might traverse some other route is when a link failure occurs, but this is already mapped in the logic for the connection. This is shown in the routing example in Figure 10-9 where the end nodes are pre-mapped in the Frame Relay switches across the network.

By using this arrangement of DLCI mapping across the network, the network can also accommodate various other types of traffic, such as IBM *Systems Network Architecture / Synchronous Data Link Control* (SNA/SDLC) traffic, which is very time-sensitive and times out if the traffic does not arrive in time. Moreover, if a customer is using an older form of interactive terminal traffic using some older 2770/3770 bisynchronous protocols, these can be placed into the frame.

The concept of placing the traffic inside these frames is called tunneling. Clearly what happens is the traffic is sent inside the frame transparent to the network. No checks or validations are made on the data through the network, reducing some of the delays of data handling. Also, while tunneling through the network, the data is actually encapsulated inside the Frame Relay frame, so the only place where the data is actually enacted on is at the two ends. There is some minor overhead associated with the use of these other protocols, called *subnet access protocols* (SNAP), but the overhead is minimized. This tunneling concept is shown in Figure 10-10 where the data is encapsulated inside the frame (tunneled).

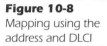

Figure 10-8

Mapping using the address and DLCI

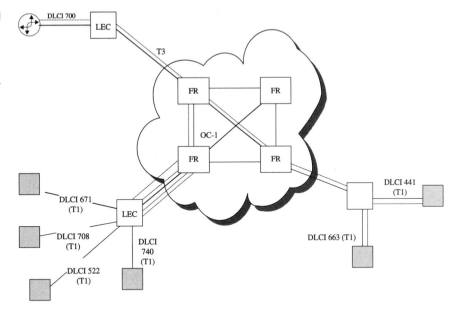

Figure 10-9
A typical mapping of the DLCI in a Fame Relay network

Figure 10-10
Traffic is tunneled in the Frame Relay frame using a small amount of overhead.

The term tunneling is getting a lot of press these days because of the Internet and the *Internet Protocols* (IP) using *Virtual Private Networking* (VPN) by tunneling through the Internet with a private link. This will be discussed in a later chapter in greater detail but is shown in Figure 10-11 with TCP/IP tunneled into a Frame Relay connection.

Frame Relay Speeds

As a means of keeping everything in order, it is appropriate to discuss the speed that can be achieved with the use of Frame Relay. Although in the beginning of this chapter, it was stated that Frame Relay was designed for speeds up to T1/E1 (1.544–2.048 Mbps); it later evolved to speeds of up to 50 Mbps. Actually, few end users have ever implemented Frame Relay at the higher speeds; this is more of a speed for the carrier community, but the need for stepped increments has always been a requirement for data transmission. Therefore, Table 10-2 is used to show some of the speed increments typically used. Other rates are possible in 4 Kbps increments, but implementations are normally done at the speeds shown in Table 10-2.

Figure 10-11
TCP/IP traffic is tunneled into a frame.

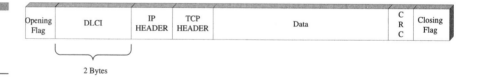

| Opening Flag | DLCI | IP HEADER | TCP HEADER | Data | C R C | Closing Flag |

2 Bytes

Table 10-2

Typical speeds used in Frame Relay

Frame Relay Access	Typical Committed Speed	Average additional Burst Speeds
56 Kbps (DS0 or ISDN)	32 Kbps	24 Kbps
64 Kbps (Clear channel DS0 or ISDN)	32 Kbps	32 Kbps
128 Kbps (ISDN)	64 Kbps	64 Kbps
128 Kbps (ISDN)	128 Kbps	0
256 Kbps	128 Kbps	128 Kbps
256 Kbps	192 Kbps	64 Kbps
384 Kbps	256 Kbps	128 Kbps
512 Kbps	384 Kbps	128 Kbps

Frame Relay Access	Typical Committed Speed	Average additional Burst Speeds
1.544 Mbps	512 Kbps	256 Kbps
1.544 Mbps	1.024 Mbps	512 Kbps
2.048 Mbps[1]	1.024 Mbps	1.024 Mbps

[1] In most implementations, when a customer exceeds 256 Kbps access, the normal installed link for access is a T1 in North America at 1.544 Mbps. This is a pricing and an availability situation.

Frame Relay Access

A link is installed between the end-user location and the network carrier's node. The normal link speed is T1, although many locations can and do use ISDN or leased lines at lower rates. Some customers may choose to install a local loop at speeds up to T3 (45 Mbps approximately) to support higher speed access and faster data throughput. (In most implementations, when a customer exceeds 256-Kbps access, the normal installed link for access is a T1 in North America at 1.544 Mbps. This is a pricing and an availability situation).

In many cases, the use of the T3 will also allow for consolidation on the same link. Many of the carriers (and in particular the LECs) will offer the T3 access and allow Frame Relay throughput at rates up to 37 or 42 Mbps. Now the *Incumbent Local Exchange Carriers* (ILECs) are offering flat-rate services and bundled capacities to be more attractive to their end users. The ILECs are the local telephone companies with their installed base of services and facilities. Often these ILECs are offering high-speed Frame Relay services LATA-wide (or statewide depending on the geographic topology of the state and LATA boundaries) all for competitively priced services. Better yet, some of the ILECs offer high-speed Frame Relay at speeds up to 10 Mbps across the LATA at competitive rates on either T3 or an OC-1. The 0–10 Mbps bursts of data are designed for the very large customer, but they may fit smaller organizations needing broadcast (or near broadcast) quality for voice and video applications in the future. With the pricing mechanisms in place by the ILECs, this is closer to reality than one might believe (possibly pre-2000).

Figure 10-12 is an example of the connection installed at a large organization yet fed across the network by lower-speed feeds from branch offices at T1 and lower rates. This scenario is likely the most common implementation for the near term but will shift as the pricing model becomes more conducive to the smaller organization.

Figure 10-12
Examples of
connections at
higher speeds

Overall Frame Relay Core Protocols

When the Frame Relay specification was developed, the primary goal was to carry data over the WAN. To handle this form of wide area communications, the core protocols for Frame Relay were established using the revised version of the data link protocols. Instead of using the network layer protocols, Layer 3 was gleaned down to efficiently carry the traffic while performing the same form of function as the network layer. Moreover, the data link layer was also streamlined to offer less overhead and processing on a link-by-link basis. Because the circuits across the wide area are much more reliable and error-free (thanks to fiber optics), the ACK and NAK functions can be eliminated. Furthermore, the use of PVCs in the connection eliminates the need for the sequence numbering on the link. One can assume that if we send multiple frames onto a circuit between two end points (even if it is a virtual circuit) the data will come out in the same sequence that it went in on the other end. Unless a frame is discarded, there should be no way that the data will arrive out of sequence. Because the data should not arrive out of sequence, there should not be a need to do the counting.

If, however, something goes wrong on the circuit, how then do we recover? The answer is that we rely on the upper layer protocols on both ends of the circuit (the transport layer) to recognize if the data is missing. If a frame is lost, then a transport will request a retransmission from the sender. This

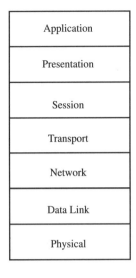

Figure 10-13
The core protocols
for Frame Relay
mimic the ISDN
standards.

eliminates much of the processing and checking at each node across the link. In Figure 10-13, the core protocols are shown for Frame Relay using a subset of the Q.922 data link layer for the actual link protocols.

Q.922 is an ITU-specified protocol for ISDN. Frame Relay was developed as an offshoot of the ISDN protocols because of the amount of work invested in the technologies and standards. As a result, many of the core protocols for Frame Relay mimic the ISDN standards. This is helpful in understanding why and how the networks have evolved throughout the past 10 years to improve native throughput and increase the acceptance of the standards.

Voice Over Frame Relay

Just as the industry was getting used to the idea of reduced overhead for data transmission, some radical thoughts began to surface in the industry. In the past, all data ran over voice networks, adjusted and accommodated according to voice standards. But what if voice could run over a data network instead? By using the capability of reduced overhead, more reliable circuits, and faster throughput, the network could be tuned to accommodate voice in the form of packets of data.

Much of the pressure for voice over any data technology has been based on cost in the past. Newer ideas are based on efficiency and the convergence of the network protocols and services. Voice is fairly inefficient! Actual voice traffic is carried only about 25 percent of the time we are on a connection.

The rest of the time, we are sending silence (no information). If we can integrate the two networks and carry interleaved traffic (voice or data), then we can efficiently fill the network with traffic all the time, instead of just sending idle conditions.

Thus, a new concept was born that could be accomplished through compression and interleaving the data. Therefore, on a digital circuit, data is data and voice is too! The voice is just a data stream of ones and zeros. The Frame Relay Forum has been busy defining the standards for three service offerings, one of which is the *Voice Over Frame Relay* (VoFR) specification. In conjunction with VoFR offerings, the forum was busy developing other protocols and specifications to support some of the unique challenges with real time data on a network. This includes the use of PVC fragmentation protocols and multilink Frame Relay services.

The fragmentation protocols are necessary to support the different types of delay experienced on the network for time-sensitive traffic (that is, voice). Interleaving the voice communications (using small frames) onto a high-speed data link with larger frames is one way to handle this need. This sharing of the same physical link allows both real-time and non-delay sensitive traffic to coexist yet receive separate treatment as it moves across the network. The fragmentation of the traffic allows variability, depending on the speed of the link, the congestion, local timing needs, and the type of service being used. This makes the implementation of fragmentation available at various interface points. In Figure 10-14, the use of a fragmentation procedure is shown in three different places: at the UNI, at the NNI, and on an end-to-end basis. These three ways allow enough flexibility to accommodate the different types of service being delivered across the Frame relay network.

Figure 10-14

Three places where fragmentation can take place in support of Voice over Frame Relay (VoFR).

With the implementation of VoFR, particularly with carrying international telephony traffic, the Frame Relay Forum introduced the specification FRF.11 that deals specifically with the voice side of the business. This specification goes beyond the possibilities of fragmenting the data and incorporates the necessary steps of call set-up and call teardown. The other major issues that the FRF.11 deals with are as follows:

- Analog to digital conversion
- Digital to analog conversion (back to analog)
- Compression techniques
- The sizing and transmission of a frame of traffic

Many of the specifications involve more sophisticated steps, such as handling various forms of analog traffic, including voice, fax, and compressed video communications. To accommodate that need, the FRF.11 handles a specification dealing with multi-service multiplexing. This covers the ability to multiplex multiple voice and data channels on a single-frame relay connection (or PVC). A gateway function is used in many cases to handle the various informational streams multiplexed onto a single PVC, as shown in Figure 10-15.

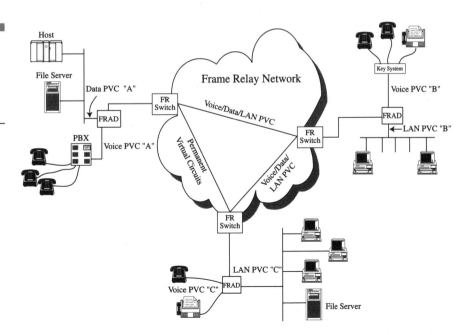

Figure 10-15
A Frame Relay gateway acts as the multiplexer of different services onto single PVC.

Compressing the Information on VoFR

By using some industry-accepted standards for compression techniques, such as *Adaptive Differential Pulse-Coded Modulation* (ADPCM) or *Code-Excited Linear Predictive Coding* (CELP), the conversation can be compressed from 64 Kbps to a data stream of 40, 32, 24 or 16 Kbps using ADPCM compression techniques, and down to 5-8 Kbps using the newer CELP standards.

Following this idea, the next step is to view how a service bureau approach might work using Frame Relay to act as an international callback service, or in the case of a corporation with multiple international locations, this service can be used for the intranet calls. One cannot underestimate the robustness and power of the Frame Relay networks. In many cases, the carriers are offering throughput across their backbone in the proximity of 99.99 percent of the *Committed Information Rate* (CIR) and with an availability of 99.5 percent or better. These two statistics absolutely beat anything we have ever seen from voice or data networks in the past. Notwithstanding the capability to achieve the throughput and availability statistics, the carriers will sign a *service-level agreement* (SLA) with these guaranteed throughput and availability characteristics without hesitation. Never before have we seen such confidence and acceptance of a single-standard interface and network topology to carry the WAN traffic.

Still, when one considers the possibility of running voice data across a network, it is uncomfortable to think that the voice networks have truly not kept pace with the developments of the data networking strategies. Do we really need another packet-switching technology, rather than improve the ones we already have in place? The answer can be found in the overall characteristics of the Frame Relay networks already discussed here.

Asynchronous Transfer Mode (ATM)

Introduction

In 1992, a group of interested parties developed a set of standards-based specifications called the *Asynchronous Transfer Mode* (ATM). This was a step at developing a single set of standards for the integration of voice, data, video, and multimedia traffic on a single backbone network. Prior to this development, the industries offered separate standards and networks for voice, others for data, and still others for video communications. Above and beyond that, data networks were also treated differently with separate data networks for point-to-point needs, dial-up data, and packet-switched data transmissions.

Beyond being expensive, this concept also proved to be confusing to users of the networks. Should separate networks be used or should an integrated approach be sought? Even beyond that, those few brave souls who tried the integrated approach were destined to more confusion because of the carrier offerings being different. What the industry needed was a single way of handling all forms of traffic and one network service that could carry the different forms of traffic. The end user was looking for something that would clear up the confusion and make life simple. Hence, ATM was born.

Unfortunately, ATM can cause as much confusion as the older techniques, strictly by virtue of its name. Why is ATM called asynchronous? Wasn't it designed to run on a synchronous networking platform and support synchronous traffic? The answer is yes! One can see that confusion reigns in this industry just because of the naming conventions used to describe the various protocols and specifications.

What Is ATM?

ATM is a member of the fast packet-switching family called *cell relay*. As part of its heritage, it is an evolution from many other sets of protocols. In fact, ATM is a statistical time-division multiplexed form of traffic that is designed to carry any form of traffic and enables the traffic to be delivered asynchronously to the network. When traffic in the form of cells arrives, theses cells are mapped onto the network and are transported to their next destination. When traffic is not available, the network will carry empty (idle) cells because the network is synchronous. In Figure 11-1, a representation of the traffic delivery is shown and Figure 11-2 shows the carrying of idle cells across the network.

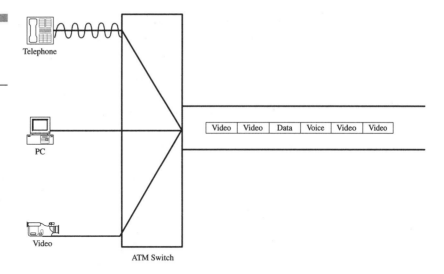

Figure 11-1
Traffic is mapped
onto the network
as it arrives.

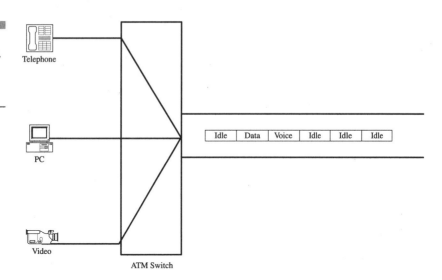

Figure 11-2
When no cells arrive,
idle cells are
transported across
the link.

Therefore, what we can derive is that the ATM technique is a combination of *Time Division Multiplexing* (TDM), with cells using preassigned slots, and Statistical TDM, with cells using whatever slots are available

or needed to handle a particular traffic flow. This offers the carrier and the network user the best of both worlds. It is also a connection-oriented protocol much the same as dial-up voice communications services, but it uses virtual circuits, such as *permanent virtual circuits* (PVC) and *switched virtual circuits* (SVC), to handle the connection. The main thrust behind the fast packet-switching arrangement is the fixed sized cell that is employed to carry the end user traffic across the various connections required.

Another area in which ATM stands out from other technologies and protocols is in the places where it is used. ATM was designed from the ground up to work across the various places where we communicate: the *Local Area Network* (LAN), the *Campus Area Network* (CAN), the *Metropolitan Area Network* (MAN), and the *Wide Area Network* (WAN). Table 11-1 summarizes the differences and locations of the capabilities of ATM in comparison to other sets of protocols.

Why the Interest in ATM?

When one considers the disappointing capacities of past technologies, we can see why there is the hype for ATM. ATM will be the basis of many of our future broadband communications systems; as such, it starts where other technologies stop. Many organizations have escalated their demands and needs for raw bandwidth, yet no single entity has emerged as a clear-cut winner to deliver the services necessary to support the demands of today's multimedia applications. Table 11-2 compares the capacities of ATM to the

Table 11-1

Summary of Where Various Protocols Are Used

Technology	LAN	CAN	MAN	WAN
Dial-up network			X	X
Leased line			X	X
Ethernet	X			
Token ring	X			
FDDI		X		
X.25 networks				X
Frame Relay				X
SMDS			X	
ATM	X	X	X	X

Table 11-2

Summary of Speeds for Various Technologies

Technology	Application	Speeds
Dial-up analog switched	Modem communications	Up to 56 Kbps circuit using current technologies. Stops at 33.6 Kbps using older modems.
Leased line	Point to Point	T1 @ 1.544 Mbps normal, T3 @ 44.736 Mbps occasionally.
Ethernet	LAN Shared/ Switched bus	10–100 Mbps normal, 1,000 Mbps emerging.
Token ring	LAN shared ring switched	4–16 Mbps normal, or 1,000 Mbps under consideration.
FDDI	Shared ring	100 Mbps normal.
X.25	Dial up or leased line	56 Kbps top end in North America, 64 Kbps top end in rest of world.
SMDS	Shared/switched bus	Up to 34 Mbps common.
Frame Relay	Point-to-point PVC	Starts at 56 Kbps normally, top end is 50 Mbps.
ATM	Switched PVC	Starts at 1.544 Mbps, top end is 622 Mbps today, 2.488 Gbps for future application.

other techniques we used in the past. This will give the reader a chance to see what the excitement is all about.

A lot of discussion is always bantered in the industry regarding the best way to handle data and voice communications. There is a continual discussion on the difference between using ATM for the higher-bandwidth intensive applications and the use of gigabit Ethernet for the same applications. One cannot predict what the outcome will ultimately be, but it keeps the end user totally confused and uncertain. What could happen is that a complement of both techniques will come rolling out to serve the application and support each other.

The use of ATM over other technologies is, however, an attractive alternative when considering the various aspects of the options. For example,

Table 11-3

Summary of ATM
Features and
Functions

One technology for voice, data, video, and multimedia
Bandwidth on demand as needed
Scalable as needs dictate
Quality of Service (QoS) is well defined
Management systems and services pre-built into ATM
Hardware-based switching instead of complicated routing and software schemes

ATM offers the following types of service, shown in Table 11-3, which may or may not be defined in other protocols.

ATM Protocols

It takes many protocols to support an ATM network, which is one of the issues that continually comes up as a negative from the supporters of the gigabit Ethernet crowd. To develop the necessary interfaces in support of the various points within a network (networks are pretty complex in themselves), different protocols are necessary. The actual protocols needed depend on where the traffic originates, what transport mechanisms must be traversed, and where the traffic will terminate. To see this in a clear picture, a summary of protocols for the ATM user is shown in Table 11-4 and also shown graphically in Figure 11-3.

One can see the reason why the gigabit Ethernet proponents support a simpler and less expansive set of protocols when dealing with LAN technologies, now moving toward a WAN protocol with Layer 3 switching being developed. The problem with ATM is that in order to support the older legacy systems, many protocol points and interfaces are necessary. To get around the problem of "fork lift" changes, the necessary protocols have been developed. Due to its protection mechanism for existing systems, ATM is its own undoing from the opponents of the technique. Too often the opponents cite the different protocols needed, instead of what they are actu-

Figure 11-3

Graphic
representation of
the ATM protocol
interfaces

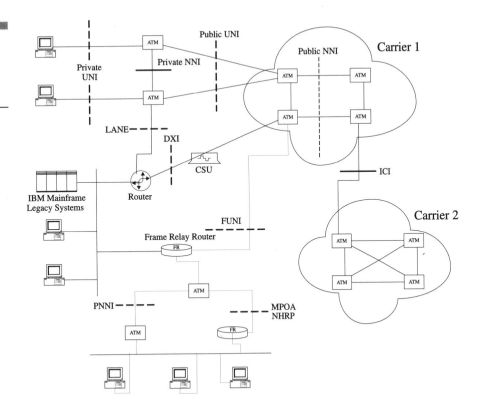

Table 11-4

Summary of Where
Protocols Are Used
for ATM

Location	Protocol/Specification
End user to LAN	Private User to Network Interface (Private UNI)
End user to WAN	Public User to Network Interface (Public UNI)
Between network nodes in a WAN	Public Network Node to Network Node Interface (Public NNI)
Between nodes in a Interprivate WAN	Private Network Node to Network Node face (Private NNI)
Between carriers in a public network	The Intercarrier Interface (ICI)
Between a Frame Relay and an ATM interface device	Frame User Network Interface (FUNI)

continues

Table 11-4

Continued

Location	Protocol/Specification
From a legacy router to the network	Data Exchange Interface (DXI)
On a LAN-to-LAN interconnection	LAN Emulation (LANE)
Between LAN nodes and other network UNIs	Next Hop Routing (Resolution) Protocol
On a LAN, PBX, or CAN interface	Multi-protocol over ATM (MPOA)

ally doing. However, the use of each of these protocols satisfies a specific need and should not be taken strictly as an overhead problem, but as an interworking function. If all one can do is complain about the various protocols, then the true benefit of ATM is lost in the mire.

Mapping Circuits Through an ATM Network

ATM uses one of two connection types. The protocol is connection-oriented, so the two choices are a PVC or a SVC. There is actually no permanency to the circuits. They are logically mapped through the network and are used when needed for PVC or dial-connected when using the SVC. In either case, the carriers promise only to make a best attempt to serve the needs of the end user when the time is appropriate. With no true guarantees, the consumer is at risk (sort of). However, the concept is that the network provider will provide a committed bandwidth available to the user on-demand whenever the user wishes to use it. This forms the basis of what ATM networks are all about: on-demand, high-speed communications networks. The connection is built into a routing table in each of the switches involved with the connection from end to end. As such, the switches only need to look up a table for the incoming port and channel, and then determine the mapping (in the same table) for the output port and channel. Using _virtual path identifiers_ (VPI) and _virtual channel identifiers_ (VCI), the carrier maps the table, as shown in Figure 11-4.

The structure of the link is shown in Figure 11-5, where a full virtual connection is mapped through the various switches across the network. Here the

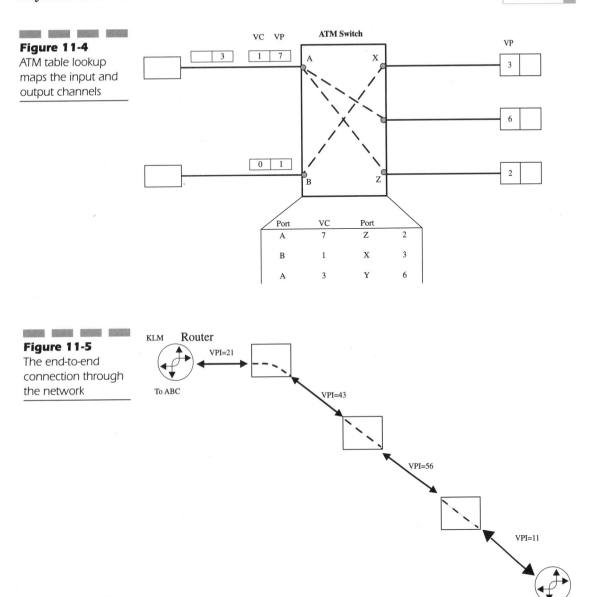

Figure 11-4
ATM table lookup maps the input and output channels

Figure 11-5
The end-to-end connection through the network

end-user device is connected across an ATM access link through a switch. The switches provide the cross connection and link to the next downstream node. Note that the connection from the end user to the network may be on a T1, T3, or OC-n. From the first switch out, the network will use SONET or *Synchronous Digital Hierarchy* (SDH) capabilities possibly mapped onto a

Figure 11-6
Virtual Path switching re-maps the path but keeps the channel the same.

Figure 11-7
The virtual path and virtual channel switches process and re-map both elements.

Dense Wave Division Multiplexer (DWDM). The network carrier will use whatever services and bandwidth is available at the connection points.

In Figure 11-6, the network switches handle the mapping on the basis of VPI switching. VPI switching means that the switches use the virtual path for mapping through the network and will re-map from one virtual path to another, while the virtual channel number is held consistent through the entire network.

A second alternative is to use VPI/VCI switching, as shown in Figure 11-7. In this case, the ATM switches along the route will switch and re-map both on a virtual path and a virtual channel. This is an installation-specific arrangement depending on how the carrier chooses to handle the switching.

The ATM Layered Architecture

Nearly all documents describe the layered architecture of every protocol against the OSI model as a reference only. The ATM architecture is no different when trying to compare what the protocol is doing. Using the OSI model as a base reference, the ATM layers fall typically in the bottom two layers (data link layer and physical layer) of the architecture. ATM has been designed to run on a physical medium such as SONET.

In Figure 11-8, the ATM layer is shown as the bottom half of Layer 2 in its equivalency. There is no real way to draw true one-to-one mapping of the ATM and OSI models, but for purposes of this document, it is done that way. Now from the bottom half of the data link layer is ATM, but below the ATM layer is the physical layer such as SONET or some other physical media dependent layer (SDH, DS3, and so on).

Moving up the architectural map, we have the upper portion of Layer 2, the LLC equivalent, called the *ATM adaptation layer*. Within this portion of the Layer 2 protocol stack, several sublayers are seen, depending on the services required. For example, using Figure 11-9 as a reference, the uppermost portion of the layer is called the *service specific convergence sublayer* (SSCS). This portion of the protocol stack is used when mapping Frame Relay, SMDS, or another protocol to the ATM adaptation process. Under the SSCS is the *common part convergence sublayer* (CPCS). The combination of the SSCS and the CPCS make up the *convergence sublayer* (CS). Conver-

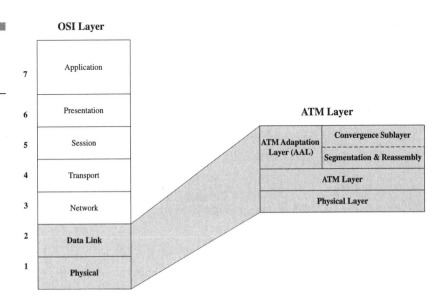

Figure 11-8
Comparing the OSI and ATM layered models

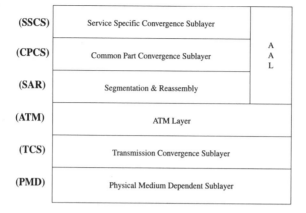

Figure 11-9

Upper layer services of ATM

(SSCS)	Service Specific Convergence Sublayer	A A L
(CPCS)	Common Part Convergence Sublayer	
(SAR)	Segmentation & Reassembly	
(ATM)	ATM Layer	
(TCS)	Transmission Convergence Sublayer	
(PMD)	Physical Medium Dependent Sublayer	

Table 11-5

Types of AAL and Services Offered

Class of Service	Type of Service			
	A	**B**	**C**	**D**
Timing	Synchronous	Synchronous	Asynchronous	Asynchronous
Bit rates	Constant	Variable	Variable	Variable
Connection Type	Connection Oriented	Connection Oriented	Connection Oriented	Connectionless Oriented
AAL	Type 1	Type 2	Type 3-4	Type 3-4-5

gence, as the name implies, is the changing and melding of the data into a common interface for the ATM networks. Following the CS portion of the upper layer is the next sublayer called the *segmentation and reassembly* (SAR). The SAR is where the data is prepared into a 48-byte payload prior to being submitted to the ATM layer below for the header. The CS and the SAR combine to form what is known as the *ATM Adaptation Layer* (AAL). Table 11-5 shows the combination of the AAL types, the services being provided, and the working relationships between the AAL and the type of service.

ATM Traffic Management

When dealing with traffic management, some of the goals of the ATM Forum and other developers include the following:

- *ATM must be flexible.* It must meet the constantly changing demands of the user population. These goals mean that the demands for traffic will rise or fall as necessary, and therefore managing this traffic is of paramount importance.

- *ATM must meet the diverse needs of the end-user population.* Many users will have varying demands for both high- and low-speed traffic across the network. Using a *quality of service* (QoS) capability throughout the ATM network, a user can determine the performance and the capabilities of how the ATM network will meet their demands. These demands must be met in terms of the delay or the actual delivery of the cells across the network. Diverse needs are always going to be different, depending on the type of service (voice, data, video, or other traffic). Meeting these diverse needs for multiple users across the single threaded network is a major goal in traffic management and traffic delivery.

- *Cost efficiency is a must.* If ATM is truly to succeed, traffic management must also include the effective usage of all of the circuitry available. ATM is designed to reduce the inefficient circuit usage by efficiently mapping cells into dead spaces, particularly when data is involved. In the past, variable amounts of data would be sent across the network. Although it is good for data because variations exist, such as the use of a fixed cell size and managing that fixed cell throughout the network in terms of its performance, buffering and delay become the crucial issues addressed.

- *Robustness in the event of failure or in the event of excess demand is a requirement of the traffic management goals.* If the network is to be readily available for all users to be able to transmit information on demand, then the network must be very robust to accommodate failures, link downtime, and so on. Through this process, the managing of traffic must accommodate such diverse needs on a WAN.

Fair allocation of traffic capacity is essential on a fair and equitable basis. The goal of the traffic management scenario is to ensure that no one user would dominate the network; rather, all users would have equal access and an equal shot at using the capacity on demand. Specific goals of delivery can be achieved through committed information capabilities, but the intent is really to fairly arbitrate the traffic capacity and divide it up among multiple users. Traffic management is a set of actions taken by network devices to monitor and control the flow of traffic on the network. A highway, for example, is built to carry a certain amount of traffic. Any more traffic at a given time (a one-hour period for example) causes congestion. Congestion causes frustration and forces some traffic to overflow to other roads. Each stream of traffic onto a network can carry a finite amount of flow. If the flow exceeds the capacity (bandwidth), then actions must be taken to minimize delays (red lights and green lights are used for flow control), control losses, redirect or discard traffic, and prevent collisions.

When using ATM networks, traffic management becomes critical. Too often the networks are built around fixed resources, which are finite and must be managed to provide equitable access and bandwidth to the end user. Network suppliers and carriers are therefore under constant pressure to get the best utilization from the networks.

Contention Management

Traffic in the form of asynchronous bursts of information (cells) enters the network at random times. This randomness is what causes the confusion and the unpredictability of the data. To manage the traffic flow, buffers are used to allow the flow and ebb of traffic volumes. Because data tends to be very bursty, it is extremely difficult to predict the demands of the network and the capacity needed at a given time. Therefore, when sufficient resources are not available, the use of buffers helps to offset the immediate demands. It is this bursty traffic that produces a contention for the network resources.

The use of "leaky buckets" in the buffering of the traffic helps to manage and control the flow of traffic onto and through the network. The leaky bucket, as the name implies, is a buffer that is constantly flowing. In Figure 11-10, a leaky bucket concept is shown with the two stages of buffering. Traffic enters into the buffers and is tagged, based on the amount of cells allowed by the carrier. If the user exceeds the amount of cell flow per increment (per second, and so on), then the buffer is filled and begins to empty out the bottom side. If more cells enter the buffer than are allowed, the cells are flagged for discard. A *first in, first out* (FIFO) concept is normally used to handle the traffic as it flows, but the end user may flag specific traffic according to application, priority, and the like.

ATM involves finding values that the network needs and making decisions about how the network will perform. These decisions involve taking action to ensure its availability to users and to ensure that all current connections are receiving adequate service based on their class of service. To facilitate these decisions, the following performances are examined:

- *The cell loss ratio.* the ratio of lost cells to the sum of the total number of lost and successfully delivered cells.

Figure 11-10
Leaky buckets allow
for buffering of the
traffic

Cells enter the network at 1/SCR

- *Cell insertion rate*   the number of cells inserted into an ATM network within a specific period of time.

- *Severely errored cell ratio.*   the ratio of severely errored cells (more than one-bit of an error in the header) to the number of successfully delivered cells.

- *Cell transfer capacity:*   the maximum number of successfully delivered cells over a specified ATM connection during a period of time (one second).

- *Cell transfer delay:*   the average delay and the arithmetic delay of a specified number of cell delays. The *cell delay variation* (CDV) is the difference between a specific cell delay and the average. This variation causes the most problems, especially with real-time voice and video.

- *Priority control:*   networks must adequately service buffers in the network nodes under all kinds of conditions. When congestion occurs (too many cells in the network), a priority mechanism can be used in the following ways:

  - *To remedy the situation.*   Some cells can be discarded under congestion circumstances.

  - *For congestion control.*   Networks must prevent congested conditions from spreading throughout the network. In this case, no sender should be allowed to overwhelm any receiver, as the network will accommodate by discarding cells.

- *As a Generic Cell Rate Algorithm (GCRA).* The GCRA is an example of a generic term. It refers to a virtual scheduling algorithm or a continuously leaky bucket and expresses a complex series of formulas. In most network implementations, it is widely known that the double leaky bucket algorithm is what we call the GCRA. Using the leaky bucket as a means of describing the GCRA is a good way to figure out exactly of what is happening. The GCRA functions like a bucket with a hole in the bottom. The bucket leaks at a steady rate, no matter when water is poured into the bucket. The bucket may be initially empty, partially full, or full to the brim. As water pours in, it may splash or completely overflow the bucket if poured in from a much larger container. Water emerges from the hole in the bucket at precisely the same rate at all times. (Of course, everyone knows there are no real "buckets" in computers and networks, although the term "bit-bucket" is constantly used.)

- *As a counter or buffer.* The simplest leaky bucket implementation is a counter. The counter has a minimum value (usually 0) and a maximum value. With the empty bucket, it holds up to 100 cells. All of a sudden, 150 cells are delivered to the network. This means that the bucket will overflow or the 50 cells will be discarded because the buffer can only hold 100. At the bottom, the cells enter the network at a ratio of 1 over the *sustained cell rate* (SCR). Cells will exit at a steady rate.

The Double Leaky Bucket

In this double leaky bucket case, which was used in most of the early implementations of ATM, cells arrive from the CPE across the UNI totally uncontrollable in time. All cells leaking into the network are sent with a *cell loss priority* (CLP) equal to zero. As long as the arriving cells do not exceed a given rate, they are admitted to the network unchanged. If the cell arrival rate exceeds the limit imposed by the network, namely the *sustained cell rate* (SCR) of the connection, the cells in excess of this limit have their loss priority bits changed to one. The cells with a CLP equal to one are subjected to another leaky bucket with another limit. Any cells sent with a CLP equal to one already set are also added to the cell stream. However, this time the limit corresponds to the maximum burst size of the connection. Cells under

Figure 11-11
A double leaky
bucket

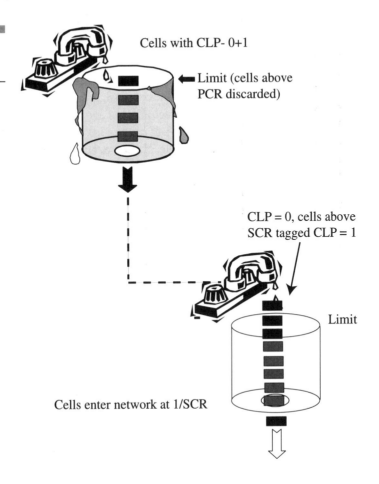

Cells with CLP- 0+1

Limit (cells above
PCR discarded)

CLP = 0, cells above
SCR tagged CLP = 1

Limit

Cells enter network at 1/SCR

the MBS are admitted onto the network, while any others are discarded.
This double leaky bucket is shown in Figure 11-11.

Here are two functions in ATM network implementation that perform
traffic control:

- *The connection administration control (CAC):* networks must set
aside the proper amount of resources to service a connection at the
time of the connection, whether it was set up as a service provision
time on a semi-permanent basis, or by means of a signaling protocol
dynamically.

- *The usage parameter control:* networks must police the *user-to-
network interface* (UNI) to make sure cell traffic volumes do not affect
overall network performance.

Traffic control and management can also be based on a credit and debit type system. As traffic enters the buffers, a certain amount of cells are authorized to enter the buffer. As the cells fill the buffers, they use up the credits allowable. When any extra cells are introduced into the buffers, then a second stage buffer can be used for the traffic as being eligible for discard. If, in fact, the eligibility for discard is exceeded, any new cells are automatically flushed away. These mechanisms are used to control the flow of data into the network. If some type of control is not used, the switches across the wide area can get into trouble. Buffer overruns and system overloads can also occur. If this continues to occur, the network suppliers cannot provide or live up to their guarantees for service levels. The continuous overruns across buffers and switches will cause severe congestion and snarl the network to a grinding halt. Data from real-time applications will be delayed, requiring a request for retransmission from the upper layers in the protocol stacks, thereby causing even more congestion. This is not a situation that can be taken lightly. To resolve the situation quickly, several traffic functions are used to manage the network and deliver the promised QoS, that elusive term everyone banters around but no one understands.

Table 11-6

Comparing the
Categories of
Service

Service Category	Equates to
Constant bit rate (CBR)	The equivalent of a dedicated point-to-point leased line. This type of service is used when a specific amount of throughput is required, but it changes infrequently. The end user can commit to a fixed bandwidth and throughput and pay accordingly for the sustained throughput demands. The user will get and use a peak cell rate (PCR) from the carrier. This type of service can be used for real-time applications such as videoconferencing, sustained file transfers, telephony applications, audio applications, and the like.
Variable bit rate in real-time (VBR-rt)	This service is a variable rate of throughput typified by some of the same real-time applications listed above, but not always the same sustained rate of throughput. This service may also apply to applications that cannot tolerate lengthy delays such as real-time SNA traffic, financial transactions, compressed and packetized video or voice applications, and some forms of multimedia. In this service, the customer will buy into a Peak Cell Rate (PCR), a Sustainable cell rate (SCR), and a Maximum Burst Size (MBS) for the traffic.

Service Category	Equates to
Variable bit rate (VBR–nrt)	This service category is characteristic of non-real-time non-real-time bursts of data but requires delivery guarantees from the carrier. This type of connection uses similar agreed-to definitions as listed in the VBR–rt mode above. This includes the PCR, SCR, and MBS modes of data transmission. The real application here may be data processing, transaction processing, banking and credit card processing, and airline reservation services. Some process control applications also fall into this category.
Unspecified bit rate (UBR)	UBR does not specify any form of QoS guarantees from the network. Applications here can handle delay and latency without too much trouble. The network provider will make a best effort to deliver the traffic within a reasonable amount of time. This can apply to applications such as e-mail, file transfers, remote printing, LAN-to-LAN interconnection, and telecommuting services for a small office/home office (SOHO).
Available bit rate (ABR)	The ABR is suited for applications that can definitely tolerate longer delays on the network. The ABR adapts to whatever resources the network has available, rather than a specified amount of throughput. Typically, the ABR also uses a minimum cell rate (MCR) as a means of controlling the traffic. The network continually feeds back information on the regular traffic and therefore allocates the MCR to hold the connection alive, while other traffic runs across the network in a priority over the ABR rate.

Categories of Service

Because of the risks associated with the congestion and the impairments to delivering the QoS, several techniques are used to shape the traffic and prevent massive congestion. The network providers offer specific service offerings based on the VPI and VCI for the end-user applications. These come in several types of services, as shown in Table 11-6 and constitute the way the traffic service offerings can be provided.

Getting to the Elusive QoS

Above and beyond the ways that the ATM networks allocate resources as shown in the categories of service described previously, other methods are used to get to the elusive quality of service guarantees. The PVCs in an ATM network can be provisioned with certain throughput parameters. The primary parameters are listed here:

- *Cell loss ratio (CLR)*
- *Maximum cell transfer delays (maxCTD)*
- *Cell delay variations (CDV)*

Each of these methods is described briefly here for the sake of knowing what the accomplished goal is.

- The cell loss ratio is a means of determining the ratio of lost cells to the total number of cells that have been transmitted. There are many reasons why cells get lost, yet the goal is to hold the amount of lost cells to a minimum. An ATM switch can discard calls that have been corrupted, especially the header. Moreover, cells can be explicitly marked (tagged) as eligible for discard if the network gets congested. The formula for the ratio is as follows:

CLR = lost cells / total cells transmitted

- The cell transfer delay is calculated as the total amount of elapsed time from when a cell enters a network (a switch) until it exits the network. This takes into account the total amount of inter-nodal processing time, buffering time, and propagation delays across whatever the medium is used.

CTD = Node Processing 1…n + buffer 1…n + propagation n

- The third formula is used to calculate the average amount of variations used in the cell delays. Often this is called jitter on the network where timing is lost or skewed a little to the left of right of the average. The mean of the variations is the overall delay variation. This is typically used for applications requiring some real-time consequence, such as packetized voice or video.

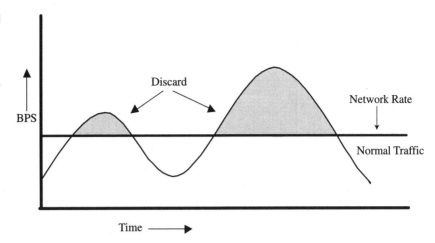

Figure 11-12
When a user exceeds
the network rate
traffic that is allowed

Shaping the Traffic

The ITU-TSS defines four possible situations when a cell enters an ATM network:

- *It will be successfully delivered.* The cell arrives at the destination with less than time T-cell delay.

- *An errored cell occurs.* A cell arrives with at least one detected bit error in the information field in the cell. Another possibility is the severely errored cell with information bits errors equal to n or n>1.

- *Lost cells.* A cell either never arrives or arrives after the time T-cell delay, in which case it is discarded at the destination.

- *An inserted cell.* A cell contains an undetected error or is misdirected by an ATM node and therefore shows up at the wrong destination.

In the bandwidth, the user is allocated a certain amount of capacity across the network. Clearly, the network will perform in different peaks and valleys, as shown in this picture. As the increase of cells per second grows and exceeds the network available rate for normal traffic over a period of time, the network will begin to discard cells. Although buffering might take place for certain periods of time, cells will likely be discarded because the network won't be able to process all of them.

If the user exceeds the network rate, the cells will be discarded and the user will therefore have to retransmit at a later time, as shown in Figure 11-12.

Figure 11-13
Cells allowed
through the network
when the user does
not exceed the
agreed upon
throughput.

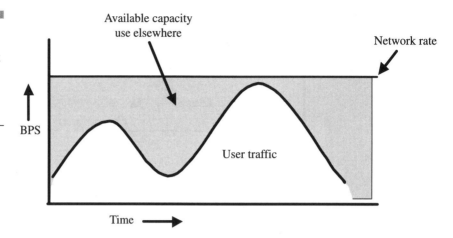

Normal Bandwidth Allocation

If, in fact, the user does NOT exceed the network rate or the agreed-to rate the network will provide, and the user traffic is shaped, under the capacity, then the available capacity left over will be provided to other users. Allocating this capacity on a fair and equitable treatment can far more efficiently use the ATM network.

Once again, as the users exceed their network thresholds, the network will discard cells, but if they do not exceed the network rate, the network should deliver all available cells delivered to the network, as shown in Figure 11-13.

What Is MPOA?

The goal of *Multi-Protocol on ATM* (MPOA) is the efficient transfer of inter-subnet unicast data in a *LAN Emulation* (LANE) environment. MPOA integrates LANE and *Next Hop Resolution Protocols* (NHRP), also known as Next Hop Routing Protocols, to preserve the benefits of LANE while allowing inter-subnet, internetwork layer protocol communication over ATM *Virtual Circuit Connections* (VCCs) without requiring routers in the data path. MPOA provides a framework for effectively synthesizing bridging and routing with ATM in an environment of diverse protocols, network technologies, and IEEE 802.1 Virtual LANs. This is designed to provide a unified paradigm for overlaying internetwork layer protocols on ATM. MPOA is capable of using both routing and bridging information to locate the optimal exit from the ATM cloud.

MPOA allows the physical separation of internetwork layer route calculation and forwarding, a technique known as virtual routing. This separation provides a number of key benefits:

- It allows efficient inter-subnet communication.
- It increases manageability by decreasing the number of devices that must be configured to perform internetwork layer route calculation.
- It increases scalability by reducing the number of devices participating in internetwork layer route calculation.
- It reduces the complexity of edge devices by eliminating the need to perform internetwork layer route calculation.

MPOA provides *MPOA Clients* (MPCs) and *MPOA Servers* (MPSs) and defines the protocols that are required for MPCs and MPSs to communicate. MPCs issue queries for shortcut ATM addresses and receive replies from the MPS using these protocols. MPOA also ensures interoperability with the existing infrastructure of routers. MPOs also make use of routers that run standard internetwork layer routing protocols, such as OSPF, providing a smooth integration with existing networks.

ATM and
Frame Relay
Internetworking

Introduction

Since 1992, two major developments have rolled into the carrier networks. The first was the implementation and rollout of Frame relay networking protocols. Frame Relay met with immediate success because of its ability to handle Wide Area Networking data traffic, replacing (or complementing) older X.25 networks. The network suppliers had to upgrade their equipment to support the newer protocols in their packet switches. Although Frame Relay met with instant success, as described in Chapter 10, it was introduced as a data-only service for the WAN. The commitment to equipment and labor to upgrade the network became a very heavy burden.

As with most carriers, network equipment has traditionally been depreciated over a 20-year period. Thus, when Frame Relay began rolling into the marketplace in 1992, the depreciation window opened with an end-date of 2022 and counting. Frame Relay was deployed in every major network around the world because of its flexibility and cost advantages over the older protocols. The result was a widely deployed and well-accepted international standard for data communications.

The second major development introduced also in 1992 was ATM. ATM is a robust set of protocols that works in more than the WAN, but is designed to work across the various platforms of network from the LAN, CAN, MAN, and WAN. Because it was designed as a transport set of protocols to work at layer 2 of the OSI equivalent model, ATM both competes with and complements the use of Frame Relay. Yet, ATM goes further than just being a data transmission set of protocols. It is designed to carry voice, data, video, and high-speed multimedia traffic. As a broadband communications set of protocols, ATM is the one set of operating protocols that meets the expectations of the end user, local and long distance carrier, and the equipment manufacturers alike. But, like all new protocols that cross boundaries (between the voice and data worlds), ATM needed some added enhancements that were not ready in 1992. Consequently, ATM did not start to catch on until late 1995 and early 1996. Moreover, ATM has been specified and studied to death, slowing its acceptance. Now, just before the new millennium, ATM has come into its own as an accepted and standardized set of protocols for the network providers in the various locations.

What remains is a problem with the rollout of the equipment and other associated interfaces. Where the carriers have endorsed and embraced Frame Relay, they now have to upgrade to an ATM backbone network. This will require significant investments. Moreover, the carriers will still have the frame relay switches in the networks that are on the books, usable and still viable traffic handling machines. To solve the problem, several techniques were developed to allow legacy systems, new systems, Frame Relay, and ATM all work together. This concept has been deemed as Frame Relay and ATM internetworking. In reality, it is a form of interworking instead of internetworking.

ATM and Frame Relay Compared

One way to understand the interworking functionality of the two sets of protocols is to compare and contrast the capabilities of the two protocols. Table 12-1 is used as an overall summary of the two techniques in use today.

What you can see from the comparison is a summary of the characteristics of the two service offerings from the carriers. This is not the whole story, but it does give the reader a visual means of seeing where and why the two techniques have become popular.

Table 12-1

Comparing Frame Relay and ATM characteristics

Features	Frame Relay	ATM
Connection Type	Connection Oriented	
Connection Mechanism	PVC	
Switched Access (SVC arrangement)	Yes but not widely implemented	Yes
Multiplexing Arrangement	Statistical TDM	
Current Speeds Available	Typically 56 Kbps – 2 Mbps, implementations up to 50 Mbps	1.544 Mbps (T1) 12.3 Mbps 25.6 Mbps 34 Mbps (E3) 45 Mbps (T3) 51 Mbps (OC1) 155 Mbps (OC3) 622 Mbps (OC12)
Area Served	WAN	LAN CAN MAN WAN
Sequencing of Data	No	No
Protocol Data Units	Variable	Fixed

continues

Table 12-1

continued

Features	Frame Relay	ATM
Protocol data Unit Size	$\leq$ 4096 bytes	53 Bytes
Flow Mechanisms	Circuit by circuit	
Traffic Congestion management	DE bit	CLP bit
Congestion Notification	FECN/BECN	Payload Type field
Bursty	Yes; defined by B^c and B^e rates	Yes, defined by PCR (maximum burst sizes)
Addressing Method	DLCI	VPI and VCI
Address Size	10 bits (normal)	24 bits
Standards based	Joint development with Frame Relay Forum and ATM Forum, ANSI specifications, ITU standards and Others	

Frame Relay Revisited

To take this a step further, Frame Relay is faster than the older networking X.25 and meets the demands of the older applications from a user's perspective. Very few applications today need a full 50 Mbps speed, but the 1.544 Mbps speeds and below are very robust and ubiquitous. This makes Frame Relay attractive also from a pricing perspective. The primary applications used with Frame Relay are SNA internetworking for mainframes and LAN to LAN connections across the WAN. Other services, such as remote access and internetworking between major corporations and their branch and small offices, are well suited for Frame Relay. The use of Frame services also fits well in the smaller corporate networks, but only scales to certain sizes before congestion and addressing become problematic. Although well deployed around the world, Frame Relay is primarily suited for the small- to mid-sized corporate network. When the network has to grow to tens of thousands of nodes, Frame Relay may run out of capacity and capability very quickly. Moreover, Frame Relay is already in place so it is used more heavily. Figure 12-1 is a graphic representation of the use of Frame Relay in the WAN by carriers and end-users alike. In this graphic, the Frame Relay switches are deployed in the carriers' networks, whereas

Figure 12-1
Frame Relay in various places

the routers on a LAN use the Frame Relay protocols. This allows the interconnectivity across the WAN in a straightforward and cost efficient manner. Frame Relay is also well suited for the bursty traffic typified across the LAN to WAN networks.

ATM Revisited

ATM, on the other hand, uses a fast relay systematic approach to data handling. By using the fixed-sized cells across the ATM switching fabric, the process of cell transmission can occur very quickly. ATM supports added functionality and capability in differing applications, such as the integration of voice and video across multiple platforms. Even though voice across ATM was slower in developing, ATM was designed to carry the various traffic types in the cells. The ATM model is shown in Figure 12-2 where ATM was designed to work. This is a quick review of the two techniques so that you will have an understanding of why they must interwork in order to achieve harmony across the networks.

What this graphic shows is the integration of the various forms of information (voice, data, and video), as well as the locations in the LAN, CAN, MAN, and WAN. You can see that the piece parts installed at the customer

Figure 12-2
ATM in various uses

locations may include an ATM router, an ATM switch, or an ATM PBX (not shown in the graphic). These changes require significant investments, not only by the consumer, but also by the carriers where each of the LEC, CLEC, and IEC interfaces use ATM switches in the backbone. ATM switches cost great amounts of money for a carrier to install and maintain.

The Frame and ATM Merger

Because of the dilemma between the two differing techniques, both the ATM and Frame Relay Forums took action to align the two standards more closely to allow for the interworking and internetworking between them. Both protocols have very defined characteristics that make them desirable. Both also have parameters that are easily matched up to the other. By aligning these parameters together, the interworking function is more easily accomplished. The similarities between the two technologies make them complements of each other without creating a direct conflict when properly planned. By combining both technologies, carriers and end users alike can protect their investments and preserve the infrastructure of the networks in place today. Without major renovations to the architecture, logical progression from frame-based services to cell-based services in the future will become far more acceptable. As a matter of fact, the combined technologies are rapidly being accepted and implemented throughout the industry.

Transparency Across the Network

The real goal is to provide for transparency across the network. End users running Frame Relay need not concern themselves with the fact that the carrier is likely running ATM in the backbone. This is shown in Figure 12-3 where the carrier is providing high-speed connectivity across the WAN, but the local attachment is using Frame Relay.

Still another way of looking at the network interworking is to have Frame Relay on one end and ATM at the other end through the network (see Figure 12-4).

Either way, this is achieved with relative ease from the perspective of the carrier and the end user. Transparency is what is necessary to make this happen. The interworking functionality is a reality today.

Figure 12-3
Frame and ATM
coming together

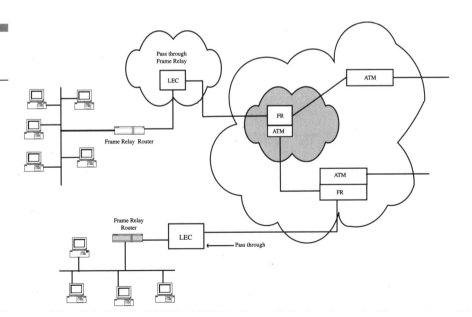

Figure 12-4
Frame Relay to ATM
conversions

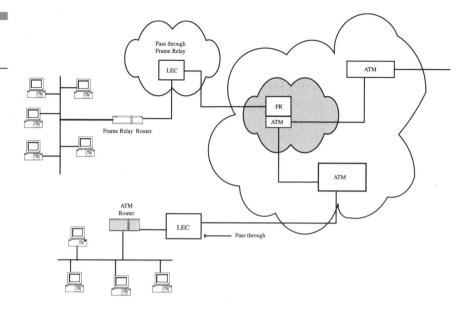

Frame User to Network Interface (FUNI)

Providing the interfaces to the internetworking services, the ATM forum specified the *Frame User to Network Interface* (FUNI) in order to allow frame-based services to access an ATM network. FUNI allows traffic to be passed from the two different networking technologies by using an industry-standardized interface. This provides the carrier and the user with a migration path for the future, while serving today's needs. The ATM forum was careful to describe the differences in two terms before specifying the protocols needed. The two terms defined are as follows:

- Interworking
- Internetworking

Although they may appear to say the same thing, subtle differences exist between the two concepts.

Interworking is a techniques that allows the two systems on the ends to run Frame Relay as shown earlier in Figure 12-3, yet run ATM across the backbone. Thus, the systems interwork with each other. Frame Relay is on both sides of the network so we can encapsulate or tunnel the traffic through the ATM network, but the traffic entering and exiting the backbone is still Frame Relay on both sides.

Internetworking, on the other hand, uses techniques to convert the traffic from one form to another, much the same as a protocol converter or a gateway. A device that sits at the edge of the network earlier performs the internetworking (or interworking) function (see Figure 12-4). The device is typically referred to as an IWF.

Taken one step further, the ATM Forum defined this interworking in two different categories:

- Service Interworking
- Network Interworking

Network interworking means that Frame is supported through the ATM backbone. The operations of the network nodes are performing actual actions on the data in the form of a convergence sublayering function. Networking interworking supports AAL 5 as a specified and optional AAL 3/4 in the performance of its interface.

FUNI is capable of handling and supporting the basic functions at the UNI such as the following:

- VPI/VCI multiplexing
- Network management
- Traffic and congestion control (shaping)
- Operation, administration, maintenance, and provisioning (OAM & P) functions

Although these functions are supported, they may be limited in their overall performance and functionality. FUNI does not support some of the traffic types (such as AAL 1 and 2), so therefore it does not support the corresponding signaling and traffic parameters for these traffic types. In addition, the FUNI is designed to support Variable Bit rate (nonreal time) and Unspecified Bit rate class of traffic.

Data Exchange Interface (DXI)

DXI is another of the interfaces used to internetwork various services across an ATM backbone. Functionally, the FUNI and DXI provide the same services with minor modifications. DXI is used with legacy routers that are not ATM equipped, but use the DXI protocols and interfaces to provide frame-based services to the *Data Service Unit* (DSU) on the circuit where the cells are then generated. The DXI uses the DSU whereas the FUNI does not.

Moreover, DXI does not use fractional T1 services, where FUNI can. A comparison of the FUNI and the DXI interfaces is shown in Figure 12-5.

When comparing the DXI and the FUNI interfaces as the means of internetworking the services, it is appropriate to view the overhead comparisons of frame-based services contrasted with the cell-based services of ATM (see Figure 12-6).

The FUNI reference model as defined by the ATM forum is shown in Figure 12-7. The interworking functionality gives an idea of the mapping of services across the FUNI to an ATM network. The upper portion of the graphic

Figure 12-5

DXI and FUNI interfaces compared

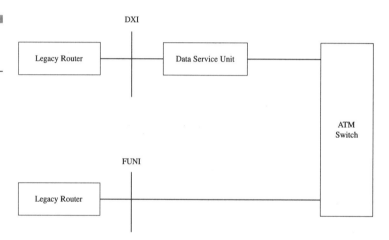

Figure 12-6

Contrast of frame- and cell-based services

Figure 12-7
FUNI reference
model

Figure 12-8
ATM Interworking
Function (IWF)

represents the block flow; whereas the lower portion of the graphic shows the protocol stacks as they compare between the interworking functions.

With this protocol analysis in mind, there are some subtleties between the Frame Relay protocol and the mapping across the ATM platform through the IWF on a network. In this case, the graphic shown in Figure 12-8 represents the interworking functions as they align to the reference model shown. Once again, the upper portion of the graphic is the block diagram of the network and the lower portion of the graphic shows the protocol stack.

What Constitutes a Frame?

The market still is unsettled. More low-end, low-cost devices are being sold every day. Users are trying to preserve their investments and hold the line on new, higher-cost investments as long as possible. With the growth projection for this market, something had to be done. Interworking is the answer. However, the frame-based services in legacy equipment and the Frame Relay devices will still exist for some time to come. The frame is therefore important to provide the interoperations and interworking between frame-based and call-based networking components. FUNI and DXI allow the frame-based access to the ATM networks, while Frame Relay allows frame-based access to a frame relay network. DXI, FUNI, and Frame Relay have similar frame structures. The headers of FUNI and DXI within the frames are identical to each other, but different from Frame Relay headers. The bit patterns fall in the same logical structure as a Frame Relay frame, so they can be mapped somewhat consistently. Figure 12-9 compares the bit patterns of the frame structures for FUNI, DXI, and Frame Relay.

Figure 12-9
Comparing the frames

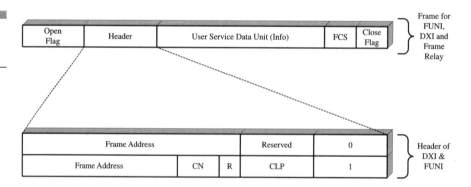

When the frame is then segmented into cells, the DXI frame address and the FUNI frame address fields can both be mapped to a VPI/VCI in ATM by using the same process. Frame Relay interworking functions can use the same addressing map. The Frame Relay to ATM interworking functions can also use other mapping of the data formats.

The next step from the original frame format is to segment the FUNI frame into ATM cells, using a mapping of the frame address into an ATM VPI/VCI. This is shown in Figure 12-10 where the actual header information is segmented into the cells and delineated into the appropriate VPI/VCI. The *Congestion Notification* (CN) bit performs much the same function as the *Forward Explicit Congestion Notification* (FECN) bit in Frame Relay. The network will set this bit during periods of network congestion in the same direction as the traffic flow where the congestion exists. The Frame Relay *Backward Explicit Congestion Notification* (BECN) bit does not have a suitable equal in the FUNI or the DXI.

FUNI Interoperability

When a FUNI device needs to establish a switched virtual circuit to some end point, it doesn't matter what the other device is (FUNI, DXI, or ATM). The use of a call set-up mechanism in signaling is the same as the ATM signaling procedure at the UNI. When FUNI traffic is delivered across the ATM networks, it doesn't matter if the data terminates at another FUNI or an ATM device. These service aspects are transparent to the network and the devices. Because FUNI is an ATM protocol, the multiprotocol encapsulation procedures at the ATM UNI are used at the FUNI, so there are no issues related to interoperability. Looking at Figure 12-11, one can see that the interoperability issues were addressed up front to prevent any downstream problems in the protocol or interoperability standards for ATM.

Figure 12-10
FUNI segmented into cells

Figure 12-11
FUNI, DXI, and ATM
interoperability

Network Interworking

The network interworking function provides the transport of Frame Relay user traffic transparently across the ATM link. It also handles the PVC signaling traffic over ATM. As already discussed, this is sometimes called tunneling through the network. Other times it is called encapsulating the traffic. Regardless of the name it is given, the function provides for the transparent movement of end user Frame Relay information across the ATM network. The benefit is that with the tunneling or encapsulation formatting, the service is as good as though the end user has a leased line service between the two end points. The benefit of this tunneling approach is connecting two Frame Relay networks across an ATM backbone. This is shown through the use of the network-interworking unit in Figure 12-12. The interworking function is shown as a separate piece of equipment between the Frame Relay and ATM network, which in some cases it is. However, newer implementations of this architecture place the network interworking function and interfaces inside an ATM switch. Regardless of where it resides, the functionality is really important, not the location of the box. The interworking function will allow for each Frame Relay PVC connection to be mapped on a one-to-one basis over an ATM PVC. In other

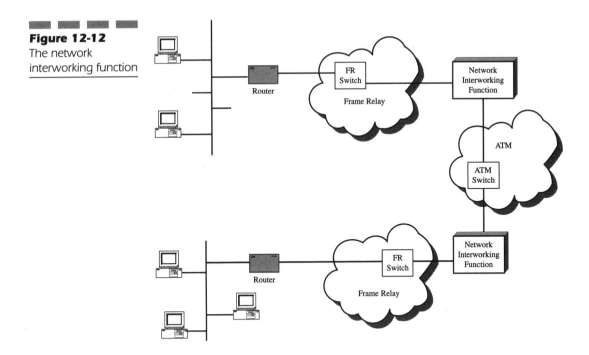

Figure 12-12
The network
interworking function

cases, many Frame Relay PVCs can be bundled together across a single, higher-speed ATM PVC. The one-to-one or the one-to-many services allow more flexibility.

Service Interworking Functions

The use of a service interworking function takes away some of the flexibility and transparency across the network. It actually acts more like a gateway (protocol converter) to facilitate the connection and communications between different disparate pieces of equipment. Figure 12-13 is a representation of the interconnection of the service interworking devices across the network. The end user actually sends traffic out across a Frame Relay network on its own PVC, and then it gets passed through the Frame Relay network to the service interworking function where the data is then mapped to an ATM PVC. The IWF functionality provides the mapping of the DLCI to the VPI/VCI, as well as other optional features. The IWF is shown as a separate box, whereas newer implementations will have the dual mode functionality inside an ATM switch.

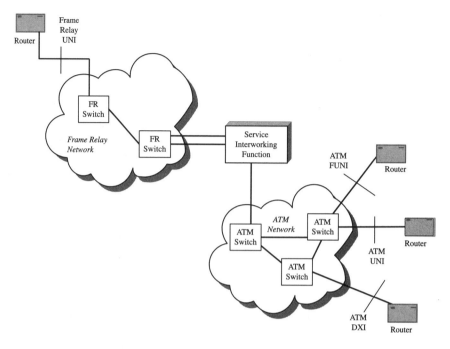

Figure 12-13
Frame/ATM service
interworking function

The DXI Interface

Several times now the additional device mentioned in the frame and ATM interworking function is the DXI interface. Using a legacy Frame Relay router, for example, one can still have access to the ATM network on an internetworking basis. This prevents the old *forklift mentality* where all users have to change hardware en masse when a new service is introduced. DXI is fairly straightforward, but does deserve some mention in its workings.

Preparing information for ATM coming from a router or a brouter DXI typically deals with a link to frame the data and prepare it for a DSU. Connecting the DXI link will be a high-speed connection at either one of the following:

- V.35
- HSSI
- EIA449/422

DXI is an open interface for the brouter to the DSU where the DSU performs all of the DXI encapsulation. When using the V.35, all that is necessary is HDLC frame-formatted data stream into the DSU from the router or brouter. The DXI transport services are provided within the DSU and then passed to the ATM layer. Several different adaptation layer processes can be used.

DXI Mode 1 A/B

Two types of DXI modes exist: mode 1 and mode 2. When using mode 1, two additional types are available: either the 1A or the 1B. When using mode 1A or 1B, a simple and efficient encapsulation of the data is provided. Data can be transparently or efficiently passed into the ATM layer.

From the DTE device a *Service Data Unit* (SDU) will be prepared and passed to the DXI data link.

- At the DXI data link layer, an HDLC frame is provided. This HDLC frame is mapped to the DXI physical interface, which is then connected to the DCE (that is, a router).

- At the DCE, the input interface from the DTE terminal devices will be at the DXI physical layer and again through a DXI data link. The DTE *service data unit* (SDU) will then be passed to the AAL5.

The AAL5 *Common Part Convergence Sub-layer* (CPCS) will then prepare the data for the SAR. From the SAR, the payload is handed to the ATM layer for the 5-byte header generation and mapped to a physical layer interface using the ATM UNI. The DXI modes 1 A and B are shown in Figure 12-14.

Figure 12-14
DXI modes 1A
and 1B

DXI mode 1 A/B simple & efficient encapsulation of SDU

- Mode 1A supports AAL5
- Mode 1B supports AAL5 plus AAL3/4

DXI Protocol Mode 1A

The DXI protocol mode 1A is synonymous to what occurs with SMDS (802.6 protocols). The DTE service unit will be 9,233 bytes of actual data. Around the data will be the typical HDLC frame formatting. The frame starts with a 1 byte opening flag. Following the opening flag, is a 2-byte DXI address. Next comes the data (up to 9,232 bytes). The frame check sequence (or CRC), which is 2 bytes long, follows. The closing flag is 1 byte long.

From the DTE, the service data unit will then be passed to a DSU through a physical interface where the ATM conversion process takes place. At the DSU network, translation information and the entire service data unit is passed through the AAL5 common part convergence sublayer. This transfer will become an AAL5 *Protocol Data Unit* (PDU). The AAL5 SDU is then broken down at the SAR layer into 48-byte payloads (SAR PDUs). These SAR PDUs will be mapped into the ATM layer where the header is generated. The DXI information in the original HDLC frame will be mapped both to and from the ATM VPI/VCI. This address mapping will help to keep everything in order. Using the mode 1A, up to 1,023 addressable devices can be used. The AAL5 layer is the easiest of the DXI protocols that can be accommodated (see Figure 12-15).

Figure 12-15
DXI protocol
mode 1A

DXI Protocol Mode 1B

As noted earlier, the DXI mode 1A supports the AAL5. The mode 1B supports AAL3/4 and AAL5 (see Figure 12-16).

The same process takes place at the DTE. First, a DTE SDU will be created. Next, the SDU will be passed into an AAL3/4 CPSC. This layer will be framed and put across the data link layer at the DXI interface and mapped onto the physical layer.

At the DCE, the data is passed to the DXI data link layer. Then it is mapped to the router as AAL3/4 or AAL5. The data will be handed down into the SAR layer, passed to ATM, and finally processed out across the UNI to the ATM network (see Figure 12-17).

DXI Mode 2

The second way of handling the DXI is to use the protocol mode 2. DXI is used to provide connectionless ATM network services, but is still important. The difference between mode 2 and mode 1A & B is that mode 2 can support up to 16,777,000 virtual connections either for AAL5 or AAL3/4 services. A DTE SDU of up to 64 Kbytes long and a frame check sequence are all that are used for the SDU. The architecture of mode 2 uses the SDU from the DTE passed directly to the AAL CPCS and then framed into a DXI data link layer and passed to the DXI physical interface. This mode supports more virtual path identifiers than the normal mode 1. Entering the router, the physical interface at the DXI level maps the SDU into the DXI data link and passes to the AAL3/4 CPCS. The data is passed to the AAL3/4 SAR.

Figure 12-16
DXI mode 1B

Mode 1B support AAL3/4 and AAL5

Figure 12-17
DXI protocol
mode 1B

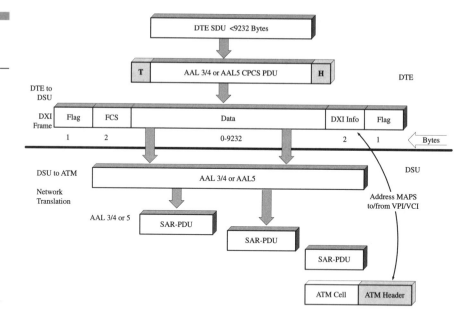

Figure 12-18
DXI mode 2

- Supports more connections (16, 777, 215 vs. 1023
- More VP (256 vs. 16)
- Larger SDU (65, 536 vs. 9232)

Now the SAR is processed to the ATM layer where the header is generated and the data is mapped onto the UNI. The mode 2 SDU supports 64 Kbytes compared to mode 1, which supports 9232 bytes. Refer to Figure 12-18 for a representation of the DXI mode 2.

DXI Protocol Mode 2

The DTE processes the protocol SDU of 64 Kbytes. The DTE will provide AAL3/4 encapsulation of the entire 64 Kbytes, adding both header and trailer information, creating the AAL CPCS PDU. Now the PDU is mapped to the DSU as a DXI frame. The AAL3/4 CPCS PDU is then encapsulated as an HDLC frame containing the DXI information field after an opening flag. Note that the DXI information field is 4 bytes long this time. Greater addressing capabilities are possible for the virtual connections. Using a 32-bit address, we can support 16,777,000 addresses.

Now the data will be framed. From there the frame check sequence and the closing flag are added. At the DSU, the protocol data unit will then be mapped using an encapsulation technique. The encapsulation is retained for the PDU and then processed into SAR PDUs and onto the ATM portion of the DSU. Once again, the DXI addressing information is mapped both to and from the VPI/VCI within the ATM header for delivery across the ATM network. This is shown in Figure 12-19 as the data is prepared in mode 2.

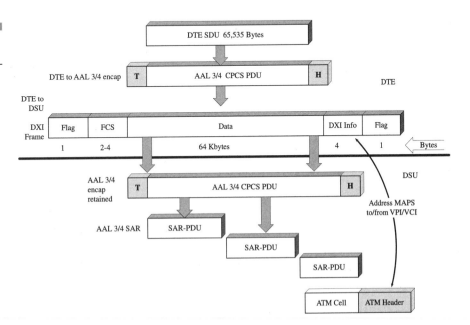

Figure 12-19
DXI protocol mode 2

Summary

Frame Relay and ATM can interwork and interoperate through several different techniques, which gives the carrier a sense of comfort. Because the legacy systems of the past can still be accommodated, the Frame Relay investments made in the early 1990s are still viable, and the ATM investments will be around for some time to come. Through the ATM protocols such as FUNI, end users have comfort knowing their networks are not obsolete. By using the DXI interfaces and protocols, the older Frame Relay routers, bridges, and brouters can still be used in an ever-changing network environment. One can now see why the interworking functions from the networking and service functionality are so important. Millions of dollars of investment can still be used and newer protocols can be deployed without making the entire network obsolete. This is what inter-networking is all about!

Cable TV Systems

Introduction

The television broadcast signal, regardless of the standard used, is one of the most complex signals used in commercial communications. The signal consists of a combination of amplitude, frequency, phase, and pulse modulation techniques all on a 6 MHz channel with a single sideband transmission process called vestigial sideband.

Cable TV appeared in the industry during the early 1960s. The initial networks installed used a basic tree architecture, in which all signals emanated from the head-end location and were distributed to individual subscribers via a series of main trunks (trees), subtrunks (branches), and feeders (twigs), as shown in Figure 13-1. This topology requires analog amplifiers to periodically boost signals to acceptable levels based on the service area being covered. However, all the benefits of solving the gain/loss problems were offset by the introduction of noise and distortion directly attributed to the amplifiers. Analog amplifiers, as noted in any communications discussion, do nothing to eliminate noise (such as cross talk, white noise, *Electromagnetic*, and *Radio Frequency Interference* [EMI/RFI]).

Figure 13-1
The CATV
architecture

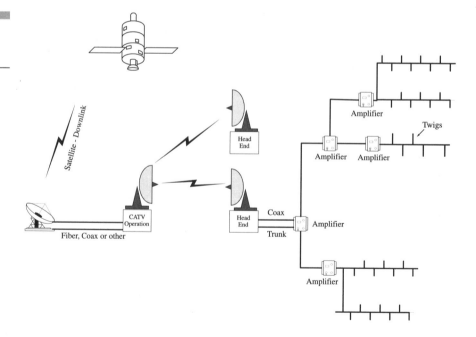

In early 1988, the CATV companies discovered that fiber optic cables could be used as a means of improving the cable infrastructure both in quality and in capacity. The initial deployments used a *Fiber Backbone* (FBB) overlay placed on top of the existing tree networks to do the following

- Improve performance
- Reduce cascading amplifier problems
- Increase reliability
- Segment systems into smaller, regional areas
- Facilitate targeted programming
- Improve upstream performance

These *Hybrid Fiber Coaxial* (HFC) networks drive the fiber closer to the consumer's door. They still use the conventional tree architecture, which branches off at the last mile (even if it is only a few hundred feet, the reference to the last mile still prevails in the communications business) from the node to the subscriber. Unfortunately, amplifiers may still be placed inefficiently, amplifier cascades are often longer than necessary, and active counts per mile may be higher than needed.

Each of these factors increases the initial investment costs without improving the reliability of the network or reducing the ongoing operating costs. Over the past decade, CATV networks have migrated from the tree architecture, to the FBB, to the current HFC platform. Comparing the number of nodes during this period of evolution, the industry has reduced node servicing from 5,000- 20,000 homes per node to approximately 500 homes. This ten-year migration illustrates the rapid advancements that have taken place in the technology/cost structure.

Many CATV providers believe a balance between cost and service capacity at 500 homes per node is acceptable. However, if subscription rates explode, a CATV company's platform must be easily adjustable to fewer nodes. In light of the initial investment costs of a system upgrade/rebuild, operators must continually consider a system that will improve reliability for the end user, while reducing the initial costs and ongoing operating costs.

In the past, the CATV industry has designed systems uniformly; whether design began at the head-end or at the node site, the coaxial portion of the system maintained the same look and feel of a tree and branch. This usually meant optimizing the design through trial-and-error methods of cascading devices to reach all areas of the system. This conventional approach to design was based on proven design techniques, which were developed prior to the broader range of design products available today.

The advent of fiber, combined with innovative new amplifier products and the need to consider future service capabilities (such as Data and Internet access) and optimal costs, is the driving force behind the development of alternative design methods.

Cable Television Transmission

Television signals that are broadcast over the air in the United States are transmitted in 6 MHz channels that are allocated to broadcasters by the *Federal Communications Commission* (FCC). The FCC regulates the location, power, and frequencies used by television stations, ensuring that stations that use the same channel are sufficiently far apart so that they do not interfere with one another in any of the areas where they may be received. But preventing interference may also require that two stations do not use adjacent channels in the same area for a more subtle reason. In the air, broadcast signal strength falls off rapidly with the distance from the transmitter, and as a result, the signal from a nearby transmitter can be several orders of magnitude stronger than that from a distant transmitter.

Because transmitters cannot contain their signal perfectly within their designated bandwidth and because receivers cannot perfectly discriminate between signals from adjacent channels, a strong nearby signal can interfere with a weak distant signal. This is known as the near-far problem, and the result is that it is not always practical to use adjacent television channels in one area.

Television signals delivered over traditional cable television networks are sent the same way as they are over the air: by dividing up the cable spectrum into 6 MHz channels of bandwidth and modulating each television signal into one channel. (This is an example of frequency division multiplexing.) But these cable systems can carry many more channels than broadcast television for two reasons.

First, the near-far problem of broadcast television is not a problem on cable systems because all channels can be transmitted at the same power level throughout the network, enabling adjacent channels to be used on the cable.

Second, cable systems are not limited to the bandwidth that is designated by the FCC for broadcast television; a cable can carry as many channels as the infrastructure will permit, which in modern systems can be a hundred channels or more.

Because the transmission, or encoding, of analog television signals is done in the same manner as it is for broadcast television, receiving these signals is straightforward. But a television that is not built specifically to receive cable will require an external receiver because a wider range of frequencies is used on cable systems and sometimes because a television receiver cannot properly discriminate between adjacent channels. This external receiver (a set-top box) retransmits a selected channel to one that the television can receive. Modern televisions, however, can often tune cable channels directly.

The Cable Infrastructure

The coaxial cable and broadband amplifier technologies define the essential capabilities of cable networks. But we also need to understand how real cable systems are actually built out of these and other components such as optical fiber, since this introduces both technological and economic constraints on using cable to support other communication applications. Although cable systems in the United States were built independently by a variety of cable companies and equipment providers, the infrastructure is similar enough from system to system so that we can talk about cable systems in general terms. Much of what we need to understand is a matter of terminology used to describe different parts of the network.

CATV has always been considered a one-way transmission system, designed to deliver TV and packaged entertainment to the residential marketplace. As the development of the architectures rolled out across the country, the CATV business, like other utility functions, became a monopoly. The industry enjoyed the freedom to offer products and services packaged to serve a local community. As a utility, the CATV companies were granted exclusivity. There were few demands placed on the cable systems because the providers delivered a sufficient number of channels to the consumer based on the coaxial technology used. The typical operator uses a 6 MHz channel capacity that is then filtered to prevent overlapping signals from interfering with other TV channels as shown in Figure 13-2. The channel filters then limit the overall capacity to approximately 4.2-4.5 MHz to carry the TV signal. This is sufficient to deliver CATV broadcast quality signals to the consumers' door for a fee.

In the event the operator runs out of capacity, new techniques exist to offer as many as three or four channels of TV on a single CATV 6 MHz capacity through the use of MPEG 2 standard compression. With a three- to four-fold increase in the channel carrying capacity on the same cable, the operators find this rewarding. This is especially true when the new MPEG standard allows the increased capacity without requiring the CATV operator to trench new fibers or coaxial cables throughout the neighborhoods.

■■■ ■■■ ■■■ ■■■

Figure 13-2
Coaxial channel
capacities at 6 MHz

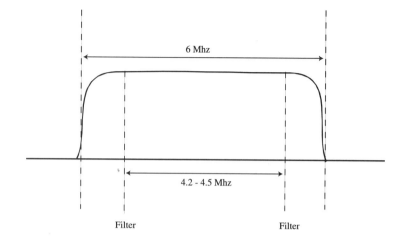

■■■ ■■■ ■■■ ■■■

Figure 13-3
Guard bands prevent
the channels from
interfering with each
other.

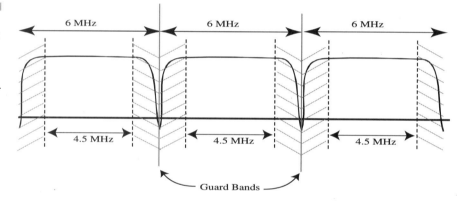

Once the cable is channeled into the current 6 MHz techniques with the bandpass filters allowing a 4.5 MHz capacity, the carriers use the sideband capacities for other operations.

1. Filter guardbands (see Figure 13-3) can prevent two channels from interfering with each other.

2. The horizontal black bar at the bottom of the TV screen, the vertical interval, usually contains test signals used by the broadcast system for online performance tests that will not interfere with the regular programming.

3. The audio information in a TV channel is a frequency-modulated carrier placed 4.5 MHz above the visual carrier at less than 1/8 of its power.

The Cable Television Distribution System

In the beginning, early television cable distribution systems were established to serve communities where residents could not receive over-the-air programming because of geographical interference. Atmospheric conditions and fading were crucial in the delivery of the airborne signals. Distances from the transmitters were also critical factors, but may have been limited by the regulatory bodies (such as the FCC and local utility commissions). When an area could not receive the signals for whatever reason, the logical distribution of a cable to these areas made prudent business sense. Regardless of the reason, CATV was used to supplement commercial TV systems whether the consumer was

- Behind a mountain
- Out of the line of sight from the Omni directional antennae
- Beyond the reception power area of a major broadcast provider

The real issue was the monopoly that was established to provide an incentive for the Cable companies to make the huge investments and attempt to service the consumer.

The term *Community Antenna Television* (CATV) has long since been extended to mean any region wired for the reception of broadcast programming, whether or not good residential antenna reception is available. Subscribers to these systems generally pay a monthly fee for the service, which usually includes increased channel selection, "pay per view," and locally originated programming. Newer services are continually being reviewed and introduced with the changes in technology and the shift in regulation.

Signal Level

The cable system requires certain parameters in order to function properly. Originally distributed via a microwave distribution (still done in some cases) and coaxial cables, the systems are under constant scrutiny and change. Throughout the CATV system, power is distributed in the form of

- TV and FM carriers
- Test signals

- Pilot tones
- DC power
- Noise

Specified levels must be maintained at each point in the system to assure good performance. The signal levels at different frequencies are just as important. Here is how they are measured. Cable TV uses frequency division multiplexing onto the coaxial cable, as shown in Figure 13-4.

This frequency division allows the provider to multiplex various channels onto a single analog carrier system and deliver the various forms of entertainment available on the cable. Not all providers are the same, but many offer more than just TV and entertainment. Some of the providers offer high quality stereo music. Others offer information such as Dow Jones or Reuters in the guard bands that travel along the cable. If the end user has a special receiver, the data are available. If the receiver is not available, then the data are not available to the individual set. Actually, the data are travelling along the cable, but fit in the black horizontal bar on the bottom of the screen. These value-added services are an additional revenue generator for the cable company. The convenience for the end user is what drives the business case. Most users who cannot receive over-the-air programming (public TV) probably have trouble receiving over-the-air radio, too!

Let's start with the cable itself. The *characteristic impedance* of distribution cable is 75 ohms (75Ω). This impedance is the amount of resistance

Figure 13-4

Frequency division multiplexing on CATV

	Channel#2	Freqency: 55.25 MHz
MUX	3	61.25
	4	67.25
	5	73.25
	6	
	7	
	8	
	9	
	10	
	27	

350-950 MHz Cable

that the cable signals "see" from the center conductor to the outer shield of the cable at the transmission frequencies. This impedance governs all the signal voltage and currents traveling the cable as covered in Ohm's law. Most CATV system measurements involve signal power differences, that is one level relative to another. Voltage differential is an awkward measure of power differential because each time a power change is measured the formula must *be calculated*. The *decibel* resolves difficulties in handling system power figures.

Digital Video on Cable TV Systems

Digital service is a key enabling technology that will allow cable systems to deliver a multitude of emerging services. High spectral efficiency, robust resistance to noise, and exceptional flexibility permit the installation of premium digital services, such as:

1. Video-on-demand
2. PCS telephony
3. Commercial data transport

Figure 13-5 is a representation of a multiservice cable operation that may well be the wave of the future for the operators.

Shown in this figure is the basic cable coming to the residence. The CATV suppliers have changed their basic architecture from a one-way cable to a two-way cable system using a *Fiber Based Backbone* (FBB). At the local hub along the route, the cable is then terminated and the HFC equipment delivers the cable to the door interface. At the home (or office), the cable serves the telephony, entertainment, and high-speed data demands of the end user. The cable operator provides the network interface, and the ancillary equipment is then hung off the cable. Subscribers already view these new capabilities with new expectations of high value and more reliable service.

As telephone companies (the LECs), long distance companies (the IECs), and the cable companies compete to deliver digital services, a key differentiation will be the quality and reliability of service. Recent acquisition and merger activity shows the IEC and CATV merger taking off as a means of getting to the consumer's door. Ensuring the quality of service requires testing digitally modulated signals. Digital services will revolutionize the way consumers view their CATV suppliers. As in an analog cable TV system, power and interference measurements are essential to maintaining digital

Figure 13-5
CATV services of
the future

cable TV services. Although the effects are different from impairments on an analog television signal, amplifier compression and spurious interference will degrade digital signals.

Forming a Digital Video Signal

In CATV, all digital modulation formats define how the data bits correspond to both carrier phase and amplitude, and how transitions between bits are made. Digital modulation formats often associated with cable systems are 16-VSB, 64 QAM, QPR, and OQPSK. However, there has been much discussion about 256 QAM for the future implementation of broadband communications services. Regardless of the technique, these formats share common characteristics in the time and frequency domains, driving common needs for measuring power and interference. Cable digital audio services are often delivered to the head-end by using the OQPSK format.

Key Features of Digital Modulation

In the frequency domain, digital modulation produces a noise-like spectrum whose bandwidth depends on the symbol rate, coding, and filtering used.

The spectrum of unfiltered *64 Quadrature Amplitude Modulation* (64 QAM) is 4.167-Msymbols per second. This produces a 25 Mbps digital data stream. The bandwidth of the filtered 64 QAM channel is approximately 4.2 MHz. Note that the broadband digital video signal is susceptible to spurious interference across the entire channel bandwidth, making control of analog and digital transmission spurs critical. Proper symbol filtering avoids spilling interference from a digital channel into adjacent video channels. 64 QAM uses six bits (2^6) and produces the technique discussed here. With 4.167 Msymbols per second and six bits per symbol, then the resultant yield is (4.167 * 6 bits = 25 Mbps).

DTV Solution Introduction

November 1, 1998 marked the beginning of a new era in the broadcast industry. After ten years of R&D and standards development, the DTV revolution entered its implementation stage. Broadcasters are still gearing to plan as well as building up their DTV facilities. The broadcasters actually completed their implementation in the ten largest cities three days early. Approximately 13,176 pioneers (first purchasers) had acquired Digital TV sets at a roaring $7,000 ± apiece. The broadcasters implemented the standard based on the digital HDTV systems developed by the Grand Alliance. The *Advanced Television Systems Committee* (ATSC) standardized this specification. It consists of three subsystems:

1. Source coding and compression

2. Service multiplex and transport

3. *Radio Frequency* (RF) transmission

Although this has met with some success, more has to be done with it. For example, antennas for DTV surfaced to the forefront of the discussion when everyone realized that Digital TV is an all or nothing proposition. Analog TV fades and gets snowy from distortion, distance, atmospheric conditions, and other issues. DTV is either there or not! If a user has a problem with reception, then a new antenna may be needed. If not enough bits make it to the set, the picture will either freeze or disappear altogether. Anyone with a satellite transmission (e.g., Direct TV) knows what this is like.

Cable TV providers should be able to overcome this problem with the cable itself, so long as their receivers are properly tuned. This may drive more customers to the CATV operators for their HDTV and DTV needs. The major networks will be broadcasting some of their programming by using

DTV channels. The cable companies are not mandated (yet) to carry DTV broadcasts. But if they choose to do so, at issue is at what resolution they will carry and deliver DTV. DTV uses an 8-VSB signal, so the QAM used by some systems may have difficulty passing through the 8-VSB. Some incompatibilities may exist for the short term.

The source coding and compression deal with bit rate reduction of video and audio. The compression layer transforms the raw video and audio samples into a coded bit stream that can be decoded by the receiver to recreate the picture and sound. The video compression syntax conforms to the MPEG-2 video standard, at a nominal data rate of approximately 18.9 Mbps. The Dolby AC-3 audio compression is used in the ATSC DTV standard to provide 5.1 channel surround-sound at a nominal rate of 384 Kbps.

The service multiplex and transport layer based on the MPEG-2 Systems Standard provides for dynamic allocation of video, audio, and auxiliary data. It utilizes a layered architecture with headers/descriptors to provide flexible operating characteristics. The flexibility of the multiplex and transport layer provides the means for multiple *Standard Definition Television* (SDTV) services. The cable operators can send a single channel of HDTV programming or use a lower resolution SDTV and split the channel to simulcast multiple programs, including data transmissions. This all depends on which digital picture format the broadcaster uses and how the station allocates bits in the data transport rate of the channel. Currently, the data rate of a 6 MHz channel is rated at 19.4 Mbps.

The flexibility of the ATSC system specifications could allow a broadcaster to mix different streams of high and low data densities and transmit them simultaneously. An example of this may be to use 8 Mbps for a sporting event and three lower rate programs at 3.8 Mbps each (such as a newscast or a soap opera program). At the same time, all four programs will be very unlikely to use all the allocated bandwidth, so a statistical TDM technique can use the remaining bits for data transmission (see Figure 13-6).

The transmission layer modulates a serial bit stream into a signal that can be transmitted over a 6 MHz television channel. The transmission system is based on a trellis-coded, 8-level *Vestigial Sideband* (VSB) modulation technique for terrestrial broadcasting.

ATSC standard-based encoding systems are one of the key elements in DTV implementation. Encoding systems will be used in the entire broadcast chain. However, not every encoder in DTV broadcast chain has to be ATSC standard-based. It should be noted that the FCC standardizes only the terrestrial-broadcasting signal. Cable operators are not required to adapt to this standard. In addition to the source coding, compression, and multiplexing, an encoding system provides ATSC standard-based systems

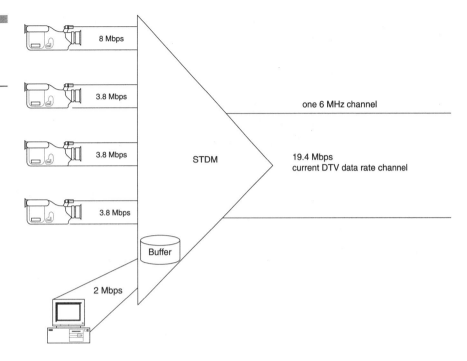

Figure 13-6
Combining multiple
streams on a single
DTV channel

information, program guide, data, and interactive services along with video
and audio.

The ATSC DTV standard, as well as MPEG 2, describes the bit stream
syntax and semantics. The standards also specify the constraints and
decoder models. However, encoding parameters are not specified by the
standards. Thus, encoder performance and systems implementation are left
to encoder designers. Thus, standard compliant does not guarantee encoder
performance.

In a standard 750 MHz system, consisting of a 5-40 MHz return bandwidth and a 52-750 MHz forward bandwidth, downstream broadcast services occupy the 52-550 MHz pass band. The remaining 200 MHz is
reserved for digital services.

The system's analog broadcast services consist of analog nonencrypted
services (basic), analog encrypted premium services, and analog pay-per-
view services. The digital broadcast services include digital audio programming, along with encrypted digitally compressed video programming.

Several different services, such as interactive video, cable modems, and
Hybrid Fiber Coax (HFC) telephony, make up the interactive services.

Cable modem equipment utilized a two-way RF system with 64 QAM
modulation in the downstream direction and QPSK modulation in the

upstream direction. Data rates for the downstream and upstream paths were 27 Mb/s and 1.7 Mb/s, respectively.

Using the system bandwidth information shown earlier, the available bandwidth in the downstream direction is 698 MHz. Both broadcast and interactive services use this bandwidth. Available upstream bandwidth is limited to 35 MHz, assuming that the entire 5-40 MHz bandpass is useable.

Migration strategies for the network must be carefully planned. Selection of downsizable architecture is crucial in this planning process. Also, conducting detailed traffic studies about these services when they are first offered in the network is crucial.

CATV operators are currently in the limelight and will remain there for some time to come. The emphasis on access to the door, deregulation and merger mania in this industry has propelled the CATV operators into one of the driver seats of technology. Whether they choose to adjust to their new-found popularity or be gobbled up by new players looking for access depends on the individual cable operator across the country (and the world).

Cable Modem Systems and Technology

Introduction

In the late 1970s a major battle arose in the communications and the computer industries. Convergence of the two industries was happening as a result of the implementation of the local area networks. In the local networking arena, users began to implement solutions to their data connectivity needs within a localized environment. Two major choices were available for their installation of wiring: baseband coaxial cable and broadband coaxial cable.

The baseband cable was based on the Ethernet developments using a 20 MHz, 50Ω coax. Designed as a half-duplex operation, Ethernet allowed the end user to transmit digital data on the cable at speeds of up to 10 Mbps. Clearly, the 10 Mbps was maximum throughput, but was attractive in comparison to the technology of twisted pair at the time (telephone wires were capable of less than 1 Mbps bursty data). Moreover, the use of the baseband technology allowed the data to be digitally applied directly onto the cable system. No analog modulation was necessary to apply the data. It was dc input placed directly onto the cable. The signal propagates to both ends of the cable before another device can transmit. This is shown as a quick review in Figure 14-1. To control the cable access, the attached devices used *Carrier Sense Multiple Access with Collision Detection* (CSMA/CD) as the access control. CSMA/CD allowed for the possibility that two devices may attempt to transmit on the cable at the same time, causing a collision and corruption of the actual data. As a result, the cable had to be very controlled, but in the late 1970s this was not a real issue.

Figure 14-1
CSMA/CD cable networks are collision domains

A second alternative was to use the broadband coaxial cable, operating at a total bandwidth of approximately 350 MHz in an analog FDM technique on a 75Ω cable. As a technology, the broadband systems were well known because they were the same as CATV, which had surfaced in the early 1960s. As a result, the technology was well deployed and commodity priced. Moreover, the 350 MHz of capacity was attractive to the computer industry and the communications industry partisans. The issues began to surface quickly regarding the benefits and losses of using each technique (see Figure 14-2).

What the issue really boiled down to was one of analog vs. digital and the baseband vs. broadband implementations to achieve this goal. One world-renowned consultant and research house in the New England area of the United States even said that people who installed a broadband cable to provide their LAN services did not know what they were doing, and were wasting their company's money! This was a hot issue throughout both industries. In reality, the issue of using a broadband cable was under the turf of the voice communications departments, whereas the baseband cables were under the primary control of the data processing/data communications departments within corporations. If one technology was chosen over another, the lines in the sand would be washed away, and the convergence of voice and data would force the convergence of the two groups.

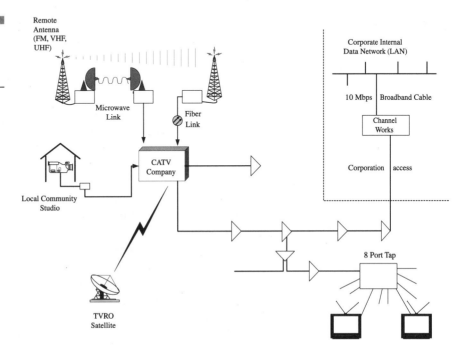

Figure 14-2
Broadband coaxial cable system from the beginning

The issue was therefore not whether to use a cable, but which type of cable to be used so that the LAN would fall under the correct jurisdictional authority within the organization. Unfortunately, control is not the goal of organizations, but access and profitability are! As an industry, too much time was wasted over semantics. However, what ultimately rolled out of the bandwidth argument was that the baseband cable systems were better for the LAN. This was the decision of the 1980s, when all traffic on the LAN was geared to data only at speeds of 10 Mbps and less.

Cable TV Technology

As discussed in Chapter 13, CATV has been around since the early 1960s. It is a proven technology. In the early days of Ethernet, *Digital Equipment Corporation* (DEC) rolled out many of their systems using baseband (Ethernet) cables. However, recognizing that some organizations needed more than just data on a large localized network, they worked with two major providers at the time to develop the interfaces for the broadband cable systems to attach an Ethernet to the CATV cable.

DEC developed several working arrangements with various suppliers to provide a *Frequency Agile Modem* (FAM) to work on the cable TV systems. The CATV companies did not necessarily own the broadband cable. Instead, this cable was locally owned in a high-rise office or a campus complex by the end user. The cable system provided a high bandwidth, but was very complex for the data and LAN departments to understand. The reason was obvious: the broadband coax operated by using frequency division multiplexing (analog techniques), which was beyond the scope of the LAN administrators and the Data Processing departments. The voice people knew of analog transmission, but had a hard time with digital transmission in those days. There was a silent department in the crux of all the arguments—the video departments within many organizations stayed out of the fight.

As DEC began to roll out various choices, the average user had to justify the connection of the analog technologies (used as a carrier) with the digital data demands of the LAN. What many organizations did on a campus was to consolidate voice, data, LAN, and video on a single cable infrastructure. What the industry came up with was a specification for 10 Broad 36 to satisfy the LAN needs over a coax cable. 10 Broad 36 stands for 10 megabits per second, on a broadband cable over 3600 meters using analog amplifiers. A classic representation of the combined services on 10 Broad 36 is shown in Figure 14-3.

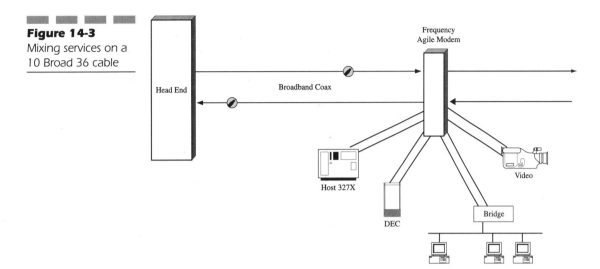

Figure 14-3

Mixing services on a 10 Broad 36 cable

The data industry was distraught because this encouraged the use of an analog carrier system to move the digital data. Over the years, however, this has been revisited several times. Wang Computer Company developed a proprietary cable system for connecting WangNet systems together by using a dual broadband coax cable. This met with only limited success because of the proprietary nature of their interfaces and the pricing model they used. Technologically, the system was sound.

Later in the evolution of this service, the term broadband LANs became popularized. Ten megabit per second Ethernet grew to 100 Mbps, and then on to gigabit Ethernet. Justifying and enabling this high-speed communications service met with some resistance until the use of the various fiber and coaxial systems emerged. By taking a quantum leap in the industry, the data and voice departments saw the benefit and need of converging the two services to the desktop in order to offer voice and video over the LAN. The 10 Mbps Ethernet and coaxial cables could not handle this offering. Enter the new technology! Hybrids of coax and fiber were introduced to the desktop. Moreover, with the access to the Internet under constant scrutiny and pressure to add speed and capacity (voice and video on the Internet), the industry began to seek a new method of bypassing the local copper loop provided by the telephone companies. The technology already at the door, of course, was the CATV. So a new idea emerged: use the CATV to support the high speed Internet access and bypass the local loop from telephone companies. Hence, cable modem technology changed the way we will do business for the future.

The New Market

The cable television companies are in the midst of a transition from their traditional core business of entertainment video programming to a position as a full-service provider of video, voice, and data telecommunications services. Among the elements that have made this transition possible are technologies such as the cable data modem. These companies have historically carried a number of data services. These services have ranged from news and weather feeds, presented in alphanumeric form on single channels or as scrolling captions, to one-way transmission of data over classic cable systems as discussed in Chapter 13.

Information providers are targeting upgraded cable network architecture as the delivery mechanism of choice for advanced high-speed data services. These changes stem from the commercial and residential data communications markets. The PC and LAN explosion in the early 1980s was rapidly followed by leaps in computer networking technology. More people now work from home, depending on connectivity from commercial online services (such as AOL, CompuServe, and Prodigy) to the global Internet.

Increased awareness has led to increasing demand for data service, and for higher speeds and enhanced levels of service. Cable is in a unique position to meet these demands. The same highly evolved platform that enables cable to provide telephony and advanced video services also supports high-speed data services. There appear to be no serious barriers to cable deployment of high-speed data transmission.

System Upgrades

The cable platform is steadily evolving into a hybrid digital and analog transmission system. Cable television systems were originally designed to optimize the one-way, analog transmission of television programming to the home. The underlying coaxial cable, however, has enough bandwidth to support the two-way transport of signals. The hybrid network is shown in Figure 14-4.

Growth in demand for Internet access and other two-way services has dovetailed with the trend within the industry to enhance existing cable systems with fiber optic technology.

Many cable companies are in the midst of the upgrade to HFC plants to improve the existing cable services and support data and other new services. Companies are taking different approaches to online service access.

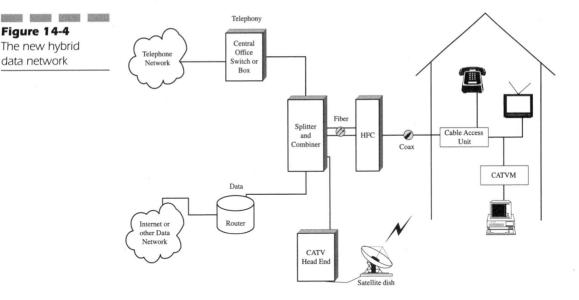

Figure 14-4
The new hybrid
data network

For some applications, customers may be accessing information stored locally at or near the cable head-end or regional hub such as the @Home®[1] services being offered in many cities. This may be temporary until wide area cable interconnections and expanded Internet backbone networks are in place to allow information access from any remote site.

Cable Modems

Digital data signals are carried over *Radio Frequency* (RF) carrier signals on a cable system. Digital data utilizes cable modems, devices that convert digital information into a modulated RF signal and convert RF signals back to digital information. The conversion is performed by a modem at the subscriber premises, and again by head-end equipment handling multiple subscribers. Look at Figure 14-5 for a block diagram of the cable modem.

A single 6 MHz channel can support multiple data streams or multiple users through the use of shared *Local Area Network* (LAN) protocols such as Ethernet, commonly used in business office LANs today. This is where the industry began in the late 1970s when Ethernet networks were applied to the broadband coaxial networks.

---

[1]@Home is a trademark of the @Home Corporation.

Figure 14-5

Block diagram of the
cable modem

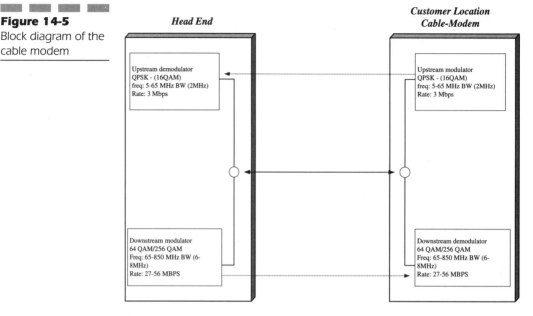

Different modulation techniques are being tried to maximize the data speed that can be transmitted through a 6-MHz channel. Modulation techniques include *Quadrature Phase Shift Keying* (QPSK), *Quadrature Amplitude Modulation* (QAM), and *Vestigial Side Band* (VSB) amplitude modulation. Comparing the data traffic rates for different types of modems shows why the cable modem is so popular under today's environment. Table 14-1 is a comparison of a file download of 500 Kbytes using different techniques.

Careful traffic engineering is being performed on cable systems so that data speeds are maximized as customers are added. Just as office LANs are routinely subdivided to provide faster service for each individual user, so too can cable data networks be custom tailored within each fiber node to meet customer demand. Multiple 6-MHz channels can be allocated to expand capacity as well.

Some manufacturers have designed modems that provide asymmetrical capabilities, using less bandwidth for outgoing signals from the subscriber. Cable systems in some locations may not have completed system upgrades, so manufacturers have built migration strategies into such modems to allow for eventual transmission of broadband return signals when the systems are ready to provide such service and customers demand it. A representative sample of the way data speeds are provided on cable modems is shown in Table 14-2.

Table 14-1

Comparison of transmission speeds

Time to Transmit a Single 500 Kbytes Image		
Telephone Modem	28.8 kbps	6–8 minutes
ISDN	64 kbps	1–1.5 minutes
Cable Modem	10 Mbps	Approximately 1 second

Source: CableLabs

Table 14-2

Representative asymmetrical data cable modem speeds

Sample Cable Modem Speeds	Upstream	Downstream
General Instrument	1.5 Mbps	30 Mbps
Hybrid/Intel	96 kbps	30 Mbps
LANcity	10 Mbps	10 Mbps
Motorola	768 kbps	30 Mbps
Zenith	4 Mbps	4 Mbps

Standards

Modems are available today from a variety of vendors, all with their own unique technical approach. These modems are making it possible for cable companies to enter the data communications market now. In the longer term, modem costs must drop and greater interoperability is desirable. Customers who buy modems that work in their current cable system need assurance that the modem will work if they move to a different geographic location served by a different cable company. Furthermore, agreement on a standard set of specifications will allow the market to enjoy economies of scale and drive down the price of each individual modem. Ultimately, those modems will be available as standard peripheral devices offered as an option to customers buying new personal computers at retail stores. The cable companies and manufacturers came together formally in December 1995 to begin working toward an open standard.

Leading U.S. and Canadian cable companies were involved in this development toward an open cable modem standard. Specifications were to be developed in three phases, and then be presented to standards-setting bodies for approval as standards. Individual vendors were free to offer their own implementations with a variety of additional, competitive features and future improvements. A data interoperability specification will comprise a number of interfaces. The resultant specification is called *Data Over Cable Service Interface Specification* (DOCSIS), which architecturally is shown in Figure 14-6 as it relates to the TCP/IP protocol stack. Note that there are several sublayers added into the DOCSIS specification at the bottom layers (for example, layer 1 and 2) of the protocol stack. This is to simplify the connection and add the dimension of security into the DOCSIS specifications.

Some interfaces reside within the cable network. Several of these system-level interfaces also will be specified in order to ensure interoperability of such important functions as authentication for login/logout, ease of installation of cable modems for reliable customer activation, and spectrum management over the cable network's hybrid fiber/coaxial plant.

Return Path

The portion of bandwidth reserved for return signals (from the customer to the cable network) is usually in the 5–40 MHz portion of the spectrum. This portion of the spectrum can be subject to ingress and other types of interference, and so cable systems offering two-way data services have been designed to operate in this environment.

Industry engineers have assembled a set of alternative strategies for return path operation. Dynamic frequency agility (shifting data from one channel to another when needed) may be designed into modems so that data signals may avoid unwanted interference as it arises. Other approaches utilize a *gate* that keeps the return path from an individual subscriber closed except for those times when the subscriber actually sends a return signal. Demarcation filters, different return laser types, and reduced node size are among the other approaches, each involving tradeoffs between capital cost and maintenance effort and cost.

Return path transmission issues have already been the subject of two years of lab and field testing and product development. The full two-way capability of the coaxial cable already passing most U.S. homes is now being utilized in many areas, and will be available in most cable systems soon.

Figure 14-6
The DOCSIS model

Figure 14-7
Frequency spectrum
allocated to the cable
modems

Full activation of the return path in any given location will depend on individual cable company circumstances, ranging from market analysis to capital availability.

The spectrum used for the forward and reverse paths is shown in Figure 14-7 as an indication of the frequencies available and the overall management of the system. This also shows that additional 6-MHz channels can be set aside to handle the data traffic on the cable modems and the cables themselves.

Applications

Cable modems open the door for customers to enjoy a range of high-speed data services, all at speeds hundreds of times faster than telephone modem calls. Subscribers can be fully connected, 24 hours a day, to services without interfering with cable television service or phone service. Among these services are

- *Information Services*—access to shopping, weather maps, household bill paying, etc.
- *Internet Access*—electronic mail, discussion groups, and the World Wide Web
- *Business Applications*—interconnecting LANs or supporting collaborative work
- *Cablecommuting*—enabling the already popular notion of working from home
- *Education*—allowing students to continue to access educational resources from home

The promises of advanced telecommunications networks, once more hype than fact, are now within reach. Cable modems and other technology are being deployed to make it happen. Regardless of the technology selected, the main goal is to get the high-speed data communications on the cable adjacent to the TV and entertainment. This gives the CATV companies the leverage to act in an arbitrage situation, competing with the local telephone companies who have dragged their feet in moving high-speed services to the consumer's door. As shown in Figure 14-8, there are several up-and-down speed capabilities that can be shared to deliver asymmetrical speeds to the consumer's door. In the particular figure, the download speed is up to 30 Mbps, whereas the upstream operates at 1.5 Mbps. For many this is sufficient based on their applications.

The Combined Corporate and End User Networking Strategies

The use of a single PC on a cable system is fine for the telecommuter (or cable commuter now), but what of the small office or home office where

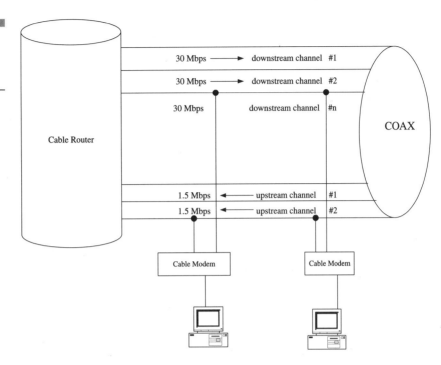

Figure 14-8

Different speeds on the up-and-down stream flows

more than a single PC is connected? In Figure 14-9 is an example of various ways the CATV connection can be accomplished. This figure uses an example of local home networking with two PCs connected to a single cable modem. Most of the providers have instructions on how to accomplish this and require a home user (or small user) to download additional software to accommodate the dual connection on a single modem. The second alternative to this is to have a router connected to the cable modem, such as a *branch office router*. The network is attached to the router, and the router is responsible to handle the dispersing of the traffic onto the cable system. This can be a very effective use of the link. Next in the figure is a connection to a hub, such as a 10 or 100 Base T connection into a LAN hub. Although the CATV providers state this is not supported and will likely not work, it has been done and works pretty well for a small office or home office connection. Using the connection directly into the hub from the cable modem makes the modem available to more users instead of just a single PC. The hub will act as a bridging function onto the modem and concentrate the traffic through the individual devices. These configurations all work, but the providers do not support problems if they arise. You are on your own if it does not work.

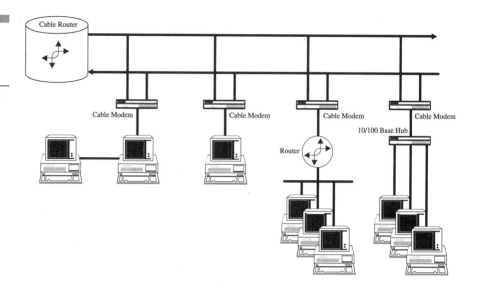

Figure 14-9
Multiple ways of
connecting to the
cable modem

A Final Thought

A final thought concerning the use of this technology is to understand the
security concerns associated with the cable modem. When the CATV systems are used, they are shared, high-speed Ethernet backbone access to the
Internet or other connections. One must be aware that on a shared cable,
the PC is a peer to all others on the same cable, even though they are in
physically different locations. In a community of 500 connections, there will
be many people who acquire the service from the CATV suppliers. The
CATV company installs according to the appropriate technology, not according to security parameters. This is okay because they are merely providing
the bandwidth to gain access. It is the end user's responsibility to turn off
all the leaks in the local system (the PC). By default, when you run
Microsoft windows environment and the appropriate networking software,
the *shares* on the PC are turned on. The end user must therefore go in and
turn them off. This means that if the shared services are not turned off, a
user down the street, or across the town, can double click on their *network
neighborhood* icon and see all the other PCs connected to the cable. Not only
can they see the devices, they can double click on the PC and see the
resources available on that PC. From there when a remote device has double clicked on your PC, they can open your drives and see your files. Unless
some provisions have been taken to block this access, the intruder (used as

a method of entry only) can read, write, edit, or delete your files. Worse yet, while the intruder is on your system, you do not even know they are there.

Many users who have cable modem service from the cable companies are not aware of the risks. Worse, the installation personnel on these systems do not totally understand or they forget to point out the risks. Therefore, users leave the PC on 100% of the time (day and night), making access to their computers totally available. The cable modem is available 100% of the time, making the computer a target for hackers and mischievous doers, without the permission or the knowledge of the penetrated computer owner. Be aware that the risk is there and find out how to shut these open doors before leaving your computer on the network. Do not assume that just because you shut your system off when you're not using it that you are protected because when you do log on, you are exposed, and the perpetrator can get on your system while you are on it, too!

Technology is a wonderful thing, but it must be controlled.

Data Over
Power Lines

Introduction

Whenever we look at the infrastructure of the Telecommunications networks, the issue of the "last mile" always crops up. For over 120 years, the Telephone Companies installed their local wires from the Central Offices to the consumers' door. This local interconnection provides all the necessary access for a customer to dial up a voice call, or a data modem call. The Telecommunications Acts being enacted around the world are privatizing and deregulating the Telephone Companies worldwide. The local loop is coming under constant barrage as new competitors form up and enter the business. The infrastructure— a pair of telephone wires to everyone's home— creates more legal suits and discussions than anything else in the industry.

As a secondary course of action, new providers are scurrying around looking for alternative methods to get to the consumer's door. Some of the discussions have centered on the use of Cable TV and wireless access. Now, a new twist has emerged! The use of the copper wires running from the electric utilities may provide an alternative route. Pilot programs and tests are being conducted around the world to see if the electric (power) wires can be an alternate delivery method. In Europe the results look favorable, which we shall see, but in North America, the studies have a long way to go.

A few years ago, the primary method of gaining new access was considered the realm of fiber optic cables to improve and provide high-speed communications. In 1996, for example, the city of Turin launched a project to link 10 million homes in Italy with fiber in an overall project costing $6.2 Billion USD.

Alternatives to fiber based technologies now include:

- xDSL
- Broadband Satellite Communications
- Cable TV
- Broadband Wireless
- New Power Line Technology

ADSL has been the "preferred"[1] method of delivering high-speed communications to the business use on the existing copper cable plant. Projections in the industry show that the ADSL growth curve from 1998 to 2002, as shown in Figure 15-1, shows a high growth area.

Looking at the three more popular solutions the use and growth figures are all quite impressive based on the use of data communications. In Figure

---

[1] Preferred method of the Telephone Companies so that they can continue to use their existing infrastructure

Figure 15-1
Projected Use of xDSL
technologies

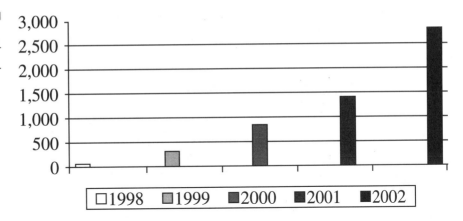

Figure 15-2
Expected volume of
three technologies to
access data

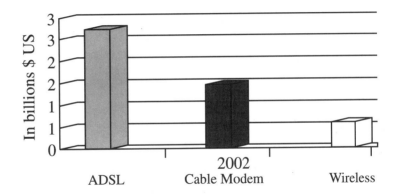

15-2 is a comparison of the expected usage in 2002 for the three compared technologies— ADSL, Cable modem and wireless technologies.

Power Line Technology

Power Line Telecommunications technology (PLT) is an emerging technology with significant business potential for both the Electric Utilities and Telecommunications companies. The deregulation of both sectors, currently underway in North America and Europe, is a major driving force in the development and exploitation of PLT. Nortel[2] is doing a lot of work and

---

[2] Nortel or Northern Telecomm is headquartered in Brampton, Ontario. Yet in Marlow, England at their European HQ, they have developed the PLT jointly with NORWEB.

research on the use of power lines to transmit data. The use in Europe is ahead more than in North America because of the way the distribution is handled. North American development will be postponed while the issues are worked out, but in Europe progress is expected rapidly.

Power Line Telecommunications (PLT) is a rapidly developing technology, targeted to use electric power lines for the transmission of data. One of the many obstacles in achieving the "Wired Society" is the investment required to build infrastructure that will reach the maximum number of information users. Current technologies (i.e. wire, coax, fiber and wireless) have not solved this problem. The result will be delays in wiring up the society for future access.

It is important to note that every building is connected to the electric power grids for commercial access to power. Moreover, every office or room has electric outlets installed. This level of wiring cannot be matched by either telephone or cable TV wiring. Thus, the emerging PLT technology opens mass production of local access at a reasonable cost. Beyond telephony, PLT can facilitate a myriad of new services normally too expensive or difficult to implement by other technologies. This includes such services as

- Remote meter reading
- Appliance control and maintenance
- Energy management
- Home automation

Technological and business impacts of PLT technology will be felt by Telecommunications and Energy providers. PLT becomes the enabling technology for utility companies to become essential providers in the "Wired Society." It can be a major factor in the mass introduction of many new data services.

Local Access

Local access is a critical infrastructure issue. It may be provided by various telecommunication technologies, of which PLT will be one. In many cases, local access infrastructure will be a hybrid. An example of how this infrastructure can look like is shown in Figure 15-3.

The first use of the power line for control messages was developed in the early 1950s. The method (known as the Ripple Control) used low frequencies (100 –900 Hz) producing a low bit rate and a demand on high transmitter power. The power often ranged in the 10^1-10^2 kW. The system provided one-

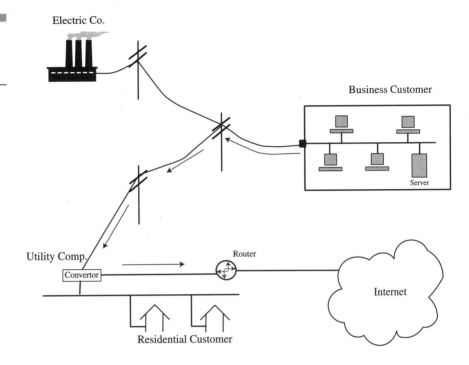

Figure 15-3
Hybrid
communications
on PLT

way communication. Primary applications include street light management, load control and rate switching. Newer systems were introduced in the early 1980s, with a slightly higher bit rate.

Studies were conducted during the mid-1980s by utility companies to determine the characteristics of the electric grid as a medium for data transfer. Frequency ranges of from 5–500 kHz were studied. The signal/noise-levels were important measurements as was attenuation of the signal. These studies were conducted both in Europe and in the US. The initial driving factor was the use of SCADA technology[3]. Two-way communications was developed in the late 1980s. Present systems came on the market during the late 1980s. The primary differences included higher frequencies and reduced signal levels.

Advanced protocol techniques are now used and systems are managed for proper data transfer. Expectations for the future will use frequencies in the GHz-range, providing broadband communications data throughput spanning multi-megabits.

---

[3] *Supervisory Control And Data Acquisition* (SCADA)

The electric power lines constitute the basis of the "Wired Society" and its economy for two separate reasons:

- *The power lines carry electricity.* The primary purpose of the wires is to supply electricity. Efficient energy supply is an essential precondition for the "Wired Society."

- *The power lines can carry data.* The last decade has brought technological innovation enabling the use of power lines as a data transmission system and information carrier in its own right.

PLT opens new business opportunities for applications and customer services, for both the Telecom and Electrical industries in Europe. There are various strong, sometimes unique, motivations to consider the PLT-enhanced power line as a very important element for the future.

- *The power grid is ubiquitous.* The low-voltage power grid constitutes an existing networked infrastructure to millions of residential customers as well as businesses. Concurrently, it is a large-scale and fine-grained infrastructure, connecting hundreds of millions of people all over the world, but also even crossing the boundaries of homes and buildings to individual wall outlets and sockets, home appliances and electrical industrial equipment.

- *The power grid offers the "last mile."* Through PLT, the power line offers the "last mile connectivity" to the individual customer. It is not the only last-mile technology, but it can be a significant competitor. Based on the individual situation, demands and capabilities, it is likely that the overall infrastructure will be a hybrid. The power line is positioned as a complement or alternative last mile. It can provide a permanent-access, two-way, always-on-line connection to the customer.

- *The power grid supports information services.* PLT technology is new and needs advances, for Telecommunications capabilities and commercial equipment cost reductions. However, the limited data rate transfer speeds are sufficient for many useful applications. A number of useful services are already feasible with data transmission in the kbps range, while current PLT speeds can handle up to 1 Mbps. Speeds up to 100 Mbps are possible within a few years.

- *The power grid is already there, so it is cost efficient.* Because the power line infrastructure is already in place, it is potentially cheaper than other forms of local telecommunications access.

Developments in PLT are driven by both business and technology factors in the telecommunications and electric sectors as follows:

1. Deregulation and privatization of the electric utilities drive investment in operating efficiencies.

2. Flat growth and margins are forcing electric utilities to focus on expansion of value-added energy service capabilities.

3. The majority of the activity in value added service development is driven by exploitation of communications technologies.

4. Deregulation of telecommunications resulted in a growing level of power utility participation, which provides significant economies of scale in developing value-added and communication-based services.

5. PLT provides a potential cost-effective solution to the "bottleneck" of local access in the telecommunications service infrastructure.

6. PLT could be deployed in less developed countries to cost-effectively raise telecommunications coverage and density.

7. PLT may provide the capability to deliver high-speed access to the home, which will accelerate the arrival of the interactive information age.

Three main categories of new Information and Communication Technology-based applications and services can be distinguished:

1. Energy services within the existing sector boundaries, but with new forms, features and scales. Examples are remote metering, remote billing, demand-side management, distribution automation and control from a distance.

2. Telecommunications services within the existing sector boundaries based on PLT technology. Examples are power line-based telephony, Internet access, voice and video transfer.

3. New categories of value-added customer services based on PLT and other ICT technologies. Examples of such services are

   - Security from a distance (in home and office)
   - Personal comfort services
   - Energy and equipment cost saving
   - "Smart" home automation

- Content-oriented electronic services (in education)
- Information
- Advice
- Entertainment
- Electronic commerce.

These services may be bundled through PLT with the conventional telecommunications and energy offerings, thus widening the service range and enhancing individual choice—features that are increasingly demanded by customers. Coupling ICT and PLT technologies gives many opportunities in applications and services. Efficient energy supply is an essential precondition.

Several alternative technologies compete for the local access infrastructure and services. Each technology has its own strengths and weaknesses. A comparison between PLT and some other relevant technologies is presented in Table 15-1.

Table 15-1

Comparison of technologies for local loop access

Technology	Strength	Weakness
Copper Wires	Mature	Slow speed
	Robust	On demand services only
	Wide variety of choices	
	Ease of use	
ISDN	Mature	Expensive for consumer
	Robust	On-demand mostly
	Available	Newer always-on not robust
Cable Modems	Excellent performance	Limited coverage (geographic)
	Mature	Many trial basis only
	Always On	Shared resource with others
	Low Cost	Slows down with contention
	Ease of installation	

Technology	Strength	Weakness
ADSL/Copper based	Great performance	Immature
	Always on	Standards not 100%
	Low cost	Different ways to implement
	Uses existing wiring	Length of wire (local loop) can be problematic
Power Line Technology	Good performance	Distance limited
	Always on	Still in development
	Coverage everywhere	Installed base limited
	Low cost increment	Possible problems with meshed networks

A technology comparison must not only take into account the state of the art today, but should also look at the ease of development and improvement in the near future. This is done in Table 15-2, which compares the upgrade paths for various Information Society infrastructure technologies.

The upgrade path for PLT is very attractive compared to other local access technologies. Other access infrastructures require significant capital investment to deliver more than one or two telephony lines. PLT systems are under development that will allow bandwidth delivery to increase incrementally, as market demand requires.

The technology comparison then shows that the major advantages of PLT technology are as follows:

■ Permanent on-line connection

■ No usage costs as with conventional switched systems

■ Good performance

■ Very good coverage

■ Symmetric, two-way communication

■ Relatively inexpensive, most of the infrastructure is already in place

■ Attractive upgrade path

■ Unique; it provides both telecommunications and energy services simultaneously

Table 15-2

Access technology upgrade path and migration

Copper Wires [1]	Cable TV Networks [2]	Wireless Local Loop [3]	PLT [4]
Broadband @> 1 Mbps			
			Requires Compression
Wideband @128 kbps- 1 Mbps			
Introduction of ADSL	Introduction of Cable Modems	Introduction of broadband radio	+
Mid-band @ up to 128 kbps			
–	–	–	+
Multi-line telephony (minimum data)			
Require additional build out	Requires additional build out	Requires additional build out	+
Telephony (minimum data)			
+	+	+	+

[1] Copper twisted pair

[2] Hybrid fiber/coax with copper termination for telecommunication

[3] NORTEL proximity21 radio system

[4] Based on an OFDM implementation protocol

PLT is a definite and significant "last-mile" local access technology, but it will not be the only one. The use of different complementary technologies is seen as more likely, since each technology offers a different compromise between bandwidth, reach, noise immunity, and cost.

The Changing Market for Legal and Technical Environments

There is no free lunch in seizing the opportunities offered by PLT technology. Current developments operate in a highly dynamic and uncertain environment. Sector boundaries are shifting, financial power is enhanced, competition is increasing, legal regulations are being fundamentally redesigned, new market entrants appear, and old paradigms of operational and business success no longer give a guarantee for the future.

Deregulation of Electricity The electrical transmission, distribution and supply businesses of Europe are confronted with an ongoing period of radical change. The growing trends of privatization, liberalization and competition have called for fundamental changes in corporate structure, business goals, organizational culture, human resources and market and customer orientation. This period of change has come at a time when the majority of these businesses also face increasing economic pressures upon fundamental elements of their operations. Annual growth in electric supply has been averaging between 1% and 2% in most of Western Europe over recent years. Margins in electric supply are classically found to be between 2% and 3%.

Telecommunications Deregulation In parallel with this period of radical change within the electric utility industry, the telecommunications industry is experiencing deregulation and privatization leading to much greater levels of competition for previous monopoly suppliers.

A common theme across Europe and much of the developed world is power utility competition within the new telecommunications operator business. This provides further significant drivers for those utilities to exploit capabilities in *Power Line Telecommunications* (PLT) technology. The power utilities not looking to enter the telecommunications services sector directly are looking to commercially exploit PLT by providing network access to telecommunications operators.

The new PLT systems and capabilities being developed have the potential to address one of the most important aspects in the future of telecommunications. For years, the *"local loop"* or *"last mile"* has been the bottleneck. The dream of the ubiquitous "information superhighway" connecting all residential customers to a high speed Internet has not been achieved. There are two main reasons for this

- *The limitations of the existing copper local loops.* The networks require massive capital investment to support high-speed interactive services (i.e. > 2 Mbps).

- *The cost of constructing alternative high-speed access networks is immense.* Cable TV businesses in the UK (owned by Telephone Companies from the USA) have spent $30 billion (ECU) building alternative TV and telephony systems to 60% of the UK residential customer base. Unfortunately, this network requires significant additional capital investments to deliver high-speed interactive services.

This is the enigma, which is preventing the infinite possibilities of the Internet in the consumer environment to be realized. To date, telecommunications operators have been unable to justify the massive investment required to deliver the "information superhighway" access networks. New interactive applications and services that would justify such investment are not being created because the access networks are not there to support such offerings. In less developed countries PLT could provide telephony access cost effectively.

Why development has been limited There are three main reasons why exploitation of PLT to date has been limited to "narrow band" applications such as telemetry, SCADA-based remote metering etc.

- *PLT is still immature.* Research into high-speed and high frequency PLT systems has been contemplated for a long time: AT&T carried out preliminary evaluations as early as 1923. Only in recent years, have the complex modulation techniques have been developed capable of managing high-speed digital services.

- *Lack of standards and interoperability.* Standards, which identify frequencies employed for PLT to 148 kHz, restrict the capability to deploy modern voice or data systems. This has further reduced the incentive to investigate potential higher value PLT systems.

- *Commercial incentives have not been clear.* For limited applications, it is not always easy to build the business case. Therefore, investigation of PLT has been limited to narrow band applications, which are primarily targeted at reducing operational costs of the power supply utility.

Given the high investments in telecommunications and power line infrastructures, it is clear (from a cost-benefit analysis) that successful adoption and penetration of PLT technology must be based on a range of simultaneous applications or service. It is also clear that PLT requires an integrated blend of disparate capabilities (telecommunications, power engineering, advanced software applications, dedicated hardware and market research).

From a general innovation perspective, the slow rate of PLT technology adoption should not come as a surprise. It is interesting to look at how fast different technologies have penetrated the market. The Internet has reached the most widespread system for telecommunications over a short period. It took five years to reach 50 million users, compared with 38 years for broadcast radio systems and 13 years for TV.

The main technological improvements needed from the infrastructure perspective are:

- *Increased PLT transfer speed and reliability.* To keep up with needs, transfer speeds > 100 Mbps are desirable. Given that the power line is a noisy medium, further reduction of PLT transfer error rates is also required.

■ *Study of PLT frequency characteristics and ranges.* PLT frequencies are currently restricted to relatively low frequencies and narrow bands.

■ *Compatibility between energy and information functions on the power line.* Energy and information are different functions working at different frequencies on the same power line infrastructure. Each function has its own requirements, however their behavior must be compatible.

■ *Hybrid infrastructures for telecommunication.* In many local access applications, there will be a need for a combination of telecommunication infrastructure.

The Converging Markets

Convergence is a popular term used by merging digital technologies such as telecommunications, information technologies, broadcasting, and multimedia. It is clear that the term Convergence can also be properly used for the fusion of energy and information systems. This new technological and commercial environment will create vast new opportunities for everybody and especially for the power utilities. A new type of *telecommunications and energy provider* is growing out of a mixture of the old type monolithic telecommunications operators and old type power companies. This new type of operator is expected to gradually replace the existing ones.

This business development implies that the integration of ICT in energy systems is a strategic technological area. Here, PLT may gain the edge. Current utility objectives are shown in Table 15-3.

	Goal	Performance Issue
Table 15-3 Comparative goals for the future with PLT	Cost Reduction	Improve operational efficiency
		Increase productivity
		Improve asset utilization
		Outsource non-core functions

continues

Table 15-3	Goal	Performance Issue
continued	Customer Retention	Improve customer services
		Retain revenue and load
		Maintain competitive position
		Gain new customers
	New Revenue Generation	Develop new offerings to new and existing customers

Enhance Services Through Communications Functionality

Today, utilities organizations are all preparing for deregulation and the open domestic market. In the domestic market, information systems will be needed to handle the massive change into new flexible billing systems and systems that provide enhanced customer service. These systems must not only provide services towards the customers. The need is also to develop and focus on the utilities' own internal efficiency. A number of ways are considered to cut management and operational costs and to shorten repair times. Over the years 1980–1990, many utilities reduced personnel and have reached the level at which only limited changes in this respect would be possible. The next step therefore is to incorporate more efficient methods through better information management and through up-grading old ICT systems. The power grid for data communication is essential for this approach. As an example, British utilities show a total predicted investment of close to $2 billion ECU in ICT systems. Half of this investment will be related to customer and market systems. The investments are to be allocated as shown in Figure 15-4.

Of the total ICT investment, the build-up of new telecommunications infrastructures is an essential part whereby the power lines will yield advantages if the data transfer is secure and efficient. From the viewpoints of both increased operational efficiency and enhanced customer service, it is evident that ICT is a strong competitive factor for tomorrow's energy and electricity industry, as it is for today's telecommunications industry.

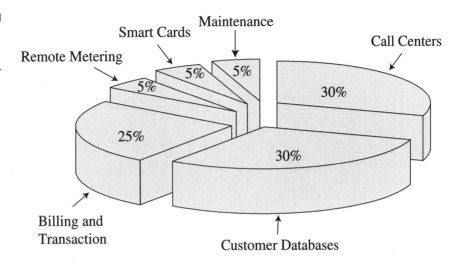

Maintenance

Smart Cards

Remote Metering

Call Centers

5%

5%

5%

30%

25%

30%

Billing and
Transaction

Customer Databases

xDSL

Introduction

One of the major problems facing the *Incumbent Local Exchange Carriers* (ILEC) is the ability to maintain and preserve their installed base. Ever since the Telecommunications Act of 1996 there has been mounting pressure on the ILECs to provide faster and more correct Internet access. In order to provide the higher speed communications capability, these carriers have continually looked for new means of providing the service.

However, the ILECs have an installed base of unshielded twisted pair in the local loop that cannot be ignored or abandoned. Therefore, a new form of communications was needed to work over the existing copper cable plant. One of the technologies selected was the use of xDSL. The DSL family includes several variations of what is known as digital subscriber line. The lower case x in front of the DSL stands for the many variations. These will include:

- *Asymmetrical Digital Subscriber Line* (ADSL)
- ISDN (like) *Digital Subscriber Line* (IDSL)
- *High Bit Rate Digital Subscriber Line* (HDSL)
- *Consumer Digital Subscriber Line* (CDSL)
- *Rate Adaptive Digital Subscriber Line* (RADSL)
- *Very High-Speed Digital Subscriber Line* (VDSL)
- *Single or Symmetric Digital Subscriber Line* (SDSL)

One can see that the variations are many. Each *Digital Subscriber Line* (DSL) capability carries with it differences in speed, throughput, and facilities used. The most popular of this family under today's technology is the use of *Asymmetrical Digital Subscriber Line* (ADSL).

ADSL is a technology being provided primarily by the ILECs because the existing cable plant can be supported and the speed throughput can vary depending on the quality of the copper. However, the most important and critical factor in dealing with ADSL technology is the ability to support speeds between 1.5 Mbps up to 8.192 Mbps. At the same time, the ILEC can also support POTS for voice or fax communications on the same line. What this means is that the ILEC does not have to install all new cabling to support high-speed communications access to the Internet.

The Asymmetrical Digital Subscriber Line (ADSL) Defined

Asymmetrical Digital Subscriber Line (ADSL) is the new modem technology to converge the existing twisted pair telephone lines into the high-speed communications access capability for various services. Most people consider ADSL as a transmission system instead of a modification to the existing transmission facilities. In reality, ADSL is a modem technology used to transmit speeds of between 1.5 Mbps and 6 Mbps under current technology. It is stated that in the future, ADSL will support speeds about 8.192 Mbps. This definition of the higher range of ADSL speeds is one that is yet to be proven, however, with changes in today's technology one can only imagine that the speeds will be achievable.

Some of the many capabilities that are being considered through the use of the DSL family are the services for converging Voice, data, multimedia, video, and Internet streaming protocols services. It is through these demands that the carriers see their future rollout of products and services to the general consuming public. In Table 16-1 various theoretical speed and distances of ADSL technologies are shown.

One should remember that the speeds and distances shown here are the theoretical limits based on good copper. If the copper has been damaged or impaired in any way, then the speed and distances will change accordingly (downward). The theoretical limits are what most engineers and providers have been touting their technologies can support. Reality is another thing, and the actual distances and speeds very likely will be less than shown

Table 16-1

Data rates for ADSL, based on installed wiring at varying gauges

Current Data Rate	Wire Gauge	Distance in K Feet	Distance in Kilometers
1.5–2.048 Mbps	24	18	5.5
1.5–2.048 Mbps	26	15	4.6
6.3 Mbps	24	12	3.7
6.3 Mbps	26	9	2.7

here. What is most important is to assume that these speeds can be established and maintained on the installed base of *Unshielded Twisted Pairs* of wire (UTP). As long as the ILEC can approximate these speeds today, the consumer will most likely not have much to complain about.

Modem Technologies

Before proceeding too far in this discussion, a review of modem technology is probably in line. Modems, or modulator/demodulator, were designed to provide for data communications across the voice dialup communications network. Through the use of modem technology introduced back in the 1960s, users were able to transmit data across the voice networks at speeds varying between 300 bps–33,600 bps. Although this may seem like high-speed communication, our demands and needs for faster communications quickly outstripped the capabilities of our current modem services, making the demand for newer services more evident. Higher speed modems could be produced, but the economics and variations on the wiring system prove this to be somewhat impractical. Instead, the providers looked for a better way to provide data communications that mimic the digital transmission speeds we are accustomed to.

Using the telephone companies' voice services, the end-user installed a modem on the local loop. This modem acts as the *Data Circuit terminating Equipment* (DCE) for the link. As shown in Figure 16-1, a modem is used on the ILEC's wires to communicate across the wide area networks such as the long distance voice networks. This figure shows that the modem is the interface to the telephone network limited to the quality of the local loop. The ILEC installs a voice grade line on the local copper cable plant and allows the end-user to connect the modem. The modem then converts the data from a computer terminal into a voice equivalent analog signal. There is no real magic in modem communications today, but in the early days of data communications, this was considered *voodoo science*. The miracle of data compression and other multimodulation techniques quickly expanded the data rates from 300 bps to today's 33.6 Kbps. Newer modems are touted to handle data at speeds of up to 56 Kbps, but there are few who actually get data across the network at these rates. So, the reality of all the pieces combined still has the consumer operating at approximately 33.6 Kbps.

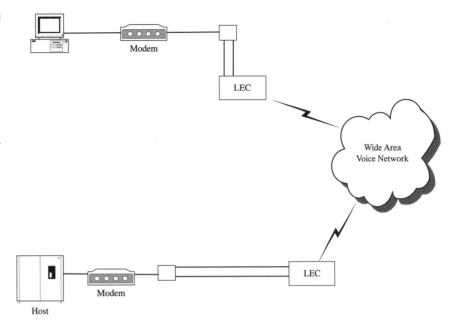

Figure 16.1
Modems are installed at the customer's location and use the existing telephone wires to transmit data across the voice network.

The Analog Modem History

In the early days of modem communications, the Bell Telephone Companies (or the Independent Telephone Companies) provided all services across North America. A customer desiring to transmit data needed only to call the local supplier who would then install the dialup telephone line, the modem, and all associated services to accommodate the desired data rates available. Leased lines were used when specific speeds or volumes were anticipated, but not guaranteed by the dialup services. Regardless of the modem and lines used, the main provider was the key ingredient. The local providers supported only what they knew they could meet, so speeds were often kept very low from a guaranteed standpoint.

If the customer had a leased line and needed better or faster data, special equipment was installed on the line to reach these goals, but at a higher monthly fee. Moreover, the technological advancement of modem technology was not a priority for the local providers because they owned the installed base.

In 1968, things began to change! With court decisions allowing the introduction of competitive devices and the connection of these devices on the regulated carrier's network, demands began to escalate. Restrictions on power output and energy levels were in place to prevent any interference from the modems on the voice network. Also, the customer-provided modems were interconnected through a data coupler (called a *Data Access Arrangement* [DAA]) provided by the local regulated carriers. This, of course, involved a fee for the connection through the telephone company that provided protection equipment.

Later, the *Federal Communications Commission* (FCC) in the US, and the *Communications Radio and Television Commission* (CRTC) in Canada allowed changes in the way the interconnection was handled. Modem manufacturers were allowed to produce their products according to a set of specifications and registrations, eliminating the need for the telephone company protection equipment and the fee associated with the monthly rentals.

Soon the market began to swell with modem products that could take advantage of the voice network to transmit data. However, limitations still existed on the speeds and services allowed by these newer devices. Most of the communications limitations came from the intent of the voice network. The telephone network was designed to carry a voice call with reasonable and reproducible voice characteristics and quality. Limitations were placed on the overall throughput of the physical wires using special filters on the wires. However, the competitive marked spurred the development of modem technologies over approximately 18 years from the old 300 bps speeds to the current speeds we now accept (28.8–33.6 Kbps). In 1997, the introduction of the 56 Kbps modem was going to revolutionize the market and speed up data transmission to meet the demands of the consumers. However, even at 56 Kbps, users were looking for more. The modems just did not satisfy the demands for higher-speed Internet access and video demands. Hence, the movement to newer techniques to provide faster data communications across the local voice telephone networks. Enter the digital subscriber line modem to meet the need.

ISDN Digital Subscriber Line (IDSL)

DSL refers to a pair of modems that are installed on the local loop (also called the last mile) to facilitate higher speeds for data transmission. Network providers do not provide a line; they use the existing lines in place and add the DSL modems to increase the throughput. DSL modems offer duplex operations—transmission in both directions at the same time. The speed of a DSL modem may be 160 Kbps on copper at distances up to 18 K' using the twisted pair wires. The bandwidth used is from 0–80 kHz, as opposed to the arbitrarily limited 0–3300 Hz on a voice line. This is the *ISDN Digital Subscriber Line* (IDSL) using the 144 Kbps full duplex, which gives us what is known as the *Basic Rate Interface* (BRI). As shown in Figure 16-2, the IDSL technique is all digital operating at two channels of 64 Kbps for voice or nonvoice operation and a 16 Kbps data channel for signaling, control, and data packets. ISDN was very slow to catch on, but the movement to the Internet created a whole new set of demands for the carriers to deal with. In fact, the carriers were caught off-guard when user demand which was moderate, escalated so quickly. Now more telephone companies (ILECs) and the newer competitive LECs (CLECs) offer ISDN services for data at a very reasonable fee. The term IDSL is new, but the gist is the same. A digital subscriber line is used to deliver ISDN services. As the deployment of IDSL was speeding up on the local loop, the providers developed a new twist, called *"always on, ISDN"* mimicking a leased set of channels that are always connected. By bonding the channels together, Internet users can *"surf the Net"* at speeds of 128 Kbps in each direction. Note this is a symmetrical digital subscriber line.

Figure 16-2
The IDSL line connection allows 128 Kbps in total simultaneously

High Speed Digital Subscriber Line (HDSL)

In 1958, the Bell Laboratories developed a voice multiplexing system that used a 64 Kbps voice modulation technique *called Pulse Coded Modulation* (PCM). Using the PCM techniques, voice calls were sampled 8,000 times per second and coded using an 8 bit encoding. These samples were then organized into a framed format using 24 time slots to bundle and multiplex 24 simultaneous conversations onto a single, four-wire circuit. Each frame carries 24 sample of 8 bits, plus one framing bit (making the frame 193 bits long), 8,000 times a second. This produces a data rate of 1.544 Mbps or what we know as a T1 (for further discussions on T1 see chapter 26). We now refer to this as a *Digital Signal Level 1* (DS–1) at the framed data rate. This rate of data transfer is used in the US, Canada, and Japan.

Throughout the rest of the world, standards were set to operate using an E1 with a signaling rate of 2.048 Mbps. The differences between the two services (T1 and E1) are significant enough to prevent the seamless integration of the two services.

However, in the digital arena, T1 required that the provider install the circuits to the customer's premises on copper (other technologies can be used but the UTP is easiest because it is already there). The local provider could install the circuit by using a four wire circuit with repeaters spaced at 3K' from the Central Office and 3 K' from the customer's entrance point. In between these two points, repeaters are used every 5–6 K'. Moreover, when installing the T1 on the copper local loop, limitations of the delivery mechanism get in the way. T1 (and E1) uses *Alternate Mark Inversion* (AMI), which demands all of the bandwidth and corrupts the cable spectrum quickly. As a result, the providers can only use a single T1 in a 50-pair cable and could not install another in adjacent cables because of the corruption. Figure 16-3 is a representation of this cable layout. This is inefficient use of the wiring to the door, making it impractical to install T1s to small office/home office and residential locations. Further limitations required the providers to remove bridge taps, clean up splices, and remove load coils from the wires to get the T1 to work.

To circumvent these cabling problems, *High Data-Rate Digital Subscriber Line* (HDSL) was developed as a more efficient way of transmitting T1 (and E1) over the existing copper wires. HDSL does not require the repeaters on a local loop of up to 12 K'. Bridge taps will not bother the service, and the splices are left in place. This means that the provider can offer HDSL as a more efficient delivery of 1.544 Mbps. The modulation rate on the HDSL service is more advanced. Sending 768 Kbps on one pair and another 768 Kbps on the second pair of wires splits the T1. This is shown in Figure 16-4.

Figure 16-3

The typical layout of the T1

Figure 16-4

HDSL is impervious to the bridge and splices. The T1 is split onto two pairs.

As already mentioned, HDSL runs at 1.544 Mbps (T1 speeds) in North America and at 2.048 Mbps (E1 speeds) in other parts of the world. Both speeds are symmetric (simultaneous in both directions). Originally, HDSL used two wire pairs at distances of up to 15 K'. HDSL at 2.048 Mbps uses three pairs of wire for the same distances, but no longer. The most recent version of HDSL uses only one pair of wire, and is expected to be more accepted by the providers. Nearly all the providers today deliver T1 capabilities on some form of HDSL.

Symmetric DSL (SDSL)

The goal of the DSL family was to continue to support and use the local copper cable plant. Therefore, the need to provide high-speed communications on a single cable pair emerged. Most local loops already employ single cable pair today, thus it is only natural to assume the providers would want this capability. SDSL was developed to provide high-speed communications on that single cable pair but at distances no greater than 10 K'. Despite this distance limitation, SDSL was designed to deliver 1.544 Mbps on the single cable pair. Typically, however, the providers provision SDSL at 768 Kbps. This creates a dilemma for the carriers because HDSL can do the same things as SDSL.

Asymmetric DSL (ADSL)

SDSL uses only one pair of wires, but is limited in its distance to provide duplex, high-speed communications. Not all users require symmetrical speeds at the same time. ADSL was therefore designed to support differing speeds in both directions over a single cable pair at distances of up to 18 K'. Because the speeds requested are typically for access to the Internet (or intranet), most users look for higher speeds in a download direction and the lower speed for an upward direction. Therefore, the asymmetrical nature of this service meets those needs.

Rate Adaptive DSL (RADSL)

Typically with equipment, installed assumptions are made based on minimum performance characteristics and speeds. In some cases special equipment is used to condition the circuit to achieve those speeds. However, if the line conditions vary, the speed will be dependent on the sensitivity of the equipment. In order to achieve variations in the throughput and be sensitive to the line conditions, *Rate Adaptive DSL* (RADSL) was developed. This gives the flexibility to adapt to the changing conditions and adjust the speeds in each direction to potentially maximize the throughput on each line. Additionally, as line conditions change, you can see the speeds changing in each direction during the transmission. Many of the ILECs have installed RADSL as their choice, given the local loop conditions. Speeds of up to 768 Kbps are the preferred rates offered by the incumbent providers.

Consumer DSL (CDSL)

Consumers are not all looking for symmetrical high-speed communication in order to achieve access to the Internet. Furthermore, the speeds of ADSL technology are more than the average consumer may be looking for. As a result, the lower speed communications capability was developed by using CDSL as the model. With other forms of DSL (such as ADSL and RADSL), splitters are used on the line to separate the voice and the data communications. CDSL does not use, nor need, a splitter on the line. Moreover, speeds of up to 1 Mbps in the download direction and 160 Kbps in the upward direction are provided. It is expected that the speeds and DSL will meet the needs of the average consumer for some time to come. As a result, a universal ADSL working group developed what is called ADSL–lite. This specification was ratified in late 1998, using the specifications from this working group for delivery to the average consumer. Because of the changes in speeds with this technique, the telephone companies are in a position to support a lower speed digital subscriber line strictly through the use of the modems without the concern for local loop. An example of this DSL-lite (G.Lite) service is provided with the Nortel 1 Mb modem.

Very High-speed DSL (VDSL)

Clearly, changes will always occur as we demand faster and more reliable communications capabilities. It was only a matter of time until some users demanded higher speed communications than was offered by the current DSL technologies. As a result, VDSL was introduced to achieve the higher speeds. If, in fact, speeds of up to 50 Mbps are demanded, then the distance limitations of the local cable plant will be a factor. In order to achieve the speeds, you can expect that a fiber feed will be used to deliver VDSL. This technique will most likely carry ATM traffic (cells) as its primary payload. The pilot program of US West communications in Gilbert, Arizona plans to deliver fiber to the door, *Fiber to the Home* (FTTH), to provide voice, data, video, and multimedia communications to the consumer. Although this pilot is still in planning, a lot of excitement has been generated over the possibilities. Another such pilot was announced in 1998 by Sprint communications with their *Integrated On-demand Networking* (ION) capability to satisfy the same communications services from the consumer's door. ION will use a combination of VDSL and ATM cells to provide integrated communications on demand. Sprint recognized after the announcement of ION that local loop, and particularly the fiber local loop, was a requirement to deliver the service. Unfortunately, many of the incumbent telephone companies do not have fiber to the door. This will cause delay until such time that the fiber can be deployed.

Table 16-2 summarizes the speeds and characteristics of the DSL technologies discussed. These are the typical installation and operational characteristics; others will certainly exist in variations of installation and implementation.

Table 16-2

Summary of DSL speeds and operations using current methods

Service	Explanation	Download	Upload	Mode of Operation
ADSL	Asymmetric DSL	1.5–8.192 Mbps	16–640 Kbps	Different up and down speeds, one pair wire
RADSL	Rate Adaptive DSL	64 Kbps–8.192 Mbps	16–768 Kbps speeds.	Different up and down. Many common operations on 768 Kbps. One pair wire.
CDSL	Consumer DSL	1 Mbps	16–160 Kbps	Now ratified as DSL-lite (G.lite). No splitters. One pair wire.
HDSL	High-data rate DSL	1.544 Mbps in North America 2.048 Mbps in rest of world	1.544 Mbps 2.048 Mbps	Symmetrical services. Two pairs of wire.
IDSL	ISDN DSL	144 Kbps (64+64+16) as BRI	144 Kbps (64+64+16) as BRI	Symmetrical operation. One pair of wire. ISDN BRI

Service	Explanation	Download	Upload	Mode of Operation
SDSL	Single DSL	1.544 Mbps 2.048 Mbps	1.544 Mbps 2–048 Mbps	Uses only 1 pair, but typically provisioned at 768 Kbps. One pair wire
VDSL	Very High data rate DSL	13–52 +/- Mbps	1.5–6.0 Mbps	Fiber needed and ATM probably used

The Hype of DSL Technologies

Why all the hype? Well the local providers are extremely excited if they can install higher-speed communications and preserve their local cable plant. No one wants to abandon the local copper loop, but getting more data reliably across the local loop is imperative. Therefore, the ability to breathe new life into the cable plant is an extension of the facilities in place.

The consumer is looking for higher speed access (primarily to access the Internet) for whatever the application. Yet, at the same time, consumers are looking for a bargain. They do not want to spend a lot of money on their communications services.

The providers are trying to bump up their revenues without major new investments. They would like to launch as many new service offerings on their existing cable plant and increase the costs to the end user. This is a business decision, not a means of trying to rake the consumer over the coals. Yet, there has to be a happy medium of providing services and generating revenues with limits on expenses. To do this, the xDSL family offers the opportunity to meet the demands while holding down investment costs. The key ingredient for success is to minimize costs and satisfy the consumer. Make no mistake, if the local provider does not offer the high-speed services, someone else will.

xDSL Coding Techniques

Many approaches were developed as a means of encoding the data onto the xDSL circuits. The more common are *Carrierless Amplitude Phase*

Modulation (CAP) and *Discreet Multitone Modulation* (DMT). *Quadrature with Phase Modulation* (QAM) has also been used, but the important part is the standardization. The industry, as a rule, selected DMT, but several developers and providers have used CAP. It is therefore appropriate to summarize both of these techniques.

Discreet Multitone

DMT uses multiple narrow band carriers, all transmitting simultaneously in a parallel transmission mode. Each of these carriers carries a portion of the information being transmitted. These multiple discrete bands, or in the world of frequency division multiplexing, subchannels, are modulated independently of each other using a carrier frequency located in the center of the frequency being used. These carriers are then processed in parallel form.

In order to process the multicarrier frequencies at the same time, a lot of digital processing is required. In the past, this was not economically feasible, but integrated circuitry has made this more feasible.

The *American National Standards Institute* (ANSI) selected DMT with the use of 256 subcarriers, each with the standard 4.3125 kHz bandwidth. These subcarriers can be independently modulated with a maximum of 15 bits/second/Hz. This allows up to 60 Kbps per tone used. Figure 16-5 shows the use of the frequency spectrum for the combination of voice and two-way data transmission. In this representation, voice is used in the normal 0–4 kHz band on the lower end of the spectrum (although the lower 20 kHz are provided). Separation is allowed between the voice channel and the upstream data communications, which operates between 20 kHz and 130 kHz. Then a separation is allowed between the upstream and the downstream channels. The downstream flow uses between 140 kHz and 1 MHz. As shown in this drawing, the separation allows for the simultaneous up and down streams and the concurrent voice channel. It is on this spectrum that the data rates are sustained. Each of the subchannels operates at approximately 4.3125 kHz and a separation of 4.3125 kHz between channels is allocated.

Figure 16-5
The ANSI DMT
specification

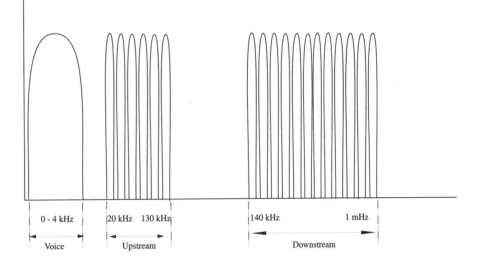

0 - 4 kHz	20 kHz 130 kHz	140 kHz 1 mHz
Voice	Upstream	Downstream

Using DMT for the Universal
ADSL service (G.Lite)

Provisions for the high-speed data rates of full ADSL are good, but not every consumer is looking for the high data rates afforded on ADSL. Therefore, the Universal ADSL Working Group decided to reevaluate the need for the end-user. What they determined is that many consumers need download speeds of up to 1.5 Mbps and upload speeds between 9.6–640 Kbps. As a result, the ADSL Lite specification was designed to accommodate these speeds as a logical stepping stone to the higher speed needs for the future. Initially introduced in early 1998, the specification was ratified in late 1998 to facilitate the lower throughput needs of the average consumer. DMT is the preferred method of delivering the G.Lite specification and service, as it is now known. This involves a slightly different method of delivering the service, but does accommodate the providers with a less expensive solution to provide full rate ADSL.

There is no way to know if the network providers can support hundreds of multimegabit ADSL up and download speeds on their existing infrastructure. But using the G.Lite specification can support lower demand users more efficiently. Similar to the DMT used in the ANSI specification, the carriers are divided as shown in Figure 16-6. Note that in this case the high end of the frequency spectrum tops out at approximately 550 kHz instead of the 1 MHz range with ADSL.

To Split or Not to Split

Another issue of using ADSL is the use of splitters on the line. In normal ADSL and Rate Adaptive DSL, the local provider uses a splitter on the line. ADSL modems usually include a *Plain Old Telephone Service* (POTS) splitter, which allows the simultaneous access to telephony applications and high-speed data access. Some vendors provision the service with an active POTS splitter device that allows the simultaneous telephone and data access. Unfortunately, with an active device, if the power or the modem fails, then the telephone also fails. This is problematic because we are accus-

Figure 16-6
The ANSI and UAWG
G.Lite spectrum

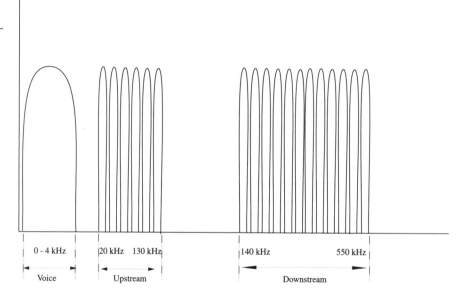

tomed to having lifeline services with our telephone systems that are always available, even if the power fails. The splitter is shown in Figure 16-7 as it is installed.

A passive POTS splitter maintains the lifeline service of telephony even if the modem fails. This is important because the telephone line is powered from the line instead of an external power source. Telephony service will be available as much as we have always expected from the normal service we have always received.

A POTS splitter is a three-pronged device, allowing telephony and simultaneous up-and-down loading data access on the same copper loop. As shown in the previous figures, the POTS service operates at the low end of the frequency spectrum. All the data signals are located in the higher frequencies. These signals start between the 20–25 kHz and above. The splitter provides a low pass filter between the copper line and the ADSL point on the modem.

One of the primary concerns is to allow a filter to block any transient noise coming from the POTS side of the line from crossing over into the ADSL side of the line. Ringing voltages can cause significant impulse noise across the line, destroying the data. ADSL modem signals are also blocked from passing onto the POTS side of the line by the filters.

Most of the POTS splitters are passive. Passive filters provide higher degrees of reliability because they donít require power, they can isolate the equipment from surges on the line, and they can arrest lightning that may

Figure 16-7
The splitter-enabled ADSL service

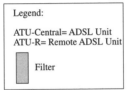

Figure 16-8
The splitterless G.Lite
installation

be coupled on the line. Active filters require power and are less adapted to suppress the surges. In many countries around the world, active filters are a requirement.

The G.Lite specification uses a splitterless device to facilitate the installation of low speed DSL services. This is shown in Figure 16-8 for implementation. Without the splitter, there is less cost for the provider and the issues of active vs. passive splitters are eliminated. This provides a different approach.

Carrierless Amplitude Phase Modulation (CAP)

CAP is closely aligned to *Quadrature Amplitude Modulation* (QAM). QAM as a technique is widely understood in the industry and well deployed in our older modems. Both CAP and QAM are a single carrier signal technique. The data rate is divided into two and modulated onto two different orthogonal carriers before being combined and transmitted. The main difference between CAP and QAM is in the way they are implemented. QAM generates two signals with a sine/cosine mixer and combines them onto the analog domain.

CAP, on the other hand, generates its two orthogonal signals and executes them digitally. Using two digital transversal bandpass filters with equal amplitude characteristics and a p/2 difference in phase response, the signals are combined and fed into a digital to analog converter. Then the data is transmitted. The advantage of CAP over QAM is that CAP is done in silicon, which is more efficient and less expensive.

CAP was one of the original proposals for use with ADSL technology. Unfortunately, this was a proprietary solution offered by a single vendor, which turned heads away from acceptance. CAP is shown in Figure 16-9 in its use of the frequency spectrum of the line. Most industry vendors agree

Figure 16-9
The spectral use
of CAP

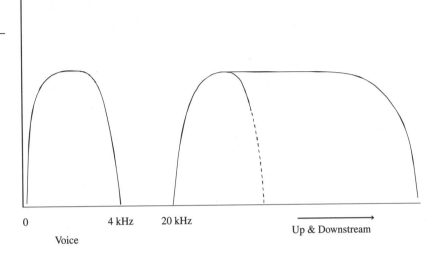

0   4 kHz   20 kHz   Up & Downstream
   Voice

that CAP has some benefits over DMT, but also that DMT has more benefits over CAP. The point here is that two differing technologies were initially rolled out for ADSL (and the other family members), which contradict each other in their implementation.

CAP uses the entire loop bandwidth (excluding the 4 kHz baseband analog voice channel) to send the bits all at once. There are no subchannels, as found in the DMT technique. The lack of subchannels removes the concern about the individual channel transmission and problems. To achieve the simultaneous send and receive capability, frequency division multiplexing is used, as is echo cancellation. Many of the *Regional Bell Operating Companies* (RBOCs) have used or tried CAP in their installations, but have also experimented with DMT.

Provisioning xDSL

In the following figures, the various architectures of the xDSL implementations are shown. The point to remember here is the goal of xDSL is to use the existing copper infrastructure and improve the speed and throughput on the installed base of wires. Consequently, the installation process attempts to minimize the added equipment (particularly at the customer's premises) and the labor required to get the equipment installed.

In Figure 16-10, the design of an ADSL model and the model components are shown. The intent of the model is to show the infrastructure of the network from the customer premises to the network provider. This model also shows the splitters in place to facilitate the ADSL model.

The next figure demonstrates the connection from the service provider to the rest of the world. In many cases, ADSL access to the local network access provider (the ILEC or other local loop provider) is then passed on to the ISP. This is designed to run over an ATM backbone, but not a firm requirement. Therefore, the NAP will assign a *DSL Access Multiplexer* card (DSLAM) and assign an ATM VPI and VCI as a default to carry the data into the *Internet Service Provider* (ISP) or other *Network Service Provider* (NSP). This is shown in Figure 16-11.

The application most commonly used is to gain high-speed access to the Internet. Many of the local service providers install the ADSL service into a single PC at the end user location, as shown in Figure 16-12. The local providers offer the customer a packaged deal with the following components:

- LAN NIC card operating at 10 Mbps
- DSL modem
- Splitter
- Management cables

The local provider will normally advise the customer that the termination must be to a single PC equipped with the NIC card. In the US, the customer owns the package when the installation is completed due to some of the regulatory constraints and the Public Utility Commission rulings. This places the burden of maintenance and diagnostics on the end user rather than the local service provider. In the case of a LAN attachment described previously, the ADSL modem is set to bridge from the LAN to the ATM network interfaces rather than route.

Figure 16-10

The ADSL model as it is laid out from the customer premises to the service provider

Central Office

Figure 16-11
Access from the NAP
to the NSP

Figure 16-12
The typical local
installation

In other cases when a LAN is present at the customer location, and the end user wants to connect all the LAN devices to the high-speed outside network for Internet access or private network access, the local carrier may suggest that a proxy server is a requirement. The proxy server (PC dedicated to act in this function) will then act as the gateway to the outside world for all devices attached to the LAN (see Figure 16-13).

An alternative to this approach is the direct connection to a LAN hub, such as that found in the telephone closet. Keeping the connection active, the carrier will normally assign an IP address, using DHCP for a contracted period of time. Normally, this is a lease period of four hours, and then the network server (outside) will renegotiate and assign a new IP address for the end user. This protects the end user network from becoming visible on the Internet and helps to prevent some of the normal security risks associated from a hard connection to the Net (see Figure 16-14).

Many of the hubs located in customer locations are 10 Base T, or 10Mbps Ethernet hubs. Occasionally, a customer may have a 10/100 auto-sensing hub or a 100 Base T hub. The local providers have been known to tell the customer that this arrangement will not work. Specific networks are already attached with direct attachment of 10/100 and 100 Base hubs with

Figure 16-13
The proxy server in lieu of a single attached PC

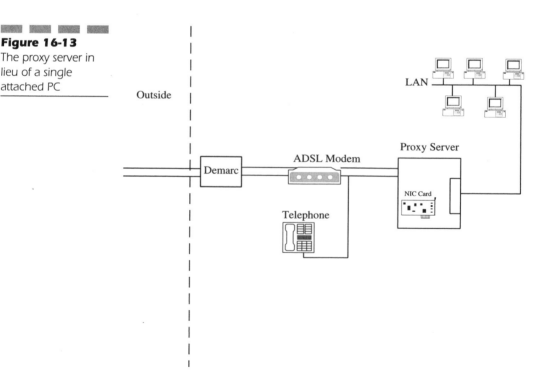

Figure 16-14
Connecting the ADSL
service to a hub

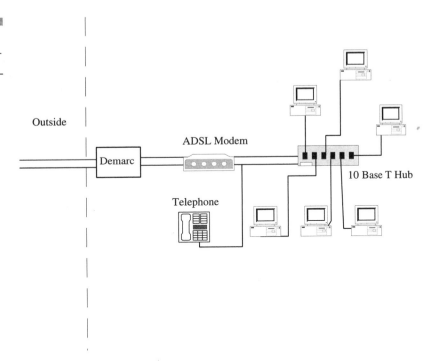

Outside

ADSL Modem

Demarc

Telephone

10 Base T Hub

no impact, as shown in Figure 16-15. The connection allows for a specific number of simultaneous connections onto the ADSL service. The local providers will always try to configure the network connections in ways they can guarantee will work and with a standard way to troubleshoot problems. By working with the variations above, the local providers still need to come up to speed on the way the data networks actually perform.

Final Comment on Deployment

The use of ADSL service is catching on. However, the local providers (ILECs and CLECs) are dragging their feet. As of late 1998, there are only about 15–20,000 total ADSL modem pairs installed in the US. In contrast, there are over 300,000 cable modems installed in residences and businesses across the country. The local owners of the copper loop have to take a more aggressive approach to delivering the high-speed services, or the consumer will go somewhere else.

Figure 16-15
Connecting the ADSL
modem directly
through a 10/100 or
a 100 Base T hub.

As the market continues to mature and standards continue to develop, the local providers must preserve their infrastructure. Consumers (small and large alike) are demanding the higher speed services. As a stepping stone for residential- and home-based businesses, the acceptance and standardization of the G.Lite specification will provide suitable transmission rates until the carriers can complete their data strategies. 1 Mbps modems, for example, giving the end user a 1 Mbps download speed and a lower 160 Kbps upload speed will suffice for many today and into the next millennium.

For the larger consumers, a full rate ADSL may be just what is needed bringing 1.544 to 8 Mbps downloads and 768 Kbps uploads to the forefront of Internet access.

Where the consumer is reluctant to proceed with ADSL, the HDSL or the SDSL services are still very attractive alternatives offering up to 1.544–2.048 Mbps symmetrical speeds or some variation as already discussed.

In the future, when high-speed media are installed to the door or to the curb, the logical stepping stone becomes the VDSL service, sometime perhaps in the year 2002–2003. Although trials are already underway, too much time passes until the results are compiled and analyzed. Therefore, the reality of VDSL for the masses is still a long way off.

Microwave-
and Radio-
Based Systems

Introduction

No one ever pays much attention to the microwave radio dishes mounted on towers, on the sides of building, or any other place. This technology has been taken for granted over the years. However, this nondescript industry has quietly grown into a $3.4 billion global business annually. Four major suppliers provide one-half of all the radio-based systems globally. Microwave has also become a vital link in the overall backbone networks over the years. Now, it has achieved new acclaim in the wireless revolution, relaying thousands of telephone conversations from place to place, bypassing the local landlines.

Microwaves are between 1 mm and 30 cm long, and operate in a frequency range from 300 MHz to 300 GHz. Microwaves were first used in the 1930s, when British scientists discovered the application in a new technology called radar.

In the 1950s, microwave radio was used extensively for long-distance telephone transmission. With the need to communicate over thousands of miles, the cost of stringing wires across the country was prohibitive. However, the equipment was both heavy and expensive. The radio equipment used vacuum tubes that were bulky, as well as highly sensitive to heat. All of that changed dramatically when integrated circuits and transistors were used in the equipment. Now the equipment is not only lightweight, but also far more economical and easy to operate. In 1950, the typical microwave radio used 2,100 watts to generate three groups of radio channels (each group consists of 12 channels), yielding 36-voice grade channel capacity. Each voice grade channel operated at the standard 4 kHz. Today, equipment from many manufacturers (and Harris/Farinon, specifically) requires only 22 watts of output to generate 2,016 voice-channels. Although there have been two orders of magnitude improvements in the quality of the voice transmission, the per-channel cost has plummeted from just over $1,000 to just under $37. This makes the transmission systems very attractive from a carrier's perspective. However, the use of private microwave radio has also blossomed over the years because of the cost and performance improvements. This is shown in the graph reflected as Figure 17-1, which details why the use of microwave has become so well accepted in the industry.

Today's microwave radios can be installed quickly and relocated easily. The major time delays are usually in getting through the regulatory process in a governmentally controlled environment. Several installations have taken over a year to be approved, only to have the radio system installed and running within a day or two. In many situations, microwave systems provide

Figure 17-1
Comparison of cost per channel over the years

more reliable service than landlines, which are vulnerable to everything including flooding, rodent damage, backhoe cuts, and vandalism. Using a radio system, a developing country without a wired communications infrastructure can install a leading-edge telecommunications system within a matter of months. For these reasons, regions with rugged terrain or without any copper landline backbone in place find it easier to leap into the wireless age and provide the infrastructure at a fraction of the cost of installing wires.

Throughout the world, government owned and controlled monopolies are being eliminated. Brazil opened its doors to international Telecom competition, allowing microwave radio systems and a mobile telephone system supplied by North American firms. In Russia, one of the leading systems manufacturers installed an integrated network along a 3,600-mile gas pipeline. Using microwave radios and digital telephone switches, this link sends data and voice from Siberia to Russia's southern border. A similar system is used for the railroad system, using a trunked radio system, microwave radio-relay, and digital telephone switches. The telephone service may serve more than the railroad, incorporating hundreds of thousands of people residing alongside the railway line.

The cellular and PCS industries invested heavily in microwave radios to interconnect the components of their networks. This is shown in Figure 17-2 where the interconnection is used in the cellular world. In addition, a

Figure 17-2
Cellular
interconnection of
microwave radio

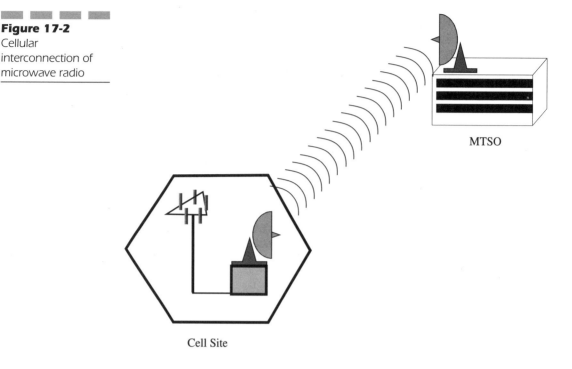

MTSO

Cell Site

new use of microwave radio, called micro/millimeter wave radio, is bringing transmission directly into buildings through a new generation of tiny receiver dishes.

WinStar Communications, a new *Competitive Local Exchange Carrier* (CLEC), is pioneering the use of micro/millimeter wave radio communications in the 30+ GHz frequency range. This allows the CLEC to deliver broadband communications to the consumer's door without the use of telephone company wires.

The new PCS industry is expected to choose microwave radio technology for the interconnection and backhaul transport on its expanding network. The PCS suppliers and the cellular suppliers do not want to pay the local telephone company for monthly T1 access lines from the cell sites to the mobile switching sites. Therefore, to eliminate the monthly recurring charges, they have installed microwave radio systems in the 18–23 GHz frequency range. Tens of thousands of new cell sites and PCS sites will be constructed over the next few years, further expanding the use of microwave radio $IPCS systems>systems in each of these sites. As third-generation, handheld devices make their way into the industry, more wireless interconnectivity will be used.

Microwave also played a very crucial part of the PCS industry as the PCS systems use the 1.9–2.3 GHz frequency band. Fixed systems operators such as police, fire, electric utilities, and some municipal organization occupied these frequencies. To accommodate the move of these users from the 2 GHz frequency band, microwave was used to relocate the users to a new band, as mandated by the FCC. One study indicated that the PCS industry would spend over $3 billion in microwave equipment and services by 2005.

Another large demand for microwave emerged in the *Competitive Access Providers'* (CAPs) market. CAPs offer long-distance access to customers at lower prices than the local telephone companies and the newer competitors. The CAPs normally install their own fiber-optic wires. However, they recognize the benefit of expanding coverage to consumer building entrances, using a wireless, high-speed connection. The CAPs are supplementing their fiber-based networks with *Wireless CAPs*. WCAPs use microwave transmission to deliver the telecom service without the need for a costly, wire-based infrastructure (see Figure 17-3).

The newer micro/millimeter-wave radios, which are smaller and usually less expensive than other microwaves, are also popular with these CAPs and PCS suppliers. They are used in urban areas to extend the fiber networks. These radio units use the high frequency (or millimeter) bandwidth

Figure 17-3
Wireless interconnection of fiber and CAPs

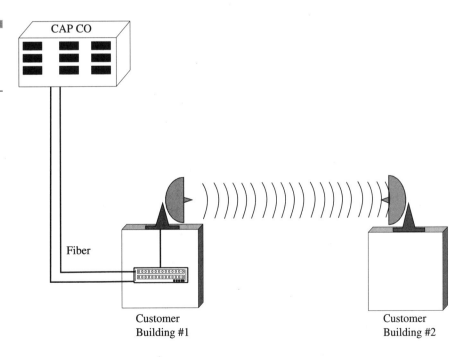

that hadn't been used before. Now, they are seen as a solution to increasing congestion in the lower frequency bands. An advantage of these systems is the small size antennas that can be hidden on rooftops without interfering with zoning ordinances or creating aesthetic controversy.

Microwave is heavily used in radio and television systems. Satellite TV relies on microwave repeaters on the satellite to retransmit TV signals to a receiving station. Microwave communication via satellite provides a more reliable signal than longer, land-based radio waves. It also improves the reception of the picture.

Some TV stations have been using microwaves to facilitate wireless communications from field cameras since 1992. What we continually hear about the "Action Cams" is a portable microwave system connected to a camera for real-time broadcast. Instead of being constrained to a fixed location, a news van can be driven and hooked up instantly as shown in Figure 17-4. The systems hooks up with a field camera with microwave units the size of a deck of cards. These can go anywhere and can operate from locations up to two miles from the van. Action and news is transmitted back to the van where it is relayed via microwave to the TV station.

Figure 17-4

Action camera and microwave systems working together

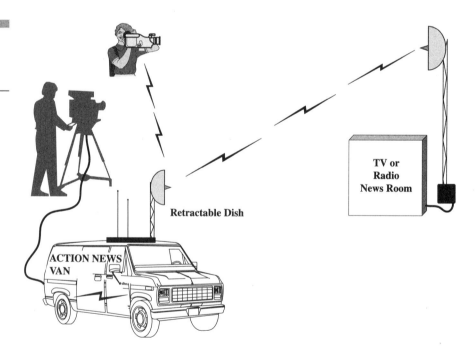

Retractable Dish

ACTION NEWS VAN

TV or Radio News Room

Other Applications

A laptop computer with a credit card-sized PRISM radio chip set can now convert incoming microwave messages into binary code for computer processing and then convert them back into microwaves for transmission (see Figure 17-5). Similarly, microwave transmission is used in Local Area Networks, on corporate or college campuses, in airports, and elsewhere. Whether it is collecting data, relaying conversations, or beaming messages from space, microwave makes the wireless revolution possible.

No one can escape the wireless hype these days. The challenge is in wading through all the confusion and misleading statements to decide whether an application fits the need. If you can make sense of it all, you may find the solution to your connectivity needs.

First, one distinction will help to narrow the playing field. In this explosion of wireless technologies, there are two major categories worth mentioning:

■ Personal wireless devices

■ Wireless devices that are used between buildings (for voice, data, and video)

Figure 17-5
Laptop computers can now send and receive microwave radio transmissions

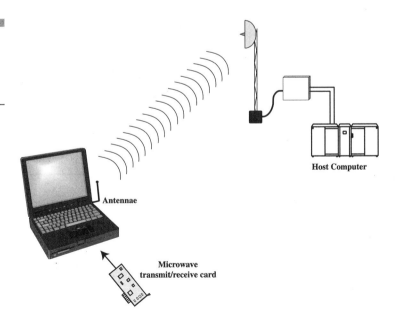

Host Computer

Antennae

Microwave
transmit/receive card

Buying the wrong personal wireless device, such as a pager or cell telephone, is annoying, but inexpensive and easily replaceable. Yet, if you're a telecommunications manager, the wrong choice to connect your sites together can have significant financial impact and your career can be shortened.

How Do You Make the Right Choices

Telecommunications and information managers have many options to connect remote sites. The options have expanded with all the wireless excitement and the expansion of VLSI integration, making the devices far more affordable. Today, there are more vendors trying to sell the end user on their products. The CLECs, CAPs, and Wireless CAPs are all vying for this portion of connectivity, but so, too, are the manufacturers.

How do you make sense of all the hype and make a decision consistent with corporate goals? That is not as complicated as it sounds. Conducting a needs assessment is critical to understanding the connectivity goal. After your needs are determined, the next step is to find a solution that works. Does the solution offer a cost justification, and it can be delivered in a reasonable time?

Complications arise when vendors don't have the actual solution needed, but try to force the solution to fit with their product. They further disparage the other vendors with fear, uncertainty, and doubt about competing products. So step one is to determine the technical requirements. Most organizations look for bandwidth and reliability. If these two requirements can be reasonably met, one need only find a vendor to deliver the following:

1. What you need
2. When you need it
3. What is reasonable financially

What about Bandwidth?

Bandwidth is always a touchy subject. It can become a "never satisfied drain" on the corporate bottom line, if due diligence is not practiced. There is a direct relationship to cost and total bandwidth. The more bandwidth needed, the greater the cost. Everyone would like as much bandwidth as possible, and at the same time wants it to be affordable. Many people make the mistake of buying more than they need, anticipating future growth. In this industry, prices keep falling as competition increases. If an organization needs an OC3 (155 Mbps) today, then laying fiber is probably the most affordable solution.

Conversely, if 10 Mbps Ethernet is the current rate of transmission, then this demand can be immediately met. Additional bandwidth can be bought later. In two to three years, the costs will plummet so that the new requirements can be met with incremental or marginal costs.

It's wiser to buy bandwidth as you need it and not before (there will be a small amount of incremental add-on, but limited). In the future, there will be the following:

- More choices
- Increased providers
- Greater availability
- Lower costs

What should be done in the interim to satisfy the need? The answer is the following:

- Lease (dark) fiber instead of paying the cost of installation
- Lease services from the ILEC or CLEC if sufficient bandwidth is available
- Buy a wireless connection such as point-to-point microwave

With a leased line (or a dark fiber) solution, the costs can be predictable, based on demand and agreed-upon bandwidth. Assuming physical facilities are available, take what is needed for the interim and order more only when necessary. This is a good intermediate step to get the bandwidth so long as the recurring charges are not exorbitant.

The alternative to leasing physical circuits from the ILEC or CLEC is wireless acquisition. For a one-time fee and limited recurring maintenance charges, bandwidth can be purchased for the immediate and future needs. If the wireless product delivers the bandwidth and reliability desired and the payback is reasonable, then wireless may be the best choice. Consider, however, the financial life and return on the investment.

It doesn't make sense to order an OC−3 (155 Mbps) connection when the need is only for 10 Mbps. Yet, there should be an upgrade path. As mentioned, leasing copper lines or fiber from the ILEC allows a migration path. Growth can be accommodated, as needed. Conversely, some wireless products limit this growth option. Some products only handle growth to a T1 or a 10 Mbps channel. Consider the expandability before buying a wireless product.

How Much Is Enough?

Risks associated with buying bandwidth fall into the two categories pointed out earlier.

1. Buying too little bandwidth will increase incremental growth costs that can add up more than buying a larger quantity at the onset.

2. Buying more bandwidth than immediately needed means paying for bandwidth that may not be required for some time, or that will be less expensive in the future.

If a T1 line slows the voice and data access to the point that users are frustrated, or unproductive, then T1 is not the solution! One year of the unproductive environment costs a fortune in lost productivity.

If a couple of Ethernet channels are needed (at 10 Mbps each), and the organization invests in a fiber optic connection, justifying the expense makes an interesting paradox.

Consider approximately how much bandwidth is needed. The answer is as much as it takes to keep the data moving, voice calls coming, and users productive enough to sell (or whatever else the mission is). There is no requirement for more, yet there must be at least that much. Meeting this equation is the one that keeps the industry guessing, including the ILECs and the CLECs, as they design their networks.

What about Reliability?

Having too much bandwidth is possible. Having too much reliability is just the opposite. Organizations lose significant amounts of money when the network connection is too slow, but far more when the link is down completely. One hour of network downtime can cost more than the profits and productivity achieved from a year of uptime. In this scenario, automatic backup is an absolute must. Buy the appropriate amount of bandwidth and make sure that the reliability is built in. Plan for the worst-case scenario! Consider an alternate backup plan. Use circuit switched or packet (frame) switched alternative connections in case of an outage.

The Choices Are Leased Lines, Fiber or Microwave

Leased lines, fiber, and microwave each have benefits and drawbacks in terms of bandwidth, reliability, price, and delivery. The tendency is usually very application-specific. Every case is different in terms of terrain, line of sight, right of way, location of a Bell central office, and so on. A __ mile fiber choice can range in cost between $20,000 and several million dollars, depending on the terrain requiring traversal. Similarly, a high-speed leased line can cost $600 a month or $20,000 a month if it crosses LATA or other rate boundaries. No one solution fits all possibilities for connectivity.

Using a wireless connection, the first, full-speed Ethernet speed solution was a major milestone. From this innovation, users had viable options, bridging the extremes between leased lines and fiber. Many users still limit their choices to the bottleneck of a T1 leased line or overpaying for a T3, which is too much bandwidth for the need.

Instead, you should find a solution that provides the needed bandwidth for a justifiable cost. All three solutions offer high reliability for the most critical connections, although a mission-critical path must be backed up.

Keep in mind also that the three choices are not mutually exclusive. They frequently work well together. Microwave handles "last mile connectivity" to a fiber backbone or serves as a lower-cost, automatic backup as insurance against "backhoe fade." You should weigh each choice, based on which offers the best cost to benefit ratio. Cost includes installation, on-going charges, upkeep, losses due to downtime, and organizational productivity.

Microwave and the Other Wireless Solutions

Prior to the 1970s, microwave was the most widely used wireless communications medium in the world. Microwave usage is making a comeback now with end users. Many user organizations were reluctant to experiment with microwave radio transmission, due to misconceptions surrounding the technology, as well as to confusion between the "wireless" products. It is important to recognize that the difference between one wireless device and another can be as different as fiber and copper wire. Both fiber and copper are "wired," but that is where the commonality ceases. The same is true between microwave and laser, spread spectrum, or cellular service.

There are even differences between one type of microwave and another. The differences are due primarily to their respective operating frequencies. Some frequencies are good for distances of 30 or 40 miles and others can barely get you across an office park. Some can only support a couple of T1s or a single video channel and others go to 10–45 megabits.

In Table 17-1, a comparison of distances and frequencies is used for representative purposes. Many times this is the best case scenario.

Microwave Radio Solutions

Private-user microwave systems are essentially the same as what the telephone company, FM radio stations, broadcasters, and fixed-site utility companies relied on before the implementation of fiber. For example, the corkscrew type antennas on news vans are shooting microwave signals back to their TV and radio stations as shown earlier. FM radio stations still rely mainly on microwave. In fact, most microwave radio uses *Frequency Modulation* (FM) radio technology.

What can microwave offer an organization? Primarily, microwave combines huge bandwidth and reliability that is better than other wireless device. In fact, microwave is typically far more reliable than the leased-line specification (99.985%) for distances across the street to twenty plus miles away. Microwave can deliver bandwidth up to 45 Mbps (and most demanded speeds in between). A properly configured system will sustain operation except in the most severe rainstorm where power and telephone line outages would be expected.

Table 17-1

Comparison of frequency bands and distances (line of sight)

Frequency	Distance
2–6 GHz	30 miles
10–12 GHz	20 miles
18 GHz	7 miles
23 GHz	5 miles
28—30 GHz	1—2 miles

The myths run rampant with radio-based systems. Despite the rumors about the various risks and perils for the radio signal, microwave usually operates 99.99+% of the time. Microwave is normally impervious to the following:

- Snow
- Sleet
- Fog
- Birds
- Pollution
- Sandstorms
- Sunspot activity

The real risk is water fade (water absorption) and multipath fade across bodies of water. These can be accommodated for the most part in design of the radio path.

A microwave link can transmit gigabytes of data without dropping a single bit (or packet when a data transmission uses packetized information). On copper wire, noise is always present. Thermal noise causes a continuous hum, white noise, and the like. A microwave path can be so clear that if no one is talking or sending data, the line is perfectly silent. This is difficult for the average lay person to understand.

Private User Microwave

Having proven that the bandwidth and reliability are readily available on a microwave system, most people look for the negative side of private microwave. The unaware consumer assumes that there must be some "gotcha" lurking in the background. One technology cannot be the most robust and reliable and yet not be the most favored. There has to be a major drawback to using microwave. The two largest drawbacks have always been the availability and the sticker price. Microwave *appears* to be the most expensive wireless option (typically, between $20,000 to $100,000 for an end-to-end link). However, if greater bandwidth and reliability are the goals, then microwave is not as expensive as it appears superficially. The initial up-front cost avoids the bandwidth and reliability tradeoff with less expensive or less robust technologies.

If a critical connection requires high bandwidth and high reliability to prevent catastrophic losses from downtime, then more money is required initially. The old saying "pay me now, or pay me later" applies here.

Another perceived "drawback" (handling over a megabit per second) is the requirement to have line-of-sight between locations. Line-of-sight issues are rarely showstoppers.

First, only a small part of the remote site needs to be visible. Even if the other site is not visible, solutions exist. Actually, this is quite common. Look for high points that can be used to get visibility between both sites. A passive or active repeater site can be implemented. Setting up a repeater is not difficult, particularly with passive repeaters. The antennas are small and lightweight enough to be placed almost anywhere; a water or radio tower, utility pole, other rooftops, and so on. An alternative is to bounce the signal off a physical obstacle (such as a mountain) and use obstacle gains the get the signal through (see Figure 17-6).

If a repeater is needed, rental space is often available on other towers. This is true in long haul but also in localized communications. Renting space on an existing tower or a rooftop is not very dramatic. Leased space typically costs $200 per month per dish. Repeaters are shown in Figure 17-7.

This assuages the fact that line-of-sight is a major drawback. Anyone who has ever negotiated right-of-way for fiber knows how tedious that can

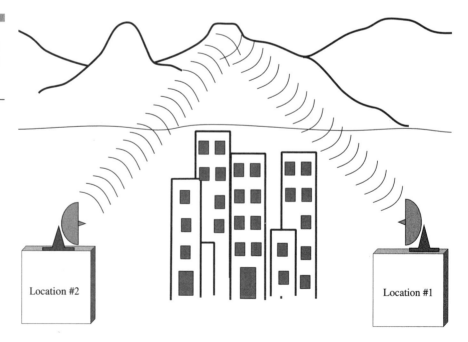

Figure 17-6
Obstacle gain uses a bounced signal off a natural obstacle, such as a mountain.

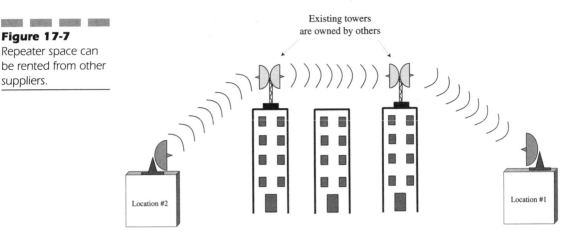

Figure 17-7
Repeater space can be rented from other suppliers.

be. With fiber, the right-of-way includes the entire physical length of the cable or conduit required for the connection. With microwave, the concern is limited to a few physical points where the dish placement will occur and a few legal issues.

Another perceived drawback of microwave, aside from price and line-of-sight, has to do with getting and maintaining the license to use. Licensing is a protection for the end user. It is a tedious process that provides structure and prevents interference and overlays in the frequency spectrum. Most organizations see this as a step for governmental control, but if a conflict arises, it can be the salvation for the licensed user. License gives you the right to use a good, clear transmission path, and that is definitely positive. The FCC is quite efficient in approving licenses for private users, and spectrum is readily available for point-to-point applications.

Licensing involves contracting someone to do a frequency search and filing a FCC license application. The frequency search is to find unused, available frequencies. Those frequencies are then reserved and filed, along with other pertinent data, on the appropriate FCC form. The process is normally completed within a few days, and it can cost $2,000 to have your vendor handle it for you. An experienced vendor can usually assure that the FCC will grant your license by avoiding amateurish mistakes.

The 23 GHz frequency band, for example, is a very common frequency band for short-haul, private-user microwave systems. Most people confuse the "band" with the actual operating frequency. The 23GHz band actually consists of 24 pairs of frequencies, ranging between 21.200 and 23.600 GHz. The number can be doubled to 48 pairs of frequencies with minor antenna changes (changing polarization from vertical to horizontal).

The radio signal is narrowly focused by the antennas at each end of the link, and transmit power is only about 60 milliwatts. These variables make it possible to use identical frequency pairs for two links originating from the same rooftop! By changing the polarization of the antennas and separating the signals, 10 degrees should provide the necessary separation and isolation.

When connecting LANs together with microwave radio systems or bridges, the important issue is selecting the right vendor. The vendor's qualifications and experience in both the LAN and microwave systems are paramount. Does the vendor possess the expertise to differentiate a network or bridge failure from a radio problem? Can the vendor provide turnkey services and assume total system responsibility. The correct vendor will ensure the successful implementation of a LAN microwave link.

Local Multipoint Distribution Services (LMDS)

Introduction

Whenever the concept of the competitive environment enters a discussion, two other discussions ensue: the wireless local loop and the use of local multipoint distribution services. This chapter will look at some of the movement in this area, so that an understanding of how and why the last mile has become so critical in meeting the demands for higher-speed broadband communications. Moreover, when looking at the incumbent Local Exchange Carriers' copper-based plant, one can only marvel at the lack of foresight in fending off the competition. The LECs have always been in control of the last mile, and invested heavily in the copper-based plant they use. Given the unshielded, twisted-pair wiring scheme and the band-limited channel capacity they deliver, one would expect them to write the cabling systems down (depreciate) as quickly as possible. Yet, to keep their costs competitively low, they have chosen the opposite route. Instead of fast depreciation, the LECs use a 20–30 year depreciation schedule on their cables. What this boils down to now is a cable plant that has not kept pace with the demands of the user and is still on the books for the provider. This is a problem for the carriers because they can ill afford to walk away from the cables in place, yet they have to breathe new life into an infrastructure that is limited in capacity. Moreover, the existing cable plant is prone to noise and disruption as a matter of course. A graphic representation of the local loop and some trouble spots is shown in Figure 18-1.

Enter the Competitive Discussion

As one discusses the possibility of the LEC having an inherent problem, there are several areas under attack at this last mile. Already discussed in earlier chapters was the use of the local cable TV operation's cable to carry voice and data communications. However, there are many areas that are being considered as competitive local loop concepts, as seen in Table 18-1.

Although many of these discussions center on new technology, the two that have gained the most momentum are the concept of the *Wireless Local Loop* (WLL) and the *Local Multipoint Distribution Services* (LMDS). The following information will consider these concepts more closely.

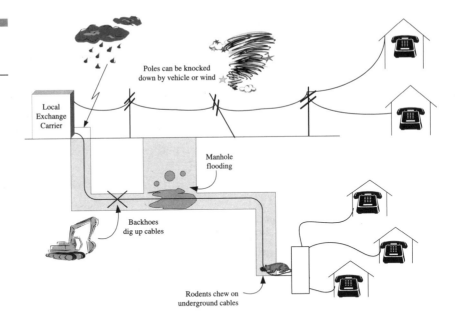

Figure 18-1

The local loop is prone to problems

Competitor	Concept	Technology
CATV Companies	Cable TV for voice and Internet access	Cable modem technology
CATV Companies	Fiber to the Curb, Coax to the door for data and voice plus entertainment	Hybrid Fiber/Coax (HFC), Fiber to the Curb (FTTC)
Cellular and PCS suppliers	Broadband PCS as a single number for all wireless voice, paging and data access	PCS on TDMA and CDMA or GSM
Local Competitors	Broadband voice and data services on wireless local loop	Local Multipoint Distribution Services (LMDS) and Microwave Multipoint Distribution Services (MMDS)
CLECs	Fiber or copper to the door	SONET and local drops on copper
New Wireless Providers	Wireless access through various methods	Wireless Local Loop (WLL)

Table 18-1

Multiple areas of competition at the local loop

Wireless Local Loop

The industry in general has placed a lot of emphasis on the wireless local loop and predicts that millions of subscribers will enjoy the benefits of untethered communications before the turn of the century. This may or may not be aggressive, but it signals the point that the competitive machine is in full swing at the last mile. Much of the growth being discussed will occur in areas where an infrastructure does not exist, such as third world countries installing the initial communications systems to the residential and business user for the first time. Many countries across the globe still do not have basic POTS, *Plain Old Telephone Service,* so it makes sense to consider a wireless connection. In some cases, the use of a *Radio in the Loop* concept (RITL) or a *Fixed Wireless Radio Access* (FWRA) concept is what the countries have dubbed the services. Countries like Brazil and China will reap many benefits from using a WLL concept, both financially and in the speed of installation. The cost of installation on a per user basis is much more favorable. One set of statistics shows the difference of the installed cable versus a wireless local access method as seen in Table 18-2.

However, the emerging underdeveloped countries are not the only places where Wireless Local Loop technology will be used. Instead, the developed countries around the world may also take advantage of the economies of scale and the financial benefits of installing the wireless local access. As a result, as many as 50 million access lines may be deployed worldwide shortly after the turn of the century and rapid growth may follow the initial installations. The day of installing copper to the door has ceased; instead, the wireless technologies may be the mode of choice for the future. No longer can the carriers afford the cost of installation and maintenance for the copper local loop.

Figure 18-2 is a representation of the overall concept of the Wireless Local Loop concept, without specific technology used but as a model for the carriers considering the use of wireless technology.

Table 18-2	**Technology Used**	**Cost per User (in US dollars)**
Cost comparison for wired versus wireless local loop	Copper Local Loop	$ 5,500.00
	Wireless Local Loop	$500.00–800.00*

*This figure will drop rapidly to approximately $200–300 per user as deployment continues and economies of scale are achieved.

Figure 18-2
Wireless Local Loop
conceptual model

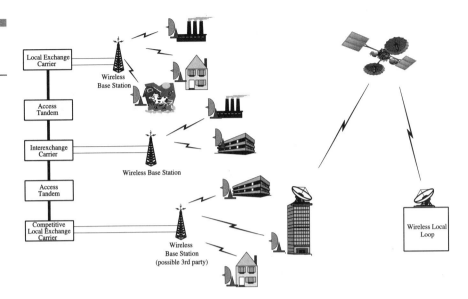

Not For Everyone

The Wireless Local Loop will encourage many new opportunists to jump into the market, but few will survive. Either the providers will be underfunded and will not survive the competition, or the larger providers looking for market share in an area of operation will gobble the smaller local providers up. In either case, the number of providers will change and the operators will continually be looking for new and competitive approaches to attract customers. Full service providers will offer the list of services as shown in Table 18-3. Others may offer pieces of these services. The point is that the end user is looking for a one-stop shopping approach, and the leverage that comes with bundled services.

Too many providers will jeopardize the success of many causing some form of shakeout, but one must consider that the end user is willing to use one or more of the providers listed in the table above. What will likely occur is a merger or a joint offering with partnering providers to get to the consumer's door. When one looks at the offerings and the carriers above, it is obvious that the services are disjointed. Some providers offer all the services, whereas others are just planning the possible services they will offer. However, the infrastructure they choose to install may have an impact on their ability to service future demands. No one answer or solution jumps out right now, but changes will occur rather quickly in this business. The economics of getting the consumer (both residential and business) to buy

Table 18-3 Summary of service offerings and providers today

Provider	Voice	Data Low Speed	Point to Point Data	CATV	Video Conferencing	Internet Access (High Speed)	Multimedia Services
				Services Offered by Carriers Today and Future Offerings			
Cable Companies	Yes	Not Avail, but possible	Not Avail, but possible	Yes	Not Avail, but possible	Yes 10 Mbps	Not Avail.
LECs	Yes	Yes	Yes	No but planned	Yes	Yes 1.5 Mbps	Limited
CLECs	Yes	Yes	Yes	No but planned	Yes/Limit	Yes 1.5 Mbps	Limited
IECs	Yes	Yes	Yes but not local loop	No merge with CATV	Yes Limited	Yes 1.5 Mbps	Limited
WLL	Yes	Yes	Limited	No, but possible	Limited	Yes 10 Mbps	Possible
Cellular Providers	Yes	Yes	No	No	No	No	No
PCS Providers	Yes	Yes	No	No	No	Limited	No

into more than one offering will set the stage for future services. One can add the numbers and see where the providers want to take this. Table 18-4 shows a summary of service offerings (on average) for the services used by the consumer. In this particular scenario, the consumer is a home office-based user or a residential user whose needs include various bundled services. If a carrier can offer the bundled services for a moderate decrease in the monthly costs, one can expect 65% of those approached to churn.

Using these bundled, one-stop pricing models, one can expect that the residential and small business customer will be tempted to use the service provider. Note that not all service providers will offer the equipment (such as the PC or the modems), whereas others may. Many of the *Competitive Local Exchange Carriers* (CLECs) are toying with this idea for their total

Table 18-4 Bundled vs. individual services plans	Service Offering	Average Monthly Price	Bundled Price
	CATV (basic cable)	$ 8.00	$ 100.00
	Extended channel services[1]	$ 23.00	
	Basic Internet service provider access	$ 20.00	
	Data access for Internet at dial-up rates	$ 25.00	
	Dial tone for voice	$ 25.00	
	High speed Internet access @ 1+ Mbps	$ 50.00	$ 35.00
	Long distance services (typical customer)	$ 25.00	$ 15.00
	Equipment costs amortized (PC, modem, telephone, etc.) monthly	$ 30.00	$ 30.00
	Cellular phone basic plan	$ 50.00	$ 30.00
	Total monthly fees	$ 256.00	$ 210.00[2]

[1] Excludes premium channel services (HBO, Showtime, and so on) and pay-per-view services

[2] The goal of the providers with one-stop shopping is to offer all the services at approximately $200.00 per month.

business provisioning. The smart provider will consider this bundled offering. As a means of meeting the customer's communications needs on a single bill, the consumer is a ready target.

However, some of the pieces may not be required. For example, many of the WLL providers include the Analog Cellular suppliers and the PCS suppliers whose recent advertisements state that consumers can remove their wired telephone and use the Cellular or PCS service for their home and business needs. This is possible and has some merit. Thus, if the consumer takes this provider up on this advertisement, the carrier loses the bundling of a $25.00 monthly dial tone service. However, the cost of the cellular plan will increase in the number of minutes used, driving that plan cost up higher. Effectively, this may become an even trade.

Another point here is the cost of the infrastructure. Once the CATV companies have delivered the basic cable services, for example, the cost of any added usage or shared bandwidth on their infrastructure is usually marginal. Thus, the profitability and mark-up is that much higher. The wired carriers understand the benefit of one-stop shopping; now the WLL carriers are learning very quickly.

What of the Bandwidth?

The bandwidth necessary for each of these services listed above changes the rules considerably. In many of the WLL providers' backbone, there is not enough bandwidth to support the number of users and the higher-speed services. For this reason, the marriage of the providers may occur sooner than expected. If a cellular provider joins forces with a WLL microwave supplier, then the bandwidth for the fixed needs at the door is assured while the cellular provider handles the demands of the roaming user. These combinations and permutations can be very complicated as the number of providers expands and the services they offer shift in any direction. The interesting point will be to see how the total market plays out with an expectation that approximately five to seven providers will dominate and the rest will be absorbed or fail.

Enter Local Multipoint Distribution Services (LMDS)

Local Multipoint Distribution Services (LMDS), as its name implies, is a broadband wireless technology that is used to deliver the multiple service offerings in a localized area. The services possible with LMDS include the following:

1. Voice dial-up services

2. Data

3. Internet access

4. Video

Just as the network providers were getting used to the battlegrounds between the *Incumbent LECs* (ILECs) and the new providers, RF spectrum was freed up around the world to support access and bypass services. Typically, the services operate in the radio frequency spectrum above 25 GHz, depending on the licenses and spectrum controlled by the regulatory bodies. This offering operates as a point-to-point, broadband wireless access method, which can provide two-way services. Because LMDS operates in the higher frequencies, the radio signals are limited to approximately five miles of point-to-point service. This makes it somewhat like a cellular operation in the way the carriers lay out their operations and "cells." An architectural concept for the LMDS operation is shown in Figure 18-3 from the perspective of the supplier to the user. This figure uses some of the premises that the service is constrained to a localized area. (Occasionally in uncongested and unpopulated areas, the signals are transmitted in much wider areas of coverage, similar to other wireless technologies. This reference to the five miles is within populated areas and the obstacles that will be encountered within the areas).

The Reasoning Behind LMDS

Point-to-point fixed microwave radio has been in use for decades in the local loop environment. Many organizations (individual businesses, utility companies, and so on) required dedicated access to their own private network facilities or to a carrier's *Point of Presence* (POP). As they would approach the ILEC, the cost to run high-speed services to the business consumer's

Figure 18-3
Typical LMDS
service areas

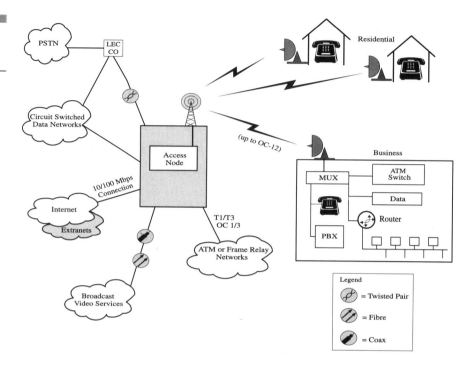

door was typically prohibitive. The monopoly owning the embedded infrastructure could literally demand any price that seemed appropriate. This
met with objections from the user, but as long as the monopoly existed,
there were few choices. The businesses therefore demanded frequency spectrum to install their own infrastructure at the last mile. The connection was
typically in a special, set-aside frequency band as shown in Figure 18-4,
using distance sensitive frequencies as listed in Table 18-5.

The problems of the fixed access methods using microwave in the past
included the following:

1. Local ordinances were unfavorable to the use of the technology

2. Local regulatory bodies had several restrictions

3. Federal authorities limited the use

4. Cost of building the tower

5. Cost of security for the site

6. Local power and other utilities might not be readily available

7. Cost of the equipment was very high

8. Maintenance costs were unnecessarily high

Figure 18-4

Typical microwave point-to-point services of the past

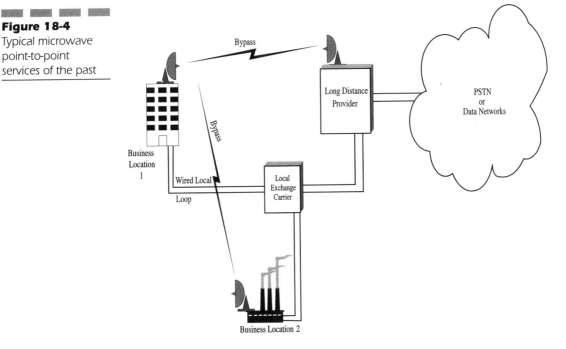

Table 18-5

Typical microwave distances, bands, and operations

Frequency Band	Distances	Use
2–6 GHz	30 miles	Commercial, utility, fixed operation, TV
10–12 GHz	20 miles	Commercial, utility, fixed operation, TV, DBS
18 GHz	7 miles	Business, limited fixed operation
23 GHz	5 miles	Business, limited fixed operation
25 GHz and above	3–5 miles	Business, bypass operation

9. FCC licensed technicians were required to do the maintenance [Reference to the US based operations. In other countries similar requirements prevailed (such as CRTC in Canada, PTT in other parts of the world)]

10. Line-of-sight frequencies might not be readily available

Each of the above areas was somewhat limiting to the demands of the end user's ability to get access to the fixed point-to-point microwave systems. The largest organizations could financially justify the use of this service because their needs were more demanding. However, smaller organizations had to rely on alternative methods or service bureaus that could provide the access at a reduced rate.

From a carrier's perspective, however, the equation changes very quickly. Using LMDS services, a new provider can install the systems more readily due to the competitive environment being introduced worldwide. The monopolies no longer mandate or dictate what the local connectivity will be like. The new providers can achieve the benefits of the LMDS world through the following means:

1. Lower cost entry into the market

2. Costs are deferred to later when services are needed. This moves the pricing model from fixed to variable costs associated with demand, as opposed to fixed size increments.

3. *Return on Investments* are achieved more quickly, encouraging the provider to enter the market

4. Less risk of customer churn, leaving the carrier stuck with large investments

5. Ease of installation and licensing makes the implementation faster

6. Standards-based services and equipment, minimizing obsolescence and proprietary solutions

The carriers seem to have found a nirvana of technology and financial benefit in a single solution. The real issues then begin to work around the need, demand, and the method of delivery. Not all systems are implemented the exact same way, so the carrier still has some choices, allowing even greater flexibility in delivering the bandwidth to the door.

Network Architectures Available to the Carriers

Various means of installing leave the carrier choices as already stated. The bulk of the carriers will likely standardize on a straight, point-to-point connectivity solution for their customers. Point-to-point TV distribution can also be provided with the LMDS offering. This increases the attractiveness of the LMDS supplier when the other services desired by the end user, such as voice, data (IP), and multimedia applications, are added to their TV distribution capability. The architecture of the LMDS will lend itself to these point-to-point services nicely. The primary pieces constituting the LMDS system are as follows:

- The *Network Operations Center* (NOC) contains all the management functions that manage all the components of a much larger infrastructure.

- The cabled infrastructure is usually fiber-based to connect the components of the LMDS to the public switched and private networks. The cabling will consist of T1/T3 or OC-1, OC-3 or OC-12 connecting to the ATM and Internet backbones.

- The Base station is where the fiber-to-radio frequency conversion takes place; the modulation of the signal across the airwaves occurs here also.

- The customer equipment, which can vary from user to user and by vendor, has to satisfy the demands of the consumer.

The architecture also varies in the modulation of the signals onto the RF (airwaves), based on the chosen strategy of the vendor. Two predominant methods of using the technology are to use an analog interface such as *Frequency Division Multiple Access* (FDMA) or a digital interface, using *Time Division Multiple Access* (TDMA). The choice will vary depending on the density of the sectors being served, the available financial choices, and the overall quality desired. The more common implementation is to use a FDMA technique to serve the customer. The use of FDMA allows better coverage and density of the applications being served, and it uses a higher modulation technique to satisfy the system demands.

Modulation Techniques

As already discussed, the modulation falls into two primary categories, FDMA and TDMA. Each of these techniques differs, but also creates other submodulation capabilities. For the broadband LMDS services, the system is usually separated into both phase and amplitude modulation of the RF. *Phase Shift Keying* (PSK) and amplitude modulation combinations have been successfully used to achieve high rates of multiplexing and carrying capacities. The options of using FDMA and TDMA are similar for the RF spectrum and will be discussed generically here.

The ultimate goal is to multiplex the most services and modulate the least amount of RF spectrum to achieve the same throughput. Table 18-6 is an example of the desired results of modulating less amount of RF to get the same amount of effective throughput (in this case a 2 Mbps data rate). The table highlights the various methods used by different vendors to achieve the data rates, such as the techniques used to modulate the signal.

We can see the capacities and differences available to modulate the signal over an FDMA technique. These variations are what will separate the supplier's ability to satisfy future demands to sustain a 2 Mbps data rate with the least amount of bandwidth.

In the TDMA alternative, time slots are more efficiently used to deliver the capacity to the end user, but the same techniques as listed previously are used, with the exception of the 64-QAM. The problem with a TDMA method is that the time slotting (fixed in most cases) uses more of the spectrum in overhead and therefore produces less efficiency in the RF side of the business. Each carrier must be aware that the extra overhead associated with this time slot usage can detract from the overall system performance.

Table 18-6

Summary of modulation techniques available for LMDS in FDMA

Technique Used	Modulation Method	Bandwidth Required to Sustain 2 Mbps Data Rate	Number of Bits Per Modulation Technique
BPSK	Binary Phase Shift Keying	2.8 MHz	1:1
DQPSK	Differential Quaternary Phase Shift Keying	1.4 MHz	2:1
QPSK	Quaternary Phase Shift Keying	1.4 MHz	2:1
8PSK	Octal Phase Shift Keying	0.8 MHz	4:1
4-QAM	Quadrature Amplitude Modulation, 4 state	1.4 MHz	2:1
16-QAM	Quadrature Amplitude Modulation, 16 states	0.6 MHz	4:1

Technique Used	Modulation Method	Bandwidth Required to Sustain 2 Mbps Data Rate	Number of Bits Per Modulation Technique
64-QAM	Quadrature Amplitude Modulation, 64 states	0.4 MHz	6:1

Two-way Service

The TDMA and FDMA modulation techniques on the LMDS network allow for the bidirectional flow between the carrier and the end user. In many cases, a different upstream is required than the downstream. The ability to modulate differently allows this to be compensated for. The more important factor is that the service will offer two-way communications. Many of the past services allowed only a one-way downstream with a dial-up upstream. This sounds incomprehensible for high-speed data, but for Internet access this method was used initially on an MMDS service and then on LMDS in the initial rollout. Later evolution of the network architectures allowed the carriers to change direction and satisfy both directions for their data needs. Moreover, for two-way voice the two-way simultaneous transmission is a must.

When lower speed users are using the system, TDMA is an effective tool for their two-way voice and data (dial up) needs. For simultaneous up and downstream services, using approximately 250 MHz in each direction, the average number of TDMA users per 5 MHz of spectrum handling use dial-up service at the DS0 rate 80. This means that a sector (or cell) using 5 separate streams of 5 MHz each, can achieve up to 4,000 simultaneous dial-up users. This is a reasonable use of the bandwidth. The overall network design on a wired world uses a 10:1 ratio of trunks to users. Using a slightly lower ratio for a wireless network connection to accommodate for fax and long-hold time traffic (i.e. Internet surfing) of approximately 8:1, the network supplier can achieve a service level of 32,000 possible customers with normal demands in a sector (cell) with 250 MHz of RF spectrum. This is a reasonable amount of traffic capacity based on standard traffic engineering design. Better ratios have been used in some networks, but the issues are coming with long hold times that exceed expectations. Using a conservative ratio of 5:1, the average network supplier can achieve service for 20,000

DS0 level users in a sector. Keep in mind that when higher-speed demands are the requirement, an FDMA arrangement will allow more flexibility. This all depends on the overall demands of the network users.

Propagation Issues

Like any radio-based system, the issue of propagation is always a concern. Like the analog cellular networks of the past, there are several factors that contribute to the quality of the signal. Many operators have to consider that at the higher frequencies (over 25 GHz), rain fade will be a critical factor. The higher the frequency, the more susceptible to rain fade than lower frequencies. One CLEC (WINSTAR) chose to use all 31 GHz radio equipment in their infrastructure to get to the customer's door. Other issues have a bearing on the design and layout of the system such as the following:

- Distance
- RF Interference (RFI)
- Electromagnetic Interference (EMI)
- Multipath fade
- Frequency reuse

In each case, the individual carrier will have to assess the overall system design specifications to meet the needs of the consumer, either residential or business. No one solution is going to satisfy all systems providers or consumers. The constant shift in network architecture will be required in a fine-tuning approach to provide the quality necessary.

Because these systems mimic the cellular network of the early 1980s, they have similar concerns. None of the concerns are insurmountable. The issue is that the carriers have a more fixed target, rather than a moving target in the wireless mobile networks. Each case, however, offers the capability to service a wide-range of needs dictated by the customer, rather than the network itself. Progress comes in many ways.

CHAPTER 19

Specialized Mobile Radio (SMR)

Introduction

The *Specialized Mobile Radio* (SMR) service was first established by the Wireless Radio Commission (part of the FCC) in 1979 to provide land/mobile communications on a commercial basis. An SMR system consists of the following:

- One or more base station transmitters
- One or more antennas
- End user radio equipment that usually consists of a mobile radio unit, either provided by the end user or obtained from the SMR operator for a fee

SMR users operate in either an "interconnected" mode or a "dispatch" mode.

1. The **interconnected mode** links the mobile radio with the *Public Switched Telephone Network* (PSTN). An end user transmits a message via the mobile radio unit to the SMR base station. The call is then routed to the local dial-up telephone network, which enables the mobile radio unit to function as a mobile telephone.

2. *Dispatch mode* allows two-way, over-the-air, voice communications between two or more mobile units or mobile units and fixed locations.

SMR customers using dispatch communications include construction companies with several trucks at different jobs or an on-the-road trucking company with a dispatch operation in a central office.

SMR systems consist of two distinct types: conventional and trunked systems. A conventional system allows the use of only one channel. If someone else is already using that end user's assigned channel, the end user must wait until the channel is available as shown in Figure 19-1. In contrast, a trunked system combines channels and contains microprocessor capabilities that automatically search for an open channel. This search capability allows more users to be served at any one time. A majority of the current SMR systems are "trunked systems."

Although SMR is primarily used for voice communications, systems are also being developed for data and facsimile services. Additionally, the development of a digital SMR marketplace allows the features shown in Table 19-1.

SMR growth is attributed to these developments and features. For example, at the end of 1998, approximately 8.8 million vehicles and portable units

Figure 19-1

Conventional radio requires users to queue on a channel

Table 19-1

Features possible with Digital SMR

Features of a Digital SMR Network
Two-way acknowledgment paging
Inventory tracking
Credit card authorization
Automatic vehicle location
Fleet management
Remote database access
Voicemail

were served by SMR systems. Several radio-based systems have been introduced and used throughout the years. Many of these operated in a very specific frequency band and in specific geographic areas of the country in which they were approved. The purpose of a radio system is to provide the wireless technology to satisfy various needs as shown in Table 19-2.

Table 19-2

Radio services
meeting the need

Needs Addressed by Radio Based Systems
Emergency operations (fire and police)
Local transportation services
Long haul communications over-the-road
Medical demands for ambulance services
Maintenance operations such as Dept. of Transportation for road systems

Although these services are old hat, they still demand the connectivity necessary to be reachable at a moment's notice. Over the years, the demands have changed and matured. The FCC and CRTC (in Canada) have issued several standards and set aside frequency bands to accommodate these services. One of the primary goals was to employ the use of *Specialized Mobile Radio* (SMR) to help meet the need.

As SMR began rolling out, the obvious players who commanded the market share developed the products and services necessary to support the use of SMR. The bigger developers and manufacturers (for example, Motorola, Ericsson, Sharp, Uniden, and others) developed products and accessories to work in the industry and meet the demand of these users. SMR also incorporated different types of service into the major systems and operation, which included a telephone, pager and short messaging service, two-way dispatch radio, and data transmission[1].

Motorola's iDEN™ technology is a classification of SMR that is based on a variety of proven RF technologies. The technology offers increased spectral efficiency and full-service integration, two of the main benefits of digital communications. The iDEN™ is also the basis for the SMR system operated by NEXTEL, a competitor in the wireless business who now touts that they have better services and coverage than most of the PCS suppliers.

Improved Spectral Efficiency

The capacity to accommodate crowded markets and worldwide growth is a critical component of iDEN™. The development of this spectrally efficient

---

[1]The data transmission services of the SMR system is limited to slow speed data, but improvements will enhance this and increase the data rate to over 33 Kbps.

technology allows multiple communications to occur over a single analog channel. This expansion of the network gives users greater access to the network and provides space for new and expanded services to be added without rebuilding the infrastructure.

iDEN™ represents a significant step toward the integration of wireless business communications systems that meet today's demands for a one-stop process. Motorola used a combination of technologies to create the increased capacities and the combination of services. Much of the enhancements and increased capacities come from Motorola's VSELP[2] vocoding technique and QAM modulation process, as well as the TDMA channel splitting process.

Motorola's VSELP-Coding Signals for Efficient Transmission

The key to the expanded capacity is the reduced transmission rate needed to send information. Motorola has developed a vocoder technology that handles the process as shown in Table 19-3.

This vocoder, known as *Vector Sum Excited Linear Predictors* (VSELP), compresses the voice signals to reduce the transmission rate needed to send information. Moreover, VSELP provides for clear voice transmission by digitizing the voice and providing high-quality audio under conditions that normally will result in a distorted analog voice. Using speech extrapolation, the VSELP decoder can "repair" the loss of a speech segment over the radio channel. The result is less distortion and interference (for example, breakup, static, and fading) as users move toward the periphery of the coverage

Table 19-3

VSELP coding process is straightforward

The VSELP Coding Process

Compresses voice signals

Creates digital packets of information (voice)

Assigns the packets a time slot

Transmits

Receives the information on the iDEN™ network

[2]VSELP is a trademark of Motorola.

area, enhancing the clarity and quality of voice communications at the out-
skirts of a cell.

QAM Modulation

While the VSELP compresses the signal and reduces the transmission rate,
Quadrature Amplitude Modulation (QAM) increases the density of the
information. QAM modulation technology was specifically designed to support the digital requirements of the iDEN™ network. Motorola's unique
QAM technology transmits information at a 64 Kbps rate. No other existing
modulation technology transmits as much information in a narrow band
channel.

Multiplied Channel Capacity

Another essential element is the *Time Division Multiple Access* (TDMA).
TDMA is a technique for dividing the wireless radio channel into multiple
communication pathways. In the iDEN™ system, each 25 kHz radio channel
is divided into six time slots. During transmission, voice and data are
divided into packets. Each packet is assigned a time slot and transmitted
over the network. At the receiving end, the packets are reassembled according to their time assignments into the original information sequence.

The Advantage of Integration

More than ever before, users are demanding multifunction devices that are
simple to use. With the iDEN™ network, users need only one telephone to
access voice dispatch, two-way telephony, short message service, and future
data transmission. This integration provides business users with flexible
communications that enable users to access information in the most efficient and convenient way, no matter where they are in the system.

The SMR systems are part of a larger family of products that are classified as trunked radio systems. Trunked radio involves a combination of
wired and wireless communication typically found in the emergency service
operation, such as fire, police, and road maintenance operations. What is
trunked radio?

A Short Overview to Trunked Radio

Historically, these systems have provided one-to-many and many-to-one voice communications service—also known as mobile dispatch services. These systems are operated by commercial entities, otherwise known as service providers that are in the business to resell their services to other entities for a profit. For over 100 years, the lines between telephone exchanges have been shared between customers. In early telephone exchanges, operators would patch calls through when a customer was located there. The operator just selected the next available circuit. Now the allocation is automatic, but the result is the same.

Radio channels were shared since the early days of radio. The operators had to listen on a frequency to determine if it was in use. Mobile telephone systems also required a customer to find an inactive channel manually. These systems were upgraded by hardware that could find a vacant mobile-telephone channel automatically and by two-way radios with subaudible (CTSS) tone equipment. This equipment was available in the 1950s.

In the 1980s, microcomputers brought a revolution in controls. A computer could be installed inside a two-way radio. The result was the development of cellular telephones and trunked radios. Both systems have a central computer managing the system. The main computer communicates with the mobile radios via an inaudible data signal.

When a trunked radio user wants to talk with someone on the same logical channel, he/she presses the microphone (Push to Talk) button. The radio sends a data signal to the controller requesting a channel. The controller responds with a physical channel number. The requesting radio switches back to receive long enough to hear this information. At the same time, all radios in the system hear the same data. These radio systems use repeaters. The base station transmits from a tall tower or building on the base frequency. The mobile units listen on that frequency, but transmits on a paired frequency (the mobile frequency).

A trunked radio system gets its name from the "trunk" used in commercial telephone systems. A "trunk" is a communications path between two or more points, typically between the telephone company central offices. The telephone trunk is shared by several different users, but each user is not aware of this sharing of lines. The caller places a call to another party, and the call is completed.

Radio communication over a trunked system is similar to the telephone system. The transmitting and receiving radio units can be thought of as the calling and called parties. Instead of telephone lines, the radio system uses radio channels to place calls. As with the telephone system, the radio users are not aware of which "trunk" or radio channel they are communicating over. Trunking a multichannel radio system increases the efficiency of the radio and the radio channels.

The concept behind trunking is very simple. Trunked radio is the pooling of several radio channels so that all the users in a given area have automatic access to any free channel. The result is a system that can provide private, wide-area, wireless communication to many different users without interference or interruption.

Trunked radio is very different from the old conventional radio where you had to contend with interference and congestion. Trunked radio, the least expensive wireless service, allows both dispatch communication and the ability to place and receive phone calls, similar to cellular.

A trunked radio system always comprises several radio channels. One channel acts as the *Control Channel* (CC), while the others carry traffic. The control channel is used for registrations, the transmission of status messages, and for call requests. This is not unlike the cellular, radio-based system using the paging and control channels in a cellular operation. The difference is that in a cellular radio network, there are several (19–21) channels set aside, whereas in trunked radio only one channel is needed (see Figure 19-2).

Upon requesting a call, a talk path is allocated on an exclusive basis to the subscriber from a pool of radio channels. The call processed on this channel. If the trunked radio system receives additional call requests, a different channel is allocated to the calling party from the pool. As soon as all channels are in use, new call requests are stored in a queue. When a channel becomes available, the requested call is switched to the first available channel on a first-in, first-out basis.

This method means that a call request need only be sent once. If the call cannot be set up immediately, the system stores the call request and processes it later.

The Control Channel (CC)

Each radio cell consists of a Trunked Radio Exchange and a *Radio Base Station* (RBS). The Trunked Radio Exchange can be used as a *Master System Controller* (MSC) or as a *Trunking System Controller* (TSC). The Trunked Radio Exchange manages the radio channels of the Radio Base Station(s). One of these channels is used as the CC. The SMR base is shown in Figure 19-3.

Figure 19-2
Control and talk
channels in a trunked
radio system

Figure 19-3
SMR base station and
radio service

When a mobile set is powered on, it automatically registers using the CC. Once the subscriber receives a positive acknowledgement from the network, the mobile is registered on the Trunked Radio System and can be used. The mobile is constantly in contact with the CC. If there is a call request, the Trunked Radio System checks whether the addressed subscriber is available. If he is not available, not registered, or engaged, this information is given to the caller. If the requested subscriber is available, the call is set up by the Trunked Radio Exchange, using a free traffic channel. Status messages and short data are submitted on the CC.

Trunked radio systems are those which share a small number of radio channels among a larger number of users. The physical channels are allocated as needed to the users who are assigned logical channels. The users only hear units on the same logical channel. This uses the available resources more efficiently since most users do not need the channel 100 percent of the time.

Service Areas and Licensing Blocks

Older 400 MHz channels are still in operation, as well as TV channel operation. Two sets of frequency bands are available for SMR operation: 800 MHz and 900 MHz. Approximately 19 MHz of spectrum is available for use by SMR operators (14 MHz in the 800 MHz band and 5 MHz in the 900 MHz band). The 800 MHz SMR systems operate on two 25 kHz channels paired, while the 900 MHz systems operate on two 12.5 kHz channels paired. Due to the different sizes of the channel bandwidths allocated for 800 MHz and 900 MHz systems, the radio equipment used for 800 MHz SMR is not compatible with the equipment used for 900 MHz SMR.

The 900 MHz SMR service was first established in 1986 and initially employed a two-phase licensing process. In Phase I, licenses were assigned in 46 *Designated Filing Areas* (DFAs), comprised of the top 50 markets. Following Phase I, the FCC envisioned licensing facilities in areas outside these markets in Phase II. Meanwhile, licensing outside the DFA was frozen while the Commission completed the Phase I process. The freeze on licensing outside DFAs continued until 1993, when Congress reclassified most SMR licensees as *Commercial Radio Service* (CMRS) providers and established the authority to use competitive bidding to select from among mutually exclusive applicants for certain licensed services. During the freeze, however, some DFA licensees elected to become licensed for secondary sites (for example, facilities that may not cause interference to primary licensees and must accept interference from primary licenses) outside their DFA to accommodate system expansion.

In response to Congress' reclassification of the SMR service in 1993, the Commission revised its Phase II proposals and established a broad outline for the completion of licensing in the 900 MHz SMR band. The 200 channel pairs in the 900 MHz service have been allocated in the 896–901 MHz and 935–940 MHz bands. Each MTA license gives the licensee the right to operate throughout the MTA on the designated channels, except where a co-channel incumbent licensee is operating already.

Frequencies have standard separation between the base and mobile pairs as shown in Table 19-4, showing the operating bands for the base and mobile radio and the separation between the channels.

Innovation and Integration

Motorola's integrated digital enhanced network technology and protocols combine dispatch radio, full-duplex telephone interconnect, short message service, and data transmission into a single integrated business communications solution. The digital technology was the result of studies indicating that a high percentage of dispatch users carried cellular telephones and 30 percent of cellular users carried pagers, along with an increasing demand for data communications. For network design efficiency, iDEN™ uses a standard seven-cell, three-sector reuse pattern.

The technology is designed to work around many SMR spectrum limitations as well. You can take individual channels and group them together to work together as a single capacity. In cellular communications, the spectrum must be contiguous. Enhanced voice places iDEN™-based services more on par with TDMA, GSM, and code division multiple access vocoders.

Table 19-4

Frequency pairing for SMR

Band	Base Station	Separation	Mobile Device
800 MHz	851–869 MHz	45 MHz lower	806–824
900 MHz	935–940 MHz	39 MHz lower	896–901 MHz
450–470 MHz	450–455 MHz	5 MHz higher	455–460 MHz
450–470 MHz	460–465	5 MHz higher	465–470
TV band	470–512 MHz	3 MHz higher	6 MHz TV channel

Although dispatch mode is simplex and not full duplex, connections are quick. It's very efficient and very fast. A typical cellular call with speed call would take seven to 10 seconds for a path to be established. With Motorola's product it takes about a second. Add in 140 character alphanumeric displays (for short message capability) and direct, circuit-switched data support, and you get innovation and integration in one neat little package.

Spectral Efficiency with Frequency Hopping

Geotek targets the same wireless niche with its digital frequency hopping, multiple access-based technology, and networks. Geotek holds a near unique position as manufacturer and spectrum owner/service operator, offering an integrated suite of mobile office solutions for dispatch, fleet management, and mobile businesses. Service offerings include dispatch, telephony, two-way messaging, automatic vehicle location, and packet data.

Frequency Hopping Multiple Access (FHMA) technology lies at the core of Geotek's networks. This TDMA and spread-spectrum derivative, originally developed by the research and development arm of the Israeli military, employs frequency hopping to achieve substantial benefits in flexibility and spectral efficiency.

FHMA achieves 25–30 times the capacity of existing analog technologies using a macrocell approach. Macrocells typically cover areas up to 70 miles in diameter with up to 10 radial slices or sectors, incorporating from five to 20 microsites.

Within sectors, FHMA implements synchronized hops from one discrete frequency to another, in a predetermined manner at both the transmitter and receiver—you stay on one piece of spectrum for only a short time.

FHMA slices information packets into pieces, shuffles, and then transmits them. Packet losses result in minimal degradation since sequential losses don't occur. In addition, the system uses two-branch diversity, incorporating two separate antennas and receivers on both ends. This "space" diversity ensures that the best incoming signals are chosen.

The system is built on TCP/IP, and every system user has an IP address. With its inherent data integration, the Internet logically becomes a larger part in Geotek's service strategy. Geotek's automatic vehicle location capability is illustrative of things to come. It has sophisticated business data applications that will drive the mobile business markets.

Digital Transition

Digital technology is an integral part of Ericsson's SMR and private radio systems (EDACS), but not as an all-or-nothing requirement. Its systems can migrate to all-digital configurations, as customer needs dictate. Ericsson's enhanced digital access communications system technology package employs standard trunked radio.

Ericsson's Aegis system, which is an all-digital system, uses adaptive multiband encoding vocoder to transform analog voice signals to digital. After the vocoding process, error protection codes are added to the digitized audio stream. This process is further augmented with synthetic audio regeneration, which replaces certain portions of the voice signal that are corrupted with noise with usable segments of speech.

EDACS embeds its control information on the channel as well. This includes unit identification or push-to-talk identification, priority scan information, and talk group segmentation. With error correction and detection added, as well as channel signaling and synchronization, the combined signal transmits at 9,600 bits per second.

The combined voice and data encoding is an EDACS feature not all systems can match. Many others require separate channels for voice, data, and control signaling. EDACS combines all three, automatically compensating for periods of high demand and with no sacrifice in capacity or reliability.

Ericsson plans system enhancement and upgrades, such as TDMA. To date, its upgrade strategy has been to protect the customers' embedded investment in private or SMR technology and provide a painless and cost-effective move to digital.

Is There Still a Benefit from Two-way Radio?

Frankly, yes! Any business with personnel in the field or any business where efficient use of resources can reduce operating costs and increase profits will benefit from two-way radio. Now that doesn't leave out too many businesses, does it? In fact, hundreds of businesses, of all sizes and types, have mobile radios in use, and the list grows with each year. Police departments, delivery services, Realtors, school systems, industrial plants, farmers, repair services, construction companies, contractors, vending companies, and many more organizations save time and money through the continued use of two-way radios.

What Kind of Savings Can Your Business Expect?

How much does it cost you to operate just one of your vehicles per hour, including labor? How many hours of vehicle time could be saved every day with better scheduling and more control of your people? When you multiply these two numbers together, you'll quickly conclude that what a two-way radio system costs doesn't even come close to what you can save.

By reducing operating costs and increasing productivity with your current workload, you may find yourself able to expand your business dramatically without investing a great deal more.

When Will You Need a Radio Service Provider?

Two-way radio as a concept is simple: wireless communication between two or more points. However, because radio waves follow a line-of-sight path, your required coverage area may require a radio service provider to satisfy the need and the demand for coverage areas greater than a local private system can meet.

A two-way radio system generally consists of three types of units: a base station at a central dispatching location, mobile stations used in vehicles, and hand-held portables utilizing battery power. In today's business, conventional radio typically will not provide the wide-area communications that most businesses require. The cost for a business to construct its own radio system consisting of multiple tower sites can be too expensive. That's why a *Specialized Mobile Radio* (SMR) system provider is used. Customers pay a small monthly fee to use the service, similar to cellular service. The difference is you only pay a flat monthly fee, which allows for unlimited communication without high monthly cellular bills! There's no per minute charge!

20

Cellular
Communications

Introduction

A lot has happened in the cellular world since its original introduction in 1984. In 1984 when cellular communications became the hot button in the industry, all systems used analog radio transmissions. Many reasons were used to justify the cellular networks. These included very limited service areas, where you just could not get service where you wanted or needed it. Poor transmission haunted the operators because of the nature of the radio systems at the time. Users experienced excessive call setup delays. Heavy demand and limited channels were some of the more common problems in an operating area.

Analog Cellular radio systems used *Frequency Division Multiple Access* (FDMA), which is an analog technique designed to support multiple users in an area with a limited number of frequencies. Analog radio systems use analog input, such as voice communications. Because these systems were designed around voice applications, no one had any thought of the future transmission of data, fax, packet data, and so on, from a vehicle.

Back then, no one was sure what the acceptance rate would be. Currently, there are approximately 63 million cellular users in the United States. Approximately 100–150,000 new users sign up every month. Estimates are that four of five new telephones sold today are wireless telephones. Therefore, acceptance has become a non-issue. The new problem is not one of acceptance, but of retaining users. The churn ratio has been as high as 15–20%.

The answer to the retention problem lies in packaging the service with the handset. By offering service plans with a usage fee of $.10-.15/minute, the acceptance rate skyrocketed. The cellular industry is still primarily running on an analog backbone. Estimates are that 90–95% of the United States has analog coverage, whereas the digital counterpart to the cellular networks only covers between 50–55%. Dual mode telephones have become the salvation for cellular providers, because they would not be able to sustain their customer base without this offering.

Coverage Areas

The cellular operators build out their networks to provide coverage in certain geographically bounded areas. This poses the following dilemmas for the providers:

- The carriers need users (more) to generate higher revenues to pay off their investment.
- They must continue the evolution from analog to digital systems, allowing more efficient bandwidth use.
- Security and protection against theft is putting pressure on the carriers and users alike (cellular carriers lost over $300 million USD in 1997 through fraud and close to $500 million in 1998).

Analog Cellular Systems

Analog systems do nothing for these needs. Using amplitude or frequency modulation techniques to transmit voice on the radio signal uses all of the available bandwidth. This means that the cellular carriers can support a single call today on a single frequency. The limitations of the systems include limited channel availability.

The analog system was designed for quick communication while on the road. Because this service could meet the needs of users on the go, the thought process regarding heavy penetration was only minimally addressed. However, as the major *Metropolitan Service Areas* (MSA) began expanding, the carriers realized that the analog systems were going to be too limiting. With only a single user on a frequency, congestion in the MSA became a tremendous problem. For example, a cellular channel uses 30 kHz of bandwidth for a single telephone call!

Cellular was designed to overcome the limitations of conventional mobile telephone. Areas of coverage are divided into honeycomb-type cells and hexagonal design of smaller sizes, as shown in Figure 20-1. The cells overlap each other at the outer boundaries. Frequencies can be divided into bands or cells to prevent interference and jamming of the neighboring cell's frequencies. The cellular system uses much less power output for transmitting. The vehicular transmitter uses 3 watts of power, while the hand-held sets use only 3/10 watts. Frequencies can be reused much more often and are closer to each other. The average cell design is approximately 3–5 miles across. The more users who subscribe to a network, the closer the transmitters are placed to each other. In rural areas, the cells are much further apart.

For normal operation, cell sites may be separated by 3–5 miles, but as more users complain of *no service* due to congestion, cell splitting occurs.

Figure 20-1
The cell patterns

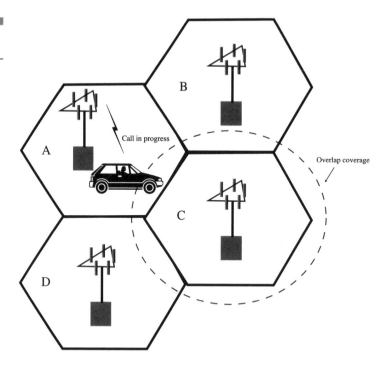

A cell can be subdivided into smaller cells, reallocating frequencies for continued use. The smaller the cell, the more equipment and other components are necessary. This places an added financial burden on the carriers as they attempt to match customer needs with returns on investments.

Log On

When the vehicle telephone powers on, it immediately logs onto the network. First, the telephone set sends a message to the *Mobile Telephone Switching Office* (MTSO). The Mobile Telephone Switching Office is the equivalent of a Class 5 Central Office. It provides all the line-in-trunk interface capabilities, much the same as the CO will do.

The information sent to the MTSO includes the electronic serial number and telephone number from the handset. These two pieces of information combined will identify the individual device.

The telephone set will use an information channel to transmit the information. Several channels are set aside specifically for the purposes of log on capabilities (see Figure 20-2).

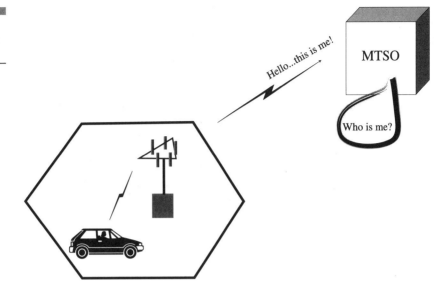

Figure 20-2
Log-on process uses
specific channels

Monitoring Control Channels

Once a telephone set has logged on, it will then scan the 21 channels set aside as control channels. Upon scanning, the telephone will lock in on the channel that it is receiving the strongest. It will then go into monitoring mode. Although the set has nothing to send, it will continue to listen to the monitored channel in the event the mobile telephone switching office has an incoming call for it. The telephone user has actually done nothing. Upon power-up, the set immediately will log onto the network, identify itself, and go into the monitoring mode.

Failing Signal

One can assume in vehicular communications that the vehicle will be in motion. As the vehicle moves from cell to cell, it will move out of range from the first site, but come into range of the second site. The received signal on the monitoring channel will begin to fail (get too weak to hear), as shown in Figure 20-3. Immediately the telephone will rescan all the monitoring channels and select a new one. After the vehicle finds a new channel, it will continue to monitor that channel until such time as it rolls out of range again. This concept of rolling from site to site allows the vehicle to be in constant touch with the mobile telephone switching office so long as the set is on.

Figure 20-3
The failing signal
procedure

Setup of a Call

When the user wishes to place a call, the steps are straightforward. The process is very similar to making a wired call.

1. Pick up the handset and dial the digits.

2. After entering the digits, press the send key.

3. The information is a dialogue between the MTSO and the handset.

4. The MTSO receives the information and begins the call setup to a trunk connection.

5. The mobile office scans the available channels in the cell and selects one.

6. The mobile office sends a message to the handset telling it which channel to use.

7. The handset then tunes its frequency to the assigned channel.

8. The mobile office connects that channel to the trunk used to set up the call.

9. The call is connected, and the user has a conversational path in both directions.

Setup of an Incoming Call

When a call is coming in from the network, things are again similar to the wireline network.

1. The mobile office receives signaling information from the network that a call is coming in.
2. The mobile office must first find the set, so it sends out a page through its network.
3. The page is sent out over the control channels.
4. Upon hearing the page, the set will respond.
5. The mobile office hears the response and assigns a channel.
6. The mobile office sends a message telling the set that it has a call and to use channel X.
7. The set immediately tunes to the channel it was assigned for the incoming call.
8. The phone rings, and the user answers.

Handoff

While the user is on the telephone, several things might happen. The first is that the vehicle is moving away from the center of the cell site. Therefore, the base station must play an active role in the process of handling the calls.

1. As the user gets closer to the boundary, the signal will get weaker.
2. The base station will recognize the loss of signal strength and send a message to the mobile office.
3. The mobile office will go into recovery mode.
4. The MTSO must determine what cell will be receiving the user.
5. The MTSO sends a message to all base stations advising them to conduct a quality of signal measurement on the channel in question.
6. Each base station determines the quality of the received signal.
7. They will advise the MTSO if the signal is strong or weak.
8. The MTSO decides which base station will host the call.

Setting Up the Handoff

After the MTSO has determined which base station will be the new host for the call, it will then select a channel and direct the new base station to set up a talk path for the call. This is all done in the background. An idle channel is set up in parallel between the base station and MTSO.

The Handoff Occurs

1. The original base station is still serving the call.
2. The new base station will host the caller.
3. The parallel channel has been set up.
4. MTSO has notified the cells to set the parallel channel in motion.
5. The MTSO sends a directive to the telephone to retune its frequency to the new one reserved for it.
6. The telephone set moves from one frequency to the new one.
7. The call is handed off from one cell to another.
8. The caller continues to converse and never knew what happened.

This procedure is shown in Figure 20-4.

Figure 20-4
The handoff process

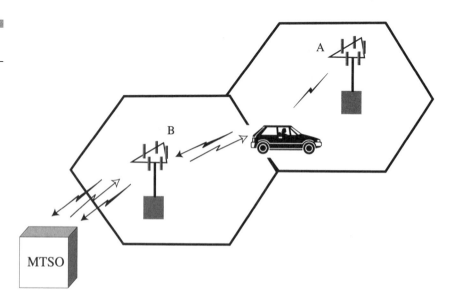

Completion of the Handoff

After the telephone has moved from one base station to the other, and one channel to another, the handoff is complete. However, the original channel is now idle, but in parallel to the original call. Therefore, the base station notifies the MTSO that the channel is now idle. The MTSO is always in control of the call. It manages the channels and the handoff mechanisms. The MTSO commands the base station to set the channel to idle and makes it available for the next call.

The Cell Site (Base Station)

The preceding discussion centered on the process of the call and referred to the base station quite a bit. The cell is comprised of a 3–5 mile radius. The base station is comprised of all the transmission and reception equipment between the base station and MTSO and the base station to the telephone. The cell has a tower with multiple antennae mounted on the top. Each cell has enough radio equipment to service approximately 45 calls simultaneously, as well as to monitor all the channels in each of the adjacent cells to it (see Figure 20-5). The equipment varies with the manufacturer and the operator, but typically, an operator will have 35–70 cells in a major location.

The Mobile Telephone Switching Office (MTSO)

The MTSO is a Class 5 Central Office equivalent. It provides the trunks and signaling interfaces to the wireline carriers. It has a full line switching component and the necessary logic to manage thousands of calls simultaneously. Like the CO infrastructure, the MTSO uses digital trunks between the MTSO and the wireline carriers (ILEC, CLEC, or IEC) either on copper, fiber, or microwave radio systems..

At the MTSO, there is a separate trunk/line interface between the MTSO and the base station. This is the line side of the switch, and it is used for the controlling call setup. Normally, the MTSO connects to the base station via a T1 operating line at 32 Kbps ADPCM. This T1 will be on copper or microwave. A MTSO is a major investment, ranging from $2 to $6 million, depending on the size and the area being served.

Figure 20-5
Seven cell pattern

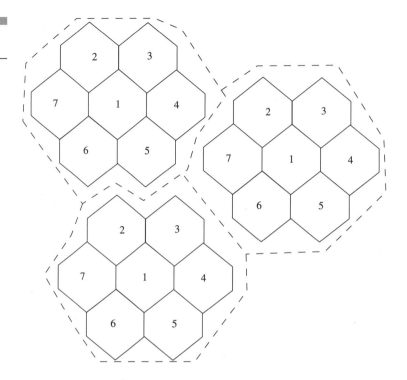

Frequency Reuse Plans and Cell Patterns

Frequency reuse is what started the cellular movement. Planning permits the efficient allocation of limited radio frequency spectrum for systems that use frequency-based channels (AMPS, DAMPS, and GSM). Frequency reuse enables increased capacity and avoids interference between sites that are sharing the frequency sets. Frequency plans exist that specify the division of channels among 3, 4, 7, and 12 cells. They define the organization of available channels into groups that maximize service and minimize interference.

As a mobile unit moves through the network, it is assigned a frequency during transit through each cell. Because each cell pattern has one low power transmitter, air interface signals are limited to the parameters of each cell. Air interface signals from nonadjacent cells do not interfere with each other. Therefore, a group of nonadjacent cells can reuse the same frequencies.

CDMA systems do not require frequency management plans because every cell operates on the same frequency. Site resources are differentiated by their PN offset (phase offset of the *Pseudorandom Noise* reference). Mobile channels are identified by a code that is used to spread across the baseband signal, and each can be reused in any cell. Using a N=7 frequency reuse pattern, all available channels are assigned to their appropriate cells. It is not necessary to deploy all radios at once, but their use has been planned ahead of time to minimize interference in the future (see Figure 20-5).

Overlapping Coverage

Each cell has its own radio equipment with an overlap into adjoining cells. This allows for the monitoring of the adjacent cells to ensure complete coverage. The cells can sense the signal strength of the mobile and hand-held units in their own areas and in the overlap areas of each adjoining cell. This is what makes the handoff and coverage areas work together. The graphic in Figure 20-6 shows the overlap coverage in the 7-cell pattern described previously.

Figure 20-6
Overlap coverage
between the cells

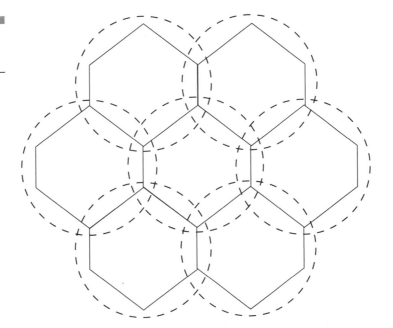

Cell Site Configurations

The mode of operation of a cell site is determined by the type of antenna used to support the air interface between the cell site and the mobile phones. When an omnidirectional antenna is used, the site serves a single 360-degree area around itself (see Figure 20-7).

The cell sites using the omnidirectional antenna are supported by a single antenna for both send and receive operations. These devices cover the full 360-degree site independently. One transmit antenna is used for each radio frame at the site (one frequency group per radio frame). Two receive antennas distribute the receive signal to every radio, providing diversity reception for every receiver at the site.

Due to the site's ability to receive signals from all directions, transmissions from neighboring sites may interfere with the site's reception. When interference reaches unacceptable levels, the site is usually sectorized to eliminate its ability to receive interfering information. Sectoring may also come into play when the site becomes so congested that the omnidirectional antennae cannot support the operations.

Figure 20-7

The omnidirectional antenna

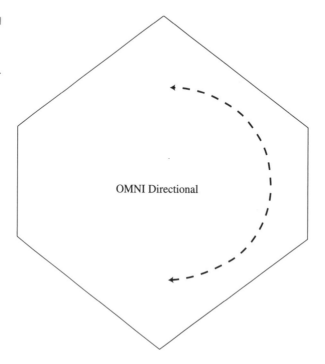

OMNI Directional

Sectorized Cell Coverage

Directional (sectorized) sites use reflectors positioned behind the antenna to focus the coverage area into a portion of a cell. Coverage areas can be customized to the needs of each site as shown in Figure 20-8, but the typical areas of coverage are as follows:

- 2 sectors using 180 degree angles
- 3 sectors using 120 degree angles
- 6 sectors using 60 degree angles

At least one transmit antenna is used in each sector (one per radio frame) and two receive antennas provide space diversity for each sector of a two- or three-sectored site. One receive antenna is used in each of the 60 degree sectors, with neighboring sectors providing the sector diversity. This is an economic issue because of the number of antennas required.

Figure 20-8
Sectorized coverage

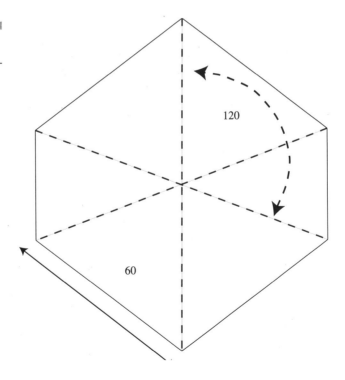

Tiered Sites

This configuration places a low power site in the same location with a high power site. Mobiles change channels as they move across the boundary between the two in order to relieve congestion in the center as shown in Figure 20-9. This configuration is used in all GSM and CDMA applications today. It is not supported on the older AMPS and Digital AMPS configurations, but may be used in newer implementations of AMPS and DAMPS. Each sector requires its own access/paging control channel to manage call setup functions. Voice traffic in each sector is supported by radios connected to antennae supporting that sector.

Reuse of Frequencies

Frequency reuse allows a particular radio channel to carry conversations in multiple locations, increasing overall capacity of the communications systems. Within a cluster, each cell uses different frequencies; however, these frequencies can be reused in cells of another cluster.

One centralized radio site with 300 channels can have 300 calls in progress at any one time. The 300 channels can be divided into four groups of 75 channels and still provide 300 calls at once. Dividing the service area into 16 sections called *cells* allows each cell to use one of the four groups of channels, increasing the call carrying capacity of the system by a value of 4 (1,200 calls at one time).

Figure 20-9
Tiered cell coverage

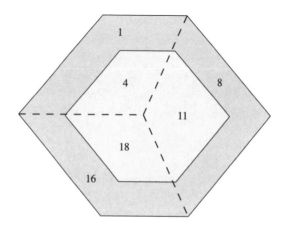

The service area can be continually divided into smaller and smaller cells to obtain greater call carrying capacity, increasing the number of calls by a factor of four with each division. The limit on how many cells can be used is determined by this information:

1. The cost infrastructure at each cell

2. Processing power of the switch that controls the system

3. Minimum power output at each site

Allocation Of Frequencies

The allocation of frequencies based on the first cellular arrangement of *Advanced Mobile Phone Services* (AMPS) was designed around 666 duplex channels. The frequency ranges were allocated in the 825–845 MHz and 870–890 MHz frequency bands. In each band, the channels use a 30 kHz separation, and 21 channels are allocated to control channels. Figure 20-10 is a representation of the channel allocation.

The FCC approved licenses for two operators of the cellular service; the wireline carrier (usually the telephone company in the area) and the non-wireline carrier (a competitor to the local telephone company). The frequencies were equally split between the wireline and non-wireline operators. This meant that only half the channels were available to each carrier and two sets of control channels were required.

Figure 20-10
Frequency allocation for cellular

Frequency 825 Mhz — 945 Mhz

| Channel | 1 | 2 | 3 | 4 | 5 | - - - - - - - - - - - - - - | 665 | 666 |

Transmit frequencies

Frequency 870 Mhz — 890 Mhz

| Channel | 1 | 2 | 3 | 4 | 5 | - - - - - - - - - - - - | 665 | 666 |

Receive frequencies

Four signaling paths are used in the cellular network to provide for signaling and control, as well as voice conversation. These can be broken into two basic function groups:

- Call setup and breakdown
- Call management and conversation

Establishing a Call from a Landline to a Mobile

From a wired telephone, the local exchange office pulses out the cellular number called to the MTSO over a special trunk connecting the telephone company to the MTSO. The MTSO then analyzes the number called and sends a data link message to all paging cell sites to locate the unit called. When the cellular unit recognizes the page, it sends a message to the nearest cell site. This cell site then sends a data link message back to the MTSO to alert the MTSO that the unit has been found. This message further notifies the MTSO which cell site will handle the call.

The MTSO next selects a cell site trunk connected to that cell and sets up a network path between the cell site and the originating CO trunk carrying the call (see Figure 20-11).

Figure 20-11
Call establishment

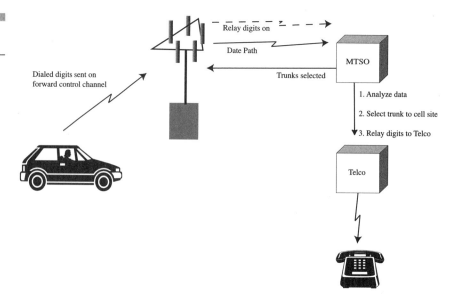

The MTSO is now also called the *Mobile Switching Center.* It is the controlling element for the entire system. The MSC is responsible for the following information:

1. All switching of calls to and from the cells
2. Blocking calls when congestion occurs
3. Providing necessary backup to the network
4. Monitoring the overall network elements
5. Handling all the test and diagnostic capabilities for the system

This is the workhorse of the cellular system. The MSC relies on two different databases within the system to keep track of the mobile stations in its area.

The first of the two databases is called the *Home Location Register* (HLR). The HLR is a database of all system devices registered on the system and owned by the operator of the MSC. These are the local devices connected to the network. The HLR keeps track of the individual device's location and stores all the necessary information about the subscriber. This includes the name, telephone number, features and functions, fiscal responsibilities, and the like.

The second database is called the *Visiting Location Register* (VLR), which is a temporary database built on roaming devices as they come into a particular MSC's area. The VLR keeps track of temporary devices while they are in an area, including the swapping of location information with the subscriber's HLR. When a subscriber logs onto a network, and it is not home to that subscriber, the VLR builds a data entry on the subscriber and tracks activity and feature usage to be consistent for the user.

Another set of databases is used by the MSC in some networks. These again are two separate and distinct functions called the *Equipment Inventory Register* (EIR) and the *Authentication Center* (AuC). These are very similar to the databases used in the GSM networks and keep track of manufacturer equipment types for consistency. The authentication center is used to authenticate the user to prevent fraudulent use of the network by a cloned device.

Global Services Mobile Communications (GSM)

Introduction

In the early 1980s, analog cellular telephony systems were rapidly growing in the European marketplace, particularly in the UK, France, Germany, and Scandinavia. In each country, the providers had developed their own internal operating systems to support this new mobile communications revolution. Unfortunately, these locally developed country systems were incompatible from system to system, country by country. Obviously, this was the least desirable of all situations for the introduction of a new system, leaving operators and users equally dissatisfied. Something had to be done to create a unified approach to the wireless networking and communications systems to bring them into harmony.

In 1982, the *Conference of European Post and Telegraph* (CEPT) created a study group to analyze what could be done. This study group was named *Groupe Spècial Mobile* (GSM), whose charter it was to develop a system that would work across the European market. The systems they proposed had to be capable of meeting certain criteria such as the following:

- Good quality speech
- Low cost for the equipment
- Efficient in the use of RF
- Capable of supporting the newer handheld telephones
- Transparent roaming capabilities
- ISDN compatible

The committees did their job well and quickly endorsed the standards and specifications to create a special mobile communications system capable of working across the international boundaries that had heretofore been blocked. These specifications were handed over to the *European Telecommunications Standards Institute* (ETSI) in late 1989. By 1990, GSM specifications were published. This turnover happened relatively quickly, considering the fact that political and economic in-fighting typifies standards setting in this industry. However, the setting of specifications led to the rollout of many GSM systems, so in 1993, 36 networks were operational in 22 countries, signifying the rapid acceptance and development of a single, standards-based network. Now over 300 operators in 133 countries have already endorsed and accepted the GSM specification for their local and national wireless networking standard.

In the graph shown in Figure 21-1, estimates highlight the growth of the GSM industry around the globe. Recently, in North America, several carriers have introduced the use of GSM in their networks, showing a unified

Figure 21-1
Subscribers on GSM
worldwide

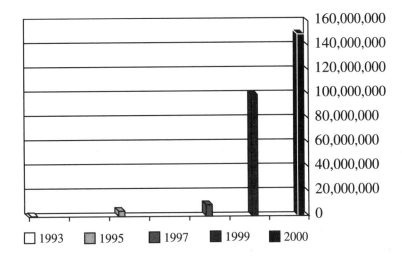

approach to the worldwide standardization. In some cases, the GSM installations in North America use a derivative of GSM called PCS 1900.

GSM amounts to approximately 38% of the world's wireless implementations spanning the 900, 1800 and 1900 MHz bands. In North America (the
United States, and Canada) there are approximately two million users
spanning 17 operators in over 1,550 cities across the continent.

GSM Concept and Services

Planning a GSM specification was not a simple task for the developers. They
initially wanted to overcome all the pitfalls of the older telecommunications
technologies and networks. One of the areas the developers sought to
address was the use of a 64 Kbps channel capacity for ISDN integration.
However, when using a wireless communications system, the in-band signaling and the robbing of bits were not conducive to delivering a full 64 Kbps
channel. The basic architecture of GSM is to support one primary service,
that being telephony from a mobile perspective. No matter what architectural model is used, speech-encoding techniques are used to transmit
the voice as a string of digital ones and zeros. GSM accommodates an emergency response system similar to the North American 911 services, whereby
the local emergency response agency is notified via a three-digit number.
Other services are automatically included in the operation of a GSM service
and network, such as:

■ Circuit switched data

■ Packet switched data

- Voice
- *Short Message Services* (SMS)
- ISDN
- Facsimile using an ITU Group 3-fax service

The GSM Architecture

The network is comprised of several components, not unlike the analog cellular networks of old. The functionality of each of the components describes the overall complexity and the degree of robustness built into the network. The pieces are shown in Figure 21-2, depicting the layout of the service elements and their interrelationship with each other. The figure shows the component, whereas the actual function that each piece performs is described below. Three major components come into play with the GSM network. These three components are the following: mobile unit, mobile base station, and mobile switching system.

- The Mobile Unit: The mobile unit, or mobile station, consists of the mobile telephone unit and typically a smart card called the subscriber interface module or *Subscriber Identity Module* (SIM). The SIM provides mobility for the individual so that a user can roam seamlessly and have all the services contracted for, regardless of the end user terminal device. By inserting the SIM into the set, the set takes on the personality of the end user. The user is then able to make and receive calls and receive the features allowed by contract, even if the set is a temporary one.

  The mobile unit has a specific identifier called the *International Mobile Equipment Identity* (IMEI). The SIM contains the *International Mobile*

Figure 21-2
Architecture of the GSM Network

Subscriber Identity (IMSI) used to identify the individual subscriber to the mobile system. Further, the IMSI contains authentication information in the form of an encrypted key. Any other pertinent information required by the systems operator is also contained in the IMSI. Because the two sets of identifiers are unique yet independent, mobility is assured for the individual.

- The Base Station Subsystem: The base station subsystem consists of two additional parts. The first part is the *Base Transceiver Station* (BTS). The second part is the *Base Station Controller* (BSC). These devices communicate between the components of the system or to disparate manufacturer's equipment through a standards-based interface (the Abis interface). The BTS is where the radio systems are located for the air interfaces to the subscriber mobile unit. Radio link protocols for GSM are used between the BTS and the mobile unit. The typical GSM air interface uses *Time Division Multiple Access* (TDMA) radio protocols.

  The BSC manages the radio resources for at least one, but possibly many, base transceiver units. The BSC is also the interface between the mobile unit and the *Mobile Switching Center* (MSC). The combination of the BTS and the BSC are defined as the *Base Station Subsystem.*

- The Network Subsystem: At the core of the Network Subsystem is the MSC. It acts like the class 5 Central Office of the *Public Switched Telephone Network* (PSTN). Moreover, the MSC provides all the necessary switching and call processing functions for the mobile unit, as well as additional functions like authentication, mobile handoff, registration onto the network, and a group of database functions. The MSC also provides the interface to the backbone signaling networks for call setup and teardown through the CCS7 (SS7) networks.

  Residing at the MSC is the *Home Location Register* (HLR), which is a database of the registered users to a specific network system. The HLR is the owner of the SIM for each subscriber of a specific network operator. When a user moves from one system to another, the *Visitor Location Register* (VLR) is enacted. The VLR is a temporary database of visiting devices in a system's area of operation. The VLR will notify the individual owner's HLR that the device is temporarily in a new location and that all requests for service can be handled by the VLR. Logically, there is only one Home Location register per GSM network, although it can be distributed across many locations. The VLR also contains selected information from the HLR, to provide all the necessary call control information for the subscriber.

It should therefore be noted that the MSC does not play into the information on specific devices because the information resides in the databases (registers) rather than in the switching system. Additional registers are used for enhanced control, such as the *Authentication Center* register (AuC) and the *Equipment Inventory Register* (EIR). The authentication center is an authentication server used to verify the user's specific information, such as password and authentication keys. The special keys and passwords are used in an over-the-air interface, so the authentication server validates the information and protects the information. The EIR is a database of the vendor specific information for all the radio sets used on the network, by manufacturer, and by IMEI. If the set is stolen or suspected of fraud, its IMEI is flagged as invalid in the database, denying a user with a clone or stolen set from using the network.

The Air and Link Interfaces

Internationally, the ITU allocated radio frequency spectrum in the 890–915 MHz band for the uplink and 935–960 MHz for the downlink in Europe. Figure 21-3 shows the frequency used in the specific up-and-down link structures. The uplink is from the mobile unit to the base station, whereas the downlink is from the base station to the mobile unit as shown in this figure. Because these same frequencies were already in use by the initial

Figure 21-3
The up-and-down link spectrum allocated

analog networks, the CEPT reserved the top 10MHz of each band for the GSM network, which was still being developed at the time. Over time the analog networks will decline, and the full 25 MHz in each band will be allotted to the GSM networks.

The Modulation Techniques Used

GSM, as stated earlier, uses TDMA in the air interface. In reality, GSM uses a combination of TDMA and *Frequency Division Multiple Access* (FDMA). Where a 25 MHz band is allotted for GSM, FDMA is used to break the 25 MHz spectrum down into a total of 124 carrier frequencies spaced at 200 kHz apart. One or more of these carrier frequencies is assigned to each of the base stations. From there, each of the carrier frequencies is subdivided in time division, using TDMA. The basic element of the TDMA is a burst of data during a burst period of 0.577 ms. There are 8 burst periods grouped into a frame. The TDMA frame is approximately 4.615 ms. This framing forms the basis of a logical channel. One physical channel is one burst period in a TDMA frame. The number and position of its corresponding burst period defines the channels. All definitions operate on a cycle of patterns that repeat every three hours.

Traffic Channel Capacities

A *Traffic Channel* (TCH) carries speech or data traffic. Traffic channels are defined by using groups of 26 TDMA frames called multiframes. A multiframe is 120 ms long (120 ms/26 frames/8 burst periods per frame). The multiframe is broken down into the following pieces:

- 24 frames carry traffic
- 1 frame carries a *Slow Associated Control Channel* (SACCH)
- 1 frame is unused

The frame breakdown is shown in Figure 21-4. The traffic channels are separated between the up-and-down link by 3 burst periods, so the mobile does not have to send and receive simultaneously.

GSM also has provisions for half-rate channels, although their implementation is not fully common. Half rate TCHs will ultimately double the capacity of a system. To do this, half rate speech coders will be used (using 7 kbps of speech instead of the normal 13 kbps).

Figure 21-4

TDMA frames and
multiframe allocation

Control Functions

Many of the aspects of the GSM use common control channels for either idle mode or dedicated mobile channels. These common channels are used by idle sets to exchange signaling information between the base and the mobile in order to change from an idle to a dedicated mode. A dedicated channel is allocated to a specific mobile while engaged in conversation, whereas an idle mode is used when the mobile is not engaged in talk. Mobiles that are in a dedicated mode monitor all the surrounding bases stations for handoff and other control information. To separate the common and dedicated channels, the common channels are defined in a 51-frame multiframe. This way a dedicated mobile can use the 26-frame multiframe for speech and traffic, yet can still listen to the monitor channels. Several different common channels are defined:

1. Broadcast Control Channel-A constant broadcast of information regarding frequencies, frequency hopping patterns, and other downlink information

2. Frequency Control and Synchronization Channel-Time synchronization patterns used to align the time slots for the cells. The cells use one of each of these channels allotted to channel slot number 0 in a TDMA frame.

3. Random Access Channel-A slotted aloha channel used by the mobile unit to request access to the network.

4. Paging Channel-The paging channel is used to signal a mobile unit of an incoming call request.

5. Access Grant Channel-This channel is used to allocate a stand-alone dedicated control channel to a mobile for signaling, usually following a request to access the network.

The Data Burst

The format of a TDMA frame is shown in Figure 21-4. Also shown in this figure is the organization of the data bursts. The normal burst carries data and signaling information. This burst lasts for 156.25 bit times, consisting of two separate 57-bit data patterns, a 26-bit training sequence for equalization, a robbed bit for forward synchronization in each block of data, 3 trailer bits at each end and 8.25 bit times guard band. The 156.25 bits are transmitted in the allotted 0.577 ms yielding a data rate of 270.833 Kbps.

Speech Coding Formats

When the speech is to be coded onto GSM, it is a digital transmission system. Therefore, the analog voice is converted to digital before transmission. The normal telephony architectures use standard PCM techniques to digitally encode a voice signal at a data rate of 64 Kbps. However, 64 Kbps is too difficult to accomplish across the radio signals. A speech compression and coding technique using a form of *linear predictive coding* (LPC) produces a 13 Kbps speech pattern. The speech is actually divided into 20 ms samples, encoded as 260 bits each, yielding the 13 Kbps speech.

The Network Structured Protocols and Interfaces

When using GSM, there is a layered protocol approach for the several interfaces and the protocols necessary to provide transparency across the medium. Figure 21-5 portrays the protocol stack at the different components of the network architecture. In the leftmost portion of the figure is the "protocols stack" for the mobile station. The TDMA protocol resides at the physical layer interface (the air interface). *Link Access Protocol Data for mobile* (LAP-Dm)

Figure 21-5
The protocol stack for
GSM interfaces

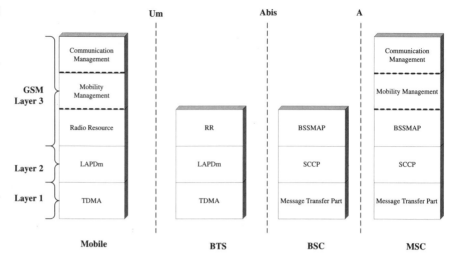

protocol, which is a derivative of the ISDN LAP-D data link layer, works at
layer 2. The GSM layer (layer 3) is subdivided into three separate sublayers;
the radio resources management sublayer, the mobility management sublayer, and the communications management sublayer.

As the information is passed between the mobile unit and the base transceiver station, the layers are similar but only used from the TDMA layer up
to the radio resource management sublayer. The upper sublayers of layer
three are not used at the user to mobile interface. This is shown in the next
part of the drawing in stack number two from the left.

Between the base transceiver systems and the Base Station Controller,
the Abis interface is used. Here at the protocol stack, the layer 1 protocol is
the *Mobile/Message Transfer Part* (MTP), which acts at the bottom three
layers of the typical OSI interfaces. The *Signal Connection Control Part*
(SCCP) and the *Base Station System Mobile Application Part* (BSSMAP)
combine to form the equivalent of an SS7 protocol stack. This is shown in
the third stack from the left.

The BSSMAP is the equivalent of a TCAP as shown in Figure 21-6,
which is merely a comparison of the SS7 protocol stack.

What is left is the stack on the right-hand side of Figure 21-5 (above),
showing the Mobile Switching Center protocols as the signal is passed to
the MSC. In this case, the Mobility Management and the Communication
Management parts are shown on the upper layers of the stack.

These additional pieces include such added features and functionality in
the protocol stack to support the following;

Figure 21-6
Comparing the TCAP
and the GSM BSSMAP

SS7

GSM-MSC

MAP	TCAP
SCCP	SCCP
MTP-3	MTP-3
MTP-2	MTP-2
MTP-1	MTP-1

CM
RR
BSSMAP
SCCP
MTP 1-3

- Radio Resource Management-Deals with link establishment and maintenance, the handoff between different base stations and the coordination of the necessary spectrum used in a particular base station operation. This is also used in setting up the necessary channel assignment when a call is incoming or a page is initiated across the wireless interface.

- Mobility Management-Deals with the constant location updates between the HLR and VLR function. Also included in the mobility management are the authentication, equipment inventory, and security capabilities.

- Communications Management-Deals with the control of call setup, routing, and teardown functions. The call control function attempts to use a point of Q.931 for routing and delivery of calls. GSM protocols are similar but unique from the standard wired telephony network. This is the equivalent of the ISUP from an SS7 standpoint.

Final Thoughts on GSM

What GSM brings to the table is a standardized platform for communications within the mobile environment. In North America, several different approaches were used to implement mobile and personal communications services. The rest of the world has lock-stepped into a global means of providing the transparency and seamless roaming capability needed to provide true communications spirit. While in Europe, a user can rent a telephone set (or buy one) and install the smart card into it. Immediately, the set takes on the identity of the individual installing the smart card. From billing information, features, and subscribed services, to credit validation and authentication, the mobile telephone is now the individual. This, in fact, is true portability, because the smart card can also carry with it the telephone number of the person it is associated with. From country to country, roaming occurs on a daily basis overseas. Yet in North America, the seamless roaming only occurs within specific carriers and networks where reciprocal agreements are in place. This means that there are still different standards being used in both parts of the world.

Most of the international countries have adopted GSM as the global platform. The North American community has many standards in place; most are not compatible with each other. This is the sad truth on a recurring basis. If a user chooses to change suppliers, then the technology in the handset may have to be changed, which is not the case in the international market.

The ITU is now working on a "third generation mobile" standard. Unfortunately, this has been somewhat of a figment of our imagination because the standards are different in countries around the world. The North American market accounts for over 60 million wireless users, who are all moving toward a broadband communications network and the personal communications services that follow. This is a fairly significant portion of the global installations (57%) for wireless communications[1]. One cannot ignore the multiple standards such as:

- GSM-TDMA
- North American-TDMA (IS−54)
- CDMA (IS−95)
- TDMA (IS−136)
- AMPS
- DECT

Many other standards have been rolled out or implemented on a sporadic basis. The one standard that has truly been implemented around the world

---

[1] As of this writing

is GSM. The "third generation mobile" standard can become a reality now that there has been some settlement between many of the players (Ericsson and Qualcomm for example), but this dream may be well into the new millennium before final realization.

Most users still think of mobile communications as simply a means of talking on the move. All that is about to change.

Work is now well advanced on the third stage of GSM standard development. Known in the business as Phase2+, the new technology will enable GSM to serve as a platform for a whole range of sophisticated services that will keep GSM operators ahead of the game. The services include features and functions such as

- Increased data transmission speeds
- High Speed Circuit Switched Data, giving a giant boost to data credibility.
- The *General Packet Radio Services* (GPRS) standard, currently under development, represents a shift to packet data, which will enable even more advanced services.
- Video conferencing and high-resolution scans are just around the corner.
- GSM will facilitate Internet access, which in turn will enable business users to share information via corporate Internets, Intranets, and extranets.
- The mobile wallet is already close to becoming a reality. Mobile phones will provide all the services currently provided by automatic teller machines and more. Customers will be able to pay bills and transfer money between checkingand savings accounts just by typing in codes on their mobile terminal. More radical still, they will be able to send electronic cash into and out of their GSM phone.
- The GSM standard is being adapted to work as a fixed access system for *Wireless Local Loop* (WLL) applications that will provide full 64kbit/s digital connections.

The technology is the richest and most flexible on the market. The future will only be constrained by the limits of our imagination.

Personal Communications Services

Introduction

There can be no doubt about the changes that have taken place in the industry since the inception of the personal communications services. The entire industry has changed, making communications an integral component of our everyday life. What was once considered as frivolous (or for the affluent only) in the past, has become commonplace today. Look around and see where the personal communications systems have penetrated. What was once a business service is now an everyday component of students, housewives, business, and regular everyday people. The personal communicator, the simple wireless telephone, has become an indispensable tool for just about every occupation. The days of having a simple pager are fast becoming obsolete as the personal communicator becomes more affordable and more competitively priced. Not that long ago, the industry was taken aback by the prospect of everyone having a personal telephone set. Today, that concept is closer to reality than ever before.

The communicator of today is not a broadband device. However, it is the precursor for the future devices. The newer ones will bring more bandwidth and more sophistication to our everyday lives. We have to merely walk before we run!

The personal communications systems have evolved from the wireless cellular and GSM networks, so there is not a lot of excitement there. However, where the original wireless networks were built on an analog networking standard, the PCS architectures are built on digital transmission systems. Therefore, several different approaches are used to deliver the capacities at the pricing models today. We shall discuss the evolution of these wireless systems and their ability to propel us into the new millennium and the communicator of the future.

Digital Systems

Because *Frequency Division Multiple Access* (FDMA) uses the full 30 kHz channels for one telephone call at a time, it is obviously wasteful (see Figure 22-1). FDMA, an analog technique, can be improved a little by using the same frequency in a *Time Division Duplex* (TDD) mechanism. Here, one channel is used and time slots are created. The conversation flows from A to and then from B to A. The use of this channel is slightly more efficient. However, when nothing is being sent, the channel remains idle. Because

Figure 22-1
FDMA slotting

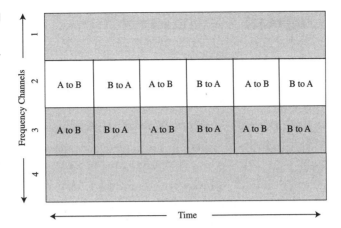

digital transmission introduces better multiplexing schemes, the carriers wanted to get more users on an already strained radio frequency spectrum.

Additional possibilities for enhancing security and reducing fraud can be addressed with digital cellular and PCS. Again, this appears to be a win-win situation for the carrier because it features:

- Fewer costs
- More users
- Better security
- Less fraud

Obviously, the carriers welcomed some of these discussions.

Once the decision was made to consider digital transmission, the major problem was how, what flavor to use, and how to seamlessly migrate the existing customer base to digital.

The digital techniques available to the carriers are the following:

- *Time Division Multiple Access* (TDMA)
- *Extended Time Division Multiple Access* (ETDMA)
- *Code Division Multiple Access* (CDMA)
- GSM (using a slightly different form of TDMA)
- *Narrowband Advanced Mobile Phone Service* (N-AMPS)

Carriers and manufacturers are using each of these systems. The various means of implementing these systems has brought about several discussions regarding the benefits and losses of each choice.

Digital Cellular Evolution

As the radio spectrum for cellular and PCS continues to become more congested, the two primary approaches in North America use derivatives of

- *Time Division Multiple Access* (TDMA)
- *Code Division Multiple Access* (CDMA)

Each of the moves to digital requires newer equipment, which means capital investments for the PCS carriers. Moreover, as these carriers compete for radio frequency spectrum, they have to make significant investments during an auction from the FCC. This places an immense financial burden on these carriers before they even begin the construction of their networks.

Time Division Multiple Access (TDMA)

TDMA uses a time division multiplexing scheme, where time slices are allocated to multiple conversations. Multiple users share a single radio frequency without interfering with each other because they are kept separate by using fixed time slots. The current standard for TDMA divides a single channel into six time slots. Then three different conversations use the time slots by allocating two slots per conversation. This provides a three-fold increase in the number of users on the same radio frequency spectrum. Although TDMA deals typically with an analog to digital conversion using a typical pulse code modulation technique, it performs differently in a radio transmission. PCM is translated into a quadrature phase-shift keying technique, thereby producing a four-phased shift, doubling the data rate for data transmission. The typical time slotting mechanism for TDMA is shown in Figure 22-2

The wireless industry began to deploy the use of TDMA back in the early 1990s when the scare radio frequency spectrum problem became most noticeable. The intent was to improve the quality of the radio transmission as well as the efficiency usage of the limited spectrum available. TDMA correctly can increase the number of users on the RF by three-fold. However, as more enhanced PCS techniques such as micro and picocells are used, the numbers can grow to as much as a forty-fold increase in the number of users on the same RF spectrum. One can see why it is so popular.

TDMA has another advantage over the older analog (FDMA) techniques. Analog transmission across the 800 MHz frequency band (in North Amer-

Figure 22-2
TDMA for PCS

ica it is 800 MHz, in much of the world it is 900 MHz) supports one primary service; voice. The TDMA architecture uses a PCM input to the RF spectrum. Therefore, TDMA can also support digital services for data in increments of 64 Kbps. The data rate can support from 64 Kbps to the hundreds of megabits/ second (120 Mbps).

The carriers like TDMA because of its ability to add data and voice across the RF spectrum and the cost associated with the migration from analog to digital. A TDMA base station costs approximately $50–80,000 to migrate to digital from FDMA. This is an attractive opportunity for the carriers developing PCS and digital cellular using the industry standards. The two standards in use today include IS−54, which is the first evolution to TDMA from FDMA. The second is IS−136, the latest and greatest technology for 800–900 and 1900 MHz TDMA service. When the IS−136 service was introduced, the addition of data and other services were introduced. These services include the *Short Message Service* (SMS), caller id display, data transmission, and other service levels. Using a TDMA approach, the carriers feel that they can meet the needs for voice, data, and video integration for the future. TDMA is the basis of the architecture for the GSM standard in Europe and the Japanese standard for PDC.

Advantages of TDMA

Unlike spread spectrum that can experience interference from other users in the same frequency, TDMA keeps all users' conversations separate. Because the conversation is limited to a timing slot, the battery on a telephone set should be extended significantly because the conversation is only taking up one-third of the time. Moreover, TDMA uses the same hierarchical structure as traditional cellular networks so it can be increased and upgraded easily.

Enhanced TDMA

Standards bodies and manufacturers all came up with variations of the use of the frequency spectrum. Nevertheless, manufacturers have been developing an *Extended-TDMA* (ETDMA), which will allow efficient use of the system. TDMA will derive a three- to five-fold increase in spectrum use, whereas ETDMA could produce ten- to fifteen-fold increases. The concept is to use a *Digital Speech Interpolation* (DSI) technique that reallocates the quiet times in normal speech, thereby assigning more conversations to fewer channels, gaining up to fifteen times over an analog channel. This is a form of statistical time division multiplexing. When a device has something to send, it places a bit in a buffer. As the sampling device sees the data in the buffer, it allocates a data channel to that device. When a device has nothing to send, then nothing is placed into the buffer. Then the sampling passes over a device with an empty buffer. Time slots are dynamically allocated based on need rather than on a fixed time slot architecture. An example of the ETDMA technique is shown in Figure 22-3

Figure 22-3

ETDMA uses a form of statistical time division multiplexing

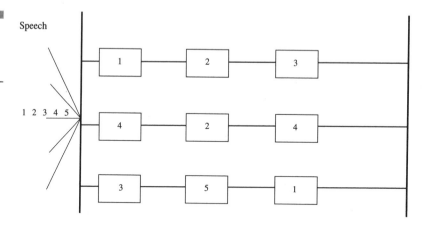

Simultaneous Voice and Data Application

    - Host Access
    - File Transfer
    - Datagram Capability (paging, credit card authentication)
    - Meter Reading
    - Fax
    - Asynchronous Terminal
    - (E-Mail)

CDMA

Spread spectrum, which was employed back in the 1920s, evolved from military security applications. Spread spectrum uses a technique of organizing the radio frequency energy over a range of frequencies, rather than a fixed frequency modulation technique. The system uses frequency hopping with time division multiplexing. At one minute the transmitter is operating at one frequency, at the next instant it is on another. The receiver is synchronized to switch frequencies in the same pattern as the transmitter. This is effective in preventing detection (interception) and jamming. Thus, additional security is derived. These techniques should produce increased capacities in ten- to twenty-fold over existing analog systems.

SPREAD SPECTRUM SERVICES

As the use of radio frequency spectrum continued to put pressure on this limited resource, the systems manufacturers and regulators were searching for some way to share spectrum among multiple users. Further, the sharing is compounded by the need to secure information while on airwaves. These pressures have led to the use of spread spectrum radio. The spreading portion of these systems using a chip set coded for your specific transmitter-to-receiver system. It can use multiple frequencies (called hopping); or it can create a coded chip set. Regardless of the operation chosen, both are designed to spread energy over a broader range of frequencies to allow for less airtime on a specific bandwidth and to ensure the integrity of the information being sent.

The technique for the spread spectrum service is called *Code Division Multiple Access* (CDMA). Many of the PCS carriers have chosen CDMA as their coding of choice. Since spread spectrum has been introduced at the commercial level, the FCC allocated spectrum in the 1.9 GHz range. This group of frequencies is heavily used by microwave users operating in the 2 GHz range. The FCC feels that the spread spectrum won't seriously impair the microwave users because it will appear only as random noise to any other frequency system operating in the same frequency ranges. However, as more users of spread spectrum are inserted into a specific frequency range, the possibility of congestion and interference exists. If you attempt to use this range of frequencies and congestion builds, then the decision could be a wrong one. Despite the benefits of frequency hopping, the benefits may be overshadowed with distance, utilization, and power constraints.

Consequently, when the FCC decided to allocate frequency spectrum to the PCS carriers, the fixed microwave users were told they had to "find a new home." This means that the operators who had radio systems in operation had to find new frequencies to use. Moreover, to facilitate the frequency change, many of the fixed operators had to upgrade their equipment to work in the new frequency band they found acceptable. Part of the auctions for the PCS frequencies included money set aside to help the fixed operators upgrade and move to a new frequency. This put added pressure on the wireless carriers to pay for the auctioned frequencies before building their infrastructure. The FCC raised approximately $13 billion in the auctions, but that placed the carriers in a negative position. They spent a lot of their capital up front to get the frequencies, before they could buy any equipment.

Seamless Networking with IS−41 and SS7

The whole intent of the IS−41 and SS7 interfaces is to allow the wireless carriers to communicate transparently and seamlessly between and among each other. Moreover, with SS7 interfaces to the wireline networks, calls can enter or exit the wireless networks flawlessly. This has been ongoing since 1994 and seems to be moving quite well. If the wireless carriers do not have physical interfaces to the telephone companies, they can use the *Independent Telecommunications Network* (ITN) to provide these services as a service bureau, for a fee. The interconnection between the networks provides industry standards-based internetworking.

Automatic Roaming

The IS−41 and SS7 provide many of the features for automatic roaming among providers, although modifications will be necessary. First, it helps to define financial responsibility between the carriers for their users. Second, it helps to provide the seamless interfaces between the wireless systems and the SS7 interfaces between the wireline and wireless carriers.

As the system operates, automatic roaming allows for the discovery of a roamer. The providing system then learns the identity of the current serving CO, or visited system and sends that information to the Home system.

A profile is established for the roamer in the visited system, allowing the network to find the end user. This, in fact, is how the wireless providers handle location portability.

Lastly, the automatic roaming allows the setup for delivery for the calls.

The wireless suppliers have been moving toward transparency through their bill and keep arrangement. In reciprocal billing, they assume that the calls cost an equal amount of money for the carriers to originate or terminate. Therefore, between wireless carriers, they do not charge each other for terminating minutes. The wireline companies on the other hand are charging for terminating minutes at approximately $.025-.03 per minute. Thus, the carriers are looking for a way to get away from the reciprocal billing and to get to a "bill and keep" arrangement with the wireline providers.

Cellular and PCS Suppliers

The cellular industry has been a success story to behold. Over the past 14 years, they have grown from nothing to over 56 million customers. With very few exceptions, the cellular carriers see themselves as a complementary carrier to the ILEC business. Some of this comes from the fact that two parts of the cellular network started when the licenses were issued: Wireline carriers meant the ILEC operated the regulated portion of the cellular network, and non-wireline meant the competition. In most cases, the ILEC who operated the wireline side of the business then spread out across the country as part of the non-wireline provider (competing for wireless services with their cousins). Therefore, the cellular industry is influenced heavily by the ILEC side of the business. The cellular providers see themselves as complementing the wireline business instead of competing with it.

PCS and SMR providers seem to be the carriers seeking a niche in the market. As they offer services to their customers they view themselves as an alternative to the ILEC wireline services and at the same time as complementary. Depending on how the customer reacts, the PCS providers will move toward one position. In checking with several PCS suppliers, the argument to make the PCS telephone number the sole number came up. The providers offer the same features and functions as the wireline service providers such as the following:

- Call forwarding (busy or don't answer)
- Voice messaging
- Three-way calling

- Caller ID
- Call transfer

The PCS providers are now saying that the customer can use the same number for their home number, their business number, and their traveling number. Why pay for two different lines and service offerings when you can do it all on one telephone (and number)? Their speech is convincing somewhat, but they fail to mention the cost of the airtime for receiving calls and the added cost for making local calls. However, they are now packaging these services in such a way that they are invisible.

Home location registers keep track of all network suppliers' users, based on their own network id. The database (HLR) has all the appropriate information to recognize the caller-by-user ESN and mobile telephone number. The database controls the features and functions the user has subscribed to. When a call comes into the network today, the called number is sent to the HLR responsible for the number dialed. These are SS7 messages. The HLR then looks into its database and determines where the caller is located or if the user is on the air.

If the user is located somewhere else, a Visiting Location Register entry exists at a remote MSC. This entry has been updated by the remote end when the user activated the telephone (powered it on and registered) or when an already powered device rolls into range of a cell site from the new location. Then an SS7 message is sent out to the HLR that the called party is now in someone else's *Automatic Line Information* (ALI) database.

If a call is coming in from the network, the SS7 inquiry comes into the HLR who then sends a redirect message to the remote MSC (the VLR) serving the end user.

Using this roaming capability, the database dips are occurring on a frequent basis. Many of the networks (wireless) are already taxing their databases. They are trying to move more information out to the STP or the local switches rather than constantly hammering into the SCP.

Final Thoughts

PCS mimics the same basic services of GSM, discussed in the preceding chapter. The primary difference is that the PCS operators are primarily North American based whereas the GSM standard has been implemented by the rest of the world. Differences exist in the overall operation, but these systems operate in similar fashion. The ability of the PCS system to learn the identity of the roaming device, is similar to the smart card used in Europe. The services allowing the Central Office equipment and databases to communicate with each other, and track the roamer simplifies the billing and the service functions. Moreover, the feature transparencies available with PCS creates a seamless network atmosphere.

Although the two differing technologies are different, they are similar at the same time. The real value is achieved when the user can operate the wireless device (the phone) transparently between supplier networks, with the same look and feel as in the home system.

Wireless Data Communications (Mobile IP)

Introduction

As the convergence of voice and data continues, a more discreet change is also coming into play. While data is considered fixed to a location, the end user is now more mobile. This opens a new set of challenges for the industry and manufacturers alike, because of the need for mobility. What once was a simple procedure of connecting the user's modem to a land line now poses the need to connect that same user to a device while mobile. Protocols need to be more flexible, accommodating the mobile user as the device is moved from location to location. Moreover, the physical devices (for example, the modems) must be moved often. In a dial-up, circuit-switched communications arrangement, this is not a major problem. The user can unplug a modem, reconnect it to a landline elsewhere, and dial from anywhere.

However, when we use IP as our network protocol, data is routed based on a network/subnetwork address. Routing tables keep track of where the user is located and route the datagrams (packets) to that subnetwork (see Figure 23-1). When a mobile user logs on and attempts to dial in to the network, the IP address is checked against a routing table and routed accordingly. Updating the tables can be extremely overhead intensive, and it can produce significant amounts of latency in the Internet or Intranet. Using an ICMP *Route Discovery Protocol* (IRDP), which is part of the TCP/IP protocol suite, helps. However, when the IRDP process updates its tables, we use a lot of bandwidth. Figure 23-2 is an example of the IRDP process where a message is generated by a host to learn all routes available to get to and through the network.

Figure 23-1
Routing table

Destination Address	Next Hop
34.1.0.0	54.34.23.12
79.2.0.0	54.34.23.12
147.9.5.0	54.32.12.10
17.12.0.0	54.32.12.10

Figure 23-2
ICMP Route Discovery
Protocol learns the
new addresses and
routers in the
network.

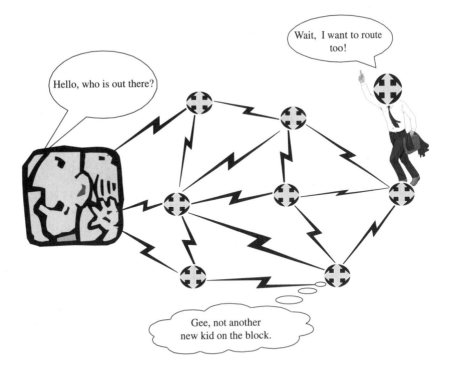

Figure 23-2
ICMP Route Discovery Protocol learns the new addresses and routers in the network.

Something needs to be done to accommodate the use of mobile IP by an escalating number of users wanting to log on anywhere. Of course one solution is to use the *Dynamic Host Configuration Protocol* (DHCP), which uses a server-based mechanism to allocate a new IP every time a user logs onto the network or assigns a static IP address to a user who may be using a diskless PC. The purpose of the DHCP was to facilitate the mobile or not permanently attached user in an ISP network where addresses are limited and casual users are the norm. So the industry had to arrive at a solution allowing casual and nomadic users the same access while they travel (roam) as when they are in there fixed office location.

Figure 23-3 shows the growth curve of wireless data users attempting to use mobile IP and wireless data communications over the past couple of years. In this graph we see that the numbers justify the concern and the effort being afforded to the problem. The number of wireless users in the world is escalating and the 500 million users shown in this graph are conservative estimates. We are living in a mobile society where users want their data, when they want it, where they want it, and how they want it! What percentage of these wireless users will want data over their connection remains to be seen. However, early estimates are that over one-fifth will want their data in a mobile environment.

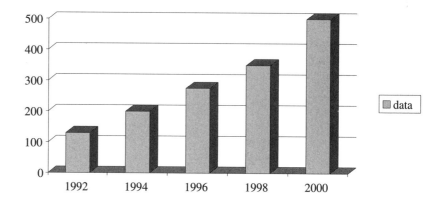

Figure 23-3
Wireless data
subscribers (in
millions)

IP Routing

When IP was first implemented, the routing scheme was based on tables that were kept in a router. The routers updated these tables periodically to keep the information current. A service advertising protocol was used to notify the downstream routers and hosts of the router's presence. As data was sent through the Internet (Intranet or private network), the IP datagrams included both the source and destination address, as shown in Figure 23-4. Routers forwarded the data on the basis of the destination address, the 32-bit address in IP version 4. Routers ignored the source address, looking only at the destination. Using the destination address, the router then concerned itself with the net hop. If a router received a datagram that was destined for a local address, it then sent the data to the physical port for the LAN sub-network it resided on. However, if the router received a datagram that was for a destination that was not local, it forwarded the datagram to its next downstream neighbor. The routing process was done on the basis of a hop-by-hop forwarding.

After receiving a datagram with an address that was local, the router would forward the data to its local physical port (the LAN). However, if a networked device was moved because the user was mobile, the router continued to send the data to the destination network in its table. A mobile user was therefore not going to receive any datagrams because IP did not guarantee delivery. This broke down the whole process because the devices were mobile, especially in global networks with mobile workers. Conceivably, a user could travel to a remote office and attempt to plug in a laptop computer with a static IP address from corporate headquarters. Because

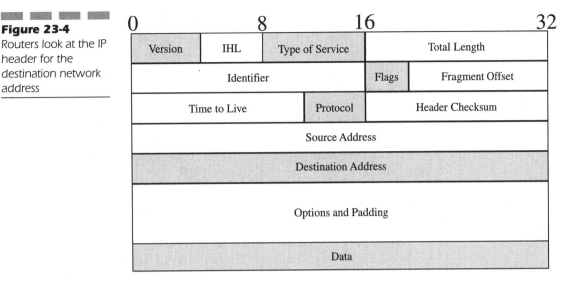

Figure 23-4
Routers look at the IP
header for the
destination network
address

the IP was static, all the IP datagrams were sent along their way to the headquarters router where the IP subnetwork address was registered.

Part of the Solution

One part of the solution is a change in IP version 4 that allows for mobile users. This mobile IP process is addressed in RFC 2002, which allows for the registration of home agents and foreign agents. A home agent is a registration process whereby a locally attached computer registers with a server or router and gets entered into the routing table. Based on RFC 2002, a mobile device (either a host or a router) can change locations from place to place without devastating effects. If mobile IP is in use by the device, the mobile device will not change its address when it moves from location to location. Instead, it will use an IP forwarding approach by using an "in-care-of" obtained from some serving device that knows of its presence. Using the home agent concept, the home agent is a server or router on the home network. The home network maintains the mapping of the mobile device's address to its current address.

Think of this in the context of the postal service. If you go on vacation for a month to Phoenix, Arizona, you still need to get your mail (for example, paychecks, bills, etc.). So, before going away, you go to the local post office and tell them to forward all your mail for the next month to Phoenix. The forwarding idea is shown in Figure 23-5. The postmaster will then make a note on the mailbox where he/she sorts your mail. This note will say to forward all mail to P.O. Box 51108, Phoenix, AZ 85076–1108. So even though people are still sending mail to your address, the post office employees (your home office agents) are handling the forwarding to you, based on the instructions you left behind.

When the vacation is over and you return to your normal domicile, you notify the local postmaster that you are back. The local folks remove the forwarding address from your normal mailbox and then resume delivery to the home address. The Phoenix postmaster, in the meantime, was responsible for knowing your temporary address while you were in his/her area. Therefore, the local postal people were handling the routing and delivery of the mail while you were in Phoenix. They were acting as your agents in the foreign location. The foreign agent received mail that was marked for your normal home address, but translated this into the current address. This combination of local (home agent) and remote (foreign agent) agents makes the process work.

Now let's assume that after vacation, you leave Phoenix, but do not go home. Instead, you go on an extended business trip to San Francisco. The

Figure 23-5
Forwarding IP traffic

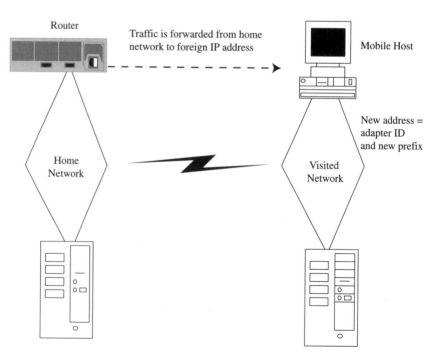

home agents then label the boxes to reflect the changes that indicate mail is now to be forwarded to the new location.

Applications that Demand Mobile IP

What are the applications that will take advantage of the mobility of the user? Well, the first one is going to follow our example for mail. But e-mail is only one of many applications that users will want to use. Other areas will include transaction processing (orders online), fax capabilities, FTP, and TELNET, Telemetry applications and multimedia are all prime candidates. This is shown in Figure 23-6 where the applications will be paramount to the ability to move around.

Applications are what the data transmission world is all about. So we need to have mobility in the IP world in order to satisfy the applications for a mobile user.

In Table 23-1 the need for these application services becomes more evident. In this table, the driving forces for the implementation of wireless data and mobility within the TCP/IP architectures are more evident.

Figure 23-6
Application of mobility in a data world

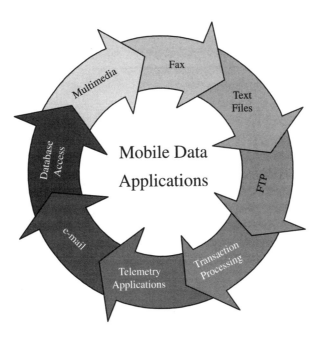

Table 23-1

Driving forces for
mobile data

Market Forces Driving Mobility	Value
Consumer Use of Computers	Increasing. In North America, the average number of households has climbed from 37% to 53% and users wired for interconnectivity.
Remote Access Needs	With more telecommuters, the need for remote access is increasing at a 10% or better ratio each year. The number of telecommuters is running at 10 million in 1999.
Users Demands for the Access They Perceive	Users want access to their data when they want it, where they want it, and how they want it. The driving force is that 150 million subscribers will be using wireless data by the end of 2000.
Wireless Innovations	A mere three years ago, wireless data was only available at 9.6 Kbps, even though it was advertised at speeds of 33.6 Kbps. In the future, mobile data and IP users will expect multimegabit speeds from their wireless data devices.
Internet	The growth on the Internet has been so dramatic that more users expect the capability to log onto the Net and get their e-mail, FTP, and transmit other forms of data anywhere and anytime. Internet growth is slotted at 200 million users in 1999–2000 period, growing to 350 million users by 2003.

Speed Isn't Everything

Some of the driving forces shown in the table couple tightly with the throughput issues of data communications. However, speed is not the only issue driving the demand for mobility. The need for speed and reliability are rapidly driving the users into a mobile environment. In the older days, the speed would hold back the use and acceptance of data transmission. Popularity in other connection devices also set the pace. An example of the devices is shown in Table 23-2. This lists the various devices people are looking to use.

Table 23-2	Devices	Expected Speeds
Speeds for wireless devices	Personal Data Assistants	Current wireless connection is 28.8 Kbps. Expect this to grow to 1 Mbps by the end of 2000.
	Laptop Computers	Current modem technology supports up to 56 Kbps downloadable and 33.6 Kbps uploadable. In the future, users will expect 2–10 Mbps.
	Palmtop Computers	Same technology as laptops.
	Mobile Telephones	Current low-speed telephony with speeds of less than 19.2 Kbps circuit-switched technology. Future will be on demand at speeds of up to 128 Kbps and then 1 Mbps.
	Paging Systems	Low speed data from short messaging systems

Variations in Data Communications (Wireless)

Several different flavors of data communications exist for wireless communications. Most people think of the services as being one type only. Earlier the discussion considered the *Short Message Services* (SMS), circuit-switched data, and packet switched data. The variations in the different technologies depend on the innovations and the air interfaces.

Regardless of the different methods, there are still many variations in using the data transmission characteristics with wireless applications and mobile users. In Table 23-3, a summary of the various methods is shown for the wireless communications techniques.

Possible Drawbacks with Wireless

Not everything with wireless data is perfect (same for wireline). Occasionally, there will be problems that must be overcome. The use of wired facilities has improved over the years with the use of fiber optics in the backbone. Newer technologies have allowed the industry to improve performance of data to a 10^{-15}–10^{-16} bit error rate. This was unheard of before the use of fiber and SONET in the backbone. The local loop (last mile) is the weakest link in the equation, producing bit error rates of 10^{-6}. Still, today

Table 23-3	Method	Air Interface
Summary of methods for data	Circuit-switched	TDMA
		CDMA
		GSM
	Packet Switched	RIM
		RAM Mobile Data
		CDPD
		IP
	SMS	Packet Switched, separate

this is not as bad as it may sound because the distances we run on copper (the local loop) are being shortened daily. The shorter the copper, the better the performance we achieve (because the weakest link is minimized, and the cables are correctable). Therefore, we see improved data performances on the local copper-data transmission systems. However, wireless (air interfaces) have been traditionally error prone, and the amount of frequencies available have always been limited. This interface is limited to the point that many people did not want to use air interfaces in the past. With the culture shift and the use of digital techniques to compound the data, air became much more acceptable.

The wireless medium is also prone to more delay and latency than the wired world. In the wired arena, the average delay for transmitting information across the nation networks is 50 msec or less. In the wired world, this delay can jump to more than 250 msec. At 250 msec, we find that echo begins to get out of control, requiring more equipment to handle this problem. However, while handling the echo control function, we introduce more latency and buffering of the real data. This almost becomes a "Catch 22" problem. The more the data is buffered and manipulated across the medium, the greater the risks of introducing errors.

Pros and Cons to Wireless

Some of the inherent problems bring other solutions and benefits to the table. For example, when using a wireless interface, the user is mobile and can go almost anywhere at whim. This allows more flexibility. However, the downside is that the use of cellular and PCS systems causes data and call handover to occur, which can be detrimental to the overall performance. Security issues cannot be ignored with data in the air. On a copper cable, data is slightly more secure because the medium is a little more difficult. The use of an interceptor in the airwaves is much simpler. Digital cellular and PCS make the data more secure than analog, but one still has to be concerned with putting anything in the airwaves.

From the perspective of financial information, the wireless data can be very expensive. In the circuit-switched arena, just as in voice, we pay the carrier a flat rate for a guaranteed amount of usage. The price is fixed whether you use the minutes or not. If you exceed the minutes of usage, you pay a premium added cost for the overage usage. Some plans do allow you to pay for only what you use; however, these plans tend to be at a higher cost per minute for all the minutes of usage. Therefore, it becomes difficult to assess the best deal, depending on variable usage. Packet switched data, such as CDPD and RAM Mobile data services, can be less expensive when small amounts of data are transferred. Yet, if the user transmits large and lengthy files, the cost of packet switched data can be as much as 500–100% higher than circuit-switched data. The variables are still less attractive to use the wireless data.

Other areas where differences exist are in the devices themselves. When dealing with a radio interface (like a cell phone), we have to be aware of the battery life of the device. Power consumption with lengthy data transfers can be critical. The use of the overall battery life is contingent upon the technology used, but current industry standards allow for 2–4 hours of talk time on a portable device (transmission time for data). Anything over that is prone to cut off and produce errors. The digital sets are better equipped to handle the data transfers, but dual mode phones can be problematic when they are in analog mode.

Satellite Communications Networking

Introduction

Satellite communications have developed through the years. The first use of satellite technology was in the military for voice communications in the early 1960s. Despite the advancements in the technology, commercial providers are prohibited from constructing and launching satellites at will. Despite these limitations, commercial satellites provide valuable information on meteorology, agriculture, forestry, geology, environmental science, and other areas.

Uses of Satellites in Agriculture

Many countries use satellite technology to improve agriculture. Satellites are used to determine productive from nonproductive yields. In many cases, they use other services such as infrared vision. Satellite-based navigation, such as *Global Positioning System* (GPS), a military invention, also assists in the process of reviewing agricultural production. Farmers can keep track of what they plant, fertilize, and spray. A farmer can now account for different kinds of soil, varying in acidity, organic content, and nitrogen levels. In addition, differences in drainage and treatment can be determined. All this improves yield and prevents the waste of valuable chemicals. This practice is well established in America. More than a dozen companies sell equipment used in agriculture.

Uses of Satellites in Oceanography

Satellites have found a niche in oceanography. In regional and local seas, satellites provide remote sensing, which gives a unique approach to monitoring the environment for the management of living resources, fishing, navigation, and coastal erosion. All maritime nations now have access to valuable information needed to manage the ocean boundaries that compromise their economic zone.

Countries use satellites in many adaptations. From military applications, agriculture and oceanography, satellites have improved the overall performance of the individual country. Each of the applications listed, however, uses significant amounts of data to perform and monitor these functions.

Commercial Providers

NASA has been working with the United States commercial satellite industry to achieve seamless interoperability of satellite and terrestrial telecommunications networks. The high latency and noise characteristics of satellite links cause inefficiencies when using *Transmission Control Protocol* (TCP) in an environment for which it was not designed. These problems are not limited to airborne communications and may appear in other high latency and noisy networks. Considerations, such as higher data rates and more users on data networks, can affect future terrestrial networks similar to those posed by satellites today.

In the future, telecommunications networks with space stations, lunar stations, and planetary stations with their inherent latency, noise, and bandwidth characteristics need to be addressed. For now, the emphasis is on earth-orbiting communications satellites.

History of Satellites

The regular use of satellite communications flourished in the 1960s. In April 1960, Tiros I was sent out to space. Tiros was the first weather satellite that sent pictures of clouds to earth. In August 1960, the United States launched Echo I, a satellite that made satellite communication possible by reflecting radio signals to earth. By 1965, more than 100 satellites were being launched into orbit each year. The 1970s produced innovation in the satellite world. Newer and more efficient instruments were developed. These innovations used computers and miniature electronics in designing and constructing satellites. During the 1980s, satellites were used to save people and other satellites. The uses of satellites in the 1990s rapidly grew for common, everyday tasks. TRW Inc. announced a plan to create a satellite system that would dominate satellite communications. This system called Odyssey would be used for the telephone business. TRW's satellites would focus on populated areas and would cover the earth uniformly. Based on current development, the satellite industry will be productive for decades to come.

How Do Satellites Work?

The basic elements of a satellite communications system are shown in Figure 24-1. The process begins at an earth station—an installation designed to transmit and receive signals from a satellite in orbit around the earth. Earth stations send information via high-powered, high-frequency (GHz range) signals to satellites, which receive and retransmit the signals back to earth where they are received by other earth stations in the coverage area of the satellite. The area that receives a signal of useful strength from the satellite is known as the satellite's footprint. The transmission system from the earth station to the satellite is called the uplink, and the system from the satellite to the earth station is called the downlink.

Figure 24-1
Satellite communication basics

Command & Data   ◯ **Pointing Control**   ● **Communications**

● **Power Supply**   ● **Mission Payload**   ◗ **Thermal Control**

Source: NASA

Satellite Frequency Bands

The three most commonly used satellite frequency bands are the C-band, Ku-band, and Ka-band. C-band and Ku-band are the two most common frequency spectrums used by today's satellites. To help understand the relationship between antenna diameter and transmission frequency, it is important to note that there is an inverse relationship between frequency and wavelength—when frequency increases, wavelength decreases. As wavelength increases, larger antennas are required to receive the signal.

C-band satellite transmissions occupy the 4 to 8 GHz frequency ranges. These relatively low frequencies translate to larger wavelengths than Ku-band or Ka-band. These larger wavelengths of the C-band mean that a larger satellite antenna is required to gather the minimum signal strength. The minimum size of an average C-band antenna is approximately 2–3 meters in diameter as shown in Figure 24-2.

Ku-band satellite transmissions occupy the 11 to 17 GHz frequency ranges. These relatively high frequency transmissions correspond to shorter wavelengths, and therefore a smaller antenna can be used to receive the minimum signal strength. Ku-band antennas can be as small as 18 inches in diameter. Figure 24-3 shows the Ku-band antenna.

Ka-band satellite transmissions occupy the 20 to 30 GHz frequency range. These very, high frequency transmissions mean very small wavelengths and very small diameter-receiving antennas.

Figure 24-2
C-band satellite
antenna

Figure 24-3
Ku-band satellite
antenna

Geosynchronous Earth Orbit (GEO) Satellites

Today, the overwhelming majority of satellites in orbit around the earth are positioned at a point 22,300 miles above the earth's equator in *Geosynchronous Earth Orbit* (GEO). At a distance of 22,300 miles, a satellite can maintain an orbit with a period of rotation around the earth exactly equal to 24 hours (see Figure 24-4). Since the satellites revolve at the same rotational speed of the earth, they appear stationary from the earth's surface. That's why most earth station antennas (satellite dishes) don't need to move after they have been properly aimed at a target satellite in the sky.

Figure 24-4
Geosynchronous
Orbit

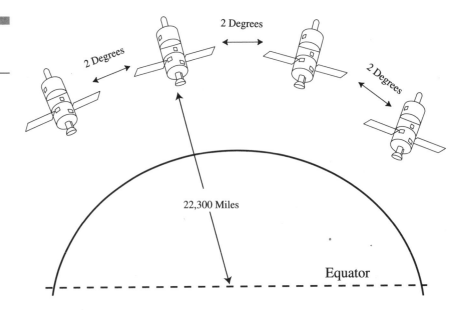

Medium Earth Orbit (MEO) Satellites

During the last few years, technological innovations in space communications have given rise to new orbits and totally new systems designs. *Medium Earth Orbit* (MEO) satellite networks have been proposed that will orbit at distances of about 8,000 miles. Signals transmitted from a MEO satellite travel a shorter distance, which translates to improved signal strength at the receiving end. This means that smaller, lighter receiving terminals can be used. Also, because the signal is traveling a shorter distance to and from the satellite, there is less transmission delay. Transmission delay is the time it takes for a signal to travel up to a satellite and back down to a receiving station. For real-time communications, the shorter the transmission delays, the better. For example, a GEO satellite requires .25 seconds for a round trip. A MEO satellite requires less than .1 seconds to complete the job. MEO operate in the 2 GHz and above frequency range.

Low Earth Orbit (LEO) Satellites

Proposed LEO satellites are divided into three categories: little LEOs, big LEOs, and Mega-LEOs. LEOs will orbit at a distance of only 500 to 1,000 miles above the earth. This relatively short distance reduces transmission delay to only .05 seconds and further reduces the need for sensitive and bulky receiving equipment. Little LEOs will operate in the 800 MHz (.8 GHz) range, big LEOs will operate in the 2 GHz or above range, and Mega-LEOs will operate in the 20–30 GHz range. The higher frequencies associated with Mega-LEOs translates into more information carrying capacity and the capability of real-time, low delay video transmission. Microsoft Corporation and McCaw Cellular (now known as AT&T Wireless Services) have partnered to deploy 288 satellites to form Teledesic, a proposed Mega-LEO satellite network. Speeds are expected at 64 Mbps downlink and 2 Mbps uplink.

Orbital Slots

With more than 200 satellites in Geosynchronous Orbit, how do we keep them from running into each other or from attempting to use the same location in space? To tackle that problem, international regulatory bodies like the ITU and national government organizations like the FCC designate the locations on the Geosynchronous Orbit, where communications satellites can be located. These locations are specified in degrees of longitude and are known as orbital slots. In response to the huge demand for orbital slots, the FCC and ITU have progressively reduced the required spacing down to only 2 degrees for C-band and Ku-band satellites.

Communications

The communications subsystem of a satellite is essential to the function of all satellites. There are many different components used in spaceborne communications' subsystems including the following:

- Special antennas
- Receivers
- Transmitters

All components must be highly reliable and low weight. Most satellites are fitted with beacons or transponders, which help with easy ground tracking. Global receive and transmit horns receive and transmit signals over wide areas on earth. Many satellites and ground stations have radio dishes that transmit and receive signals to communicate. The curved dishes reflect outgoing signals from the central horn and reflect incoming signals. A satellite's ability to receive signals is also necessary in order to trigger the return of data or to correct a malfunction if possible.

Satellite Installations

In Table 24-1, a list of the satellite installations by country is shown. The list shows the main providers and users of satellite transmission systems.

The technological and regulatory hurdles to create true high-speed satellite networks are fast becoming past tense. Low- and mid-bandwidth systems such as Motorola's Iridium and Hughes' DirecPC can handle some of the needs immediately. These systems are nothing compared to the promise of 2 Mbps, 20 Mbps, and even 155 Mbps streaming down from the sky. All that is necessary is a small antenna, a black box, and a service provider.

Table 24-1

Number of satellites in orbit by country around the world

Country	In Orbit
Argentina	1
Australia	6
Brazil	6
Bulgaria	1
Canada	16
China	15
CIS *(Former Soviet Union)*	1,322
Czechoslovakia	1
France	24
Germany	15
Hong Kong	1
India	11
Indonesia	6
Israel	1
Italy	4
Japan	55
Korea (South)	2
Luxembourg	4
Mexico	4
Portugal	1
Spain	3
Sweden	4
Thailand	2
Turkey	1
United Kingdom	18
United States	658
Arab States	3
Europe	27
International	51
NATO	8
TOTAL	2,271

This will approximate the way we buy service from an ISP today.

Is it ready for prime time yet? Not yet! Iridium's universal telephone didn't kill the cellular telephone. So broadband satellite systems won't kill terrestrial lines. Broadband satellite creators agree that broadband satellite systems will complement terrestrial networks. These satellites will provide high-speed service where terrestrial infrastructure does not exist. However, high-speed, low-cost landlines are here to stay.

Is there an application for the high-speed satellite networks? What makes them different from each other? Each of the main systems is very different. Some of the most visible ones may prove the most difficult to implement. Some of the most staid-looking systems may beat every other system to the punch.

Satellite communications are nothing new. For years, you could hook up a *Very Small Aperture Terminal* (VSAT) system and buy time on a satellite. The VSAT can deliver up to 24 Mbps in a point-to-multipoint link (for example, a multicast) and up to 1.5Mbps in a point-to-point link. These systems are fine for predictable and quantifiable data transmission, but if you need to conduct business on the fly, they can be a problem. New techniques are required to handle the on-demand usage. Primary among them are more tightly focused beams and digital signal technology, which together can increase frequency reuse (and thereby increase bandwidth) and reduce dish size from meters to centimeters. Accordingly, you also need a large unused chunk of the electromagnetic spectrum.

In 1993, NASA launched its *Advanced Communication Technology Satellite,* (ACTS). ACTS is an all-digital, Ka-band (20–30 GHz), spot-beam, *Geosynchronous Earth Orbit* (GEO) satellite system capable of delivering hundreds of megabits per second of bandwidth. NASA showed that such a system could work. The FCC has since granted orbital locations and Ka-band licenses to 13 companies including the following:

- EchoStar

- Hughes

- Loral

- Motorola

- Ka-Star

- NetSat 28

- PanAmSat

- Teledesic

All these companies aim to bring us bandwidth up to 155 Mbps to our home and office. These broadband systems are not going online before 2000.

What will we do with this speed and capacity? Anything you want! Whatever landlines can do, the satellite systems of the new millennium will also do. In Table 24-2, the list of applications covered by broadband satellite communications is shown. These are representative applications; others will be developed.

Most of the market that needs data services seems to be well served by landlines. So, why is the emphasis on the use of airborne technology? An obvious market is in places that have underdeveloped communications infrastructures. In some countries, stringing copper or fiber is out of the question because of the initial cost and the terrain where it is needed. Still, a wireless telephone has some merit. You don't need a broadband satellite to make telephone calls; Iridium and other low-earth orbit systems will likely serve that market.

So who *does* need this new class of broadband satellite communications? The first answer is the same: multinational corporations. The main problem that satellite systems solve is getting high-bandwidth access to places without a high-bandwidth infrastructure. It's unlikely that a satellite system could compete with *Digital Subscriber Line* (xDSL) to the home or fiber to the office.

Table 24-2

Applications for high-speed satellite communications

Applications
Desk-to-desk communications
Videoconferencing
High-speed Internet access
e-mail
Digital and color fax
Telemedicine
Direct TV and video
Transaction processing
Interactive distance learning
News and financial information
Teleradiology

LEO versus GEO

However, bandwidth is only half the story. The other half is latency. It's easy to talk about high-bandwidth satellite systems, but that technology has existed in VSATs for years. GEO satellites located at 22,300 miles above the equator induce 250 milliseconds of round trip delay. With that kind of latency built into the system, (not counting latency added by the various gateways and other translations), a telephone conversation is annoying. Any interactive data intense application has to be nonlatency-sensitive. Online transaction processing will have a problem using a GEO satellite system.

Moving the satellites closer to earth will help significantly. That's just what systems such as Teledesic[1], Skybridge[2], and Celestri[1] will do. With *Low Earth Orbits* (LEOs) under 1,000 miles, these systems reduce latency to .1 second. While GEOs are a well-known technology, LEOs are new. The biggest problem is that you need a lot of them to get global coverage. At one point, Teledesic planned a constellation of more than 842 satellites[3]. Until recently, the concept of launching dozens or hundreds of multimillion-dollar satellites was a pipe dream. Each of Teledesic's 288 satellites is estimated to cost $20 million apiece.

Price is only one issue. Finding a company to launch all these satellites poses another obstacle. Teledesic set an 18-month to two-year launch window to get its 288 satellites airborne. LEO system planners are talking about putting more satellites into orbit in the next five years than the world has put into orbit over the past 40 years. Once the LEO satellites are in orbit, there's an entirely new set of problems. There's the matter of space junk.

Niches in the GEO Sphere

LEOs will be good for high-speed networking, teleconferencing, Telemedicine, and interactive applications. GEOs will be better for information downloading and video distribution, such as broadcasting and multicasting. We're able to use GEO satellites to transport at least 24Mbps of broadcast

---

[1] Teledesic is a joint venture between Microsoft and Mc Caw Cellular.

[2] Skybridge is a venture of Alcatel of Belgium.

[3] Teledesic dropped the number of required LEO satellites to 288 since the first announcement.

IP data and over 2Mbps of point-to-point TCP/IP data. The latter uses technologies such as TCP spoofing. Several vendors have been using this technique for years to deliver Internet and Intranet content at high speed. Ground terminals can use similar TCP spoofing technologies. Nevertheless, there's still the 250-millisecond delay that you just can't get around. Any lossless protocol is going to have problems with this latency. Even if TCP spoofing works, given TCP's 64-Kb buffer makes this somewhat suspect, there's the matter of other protocols such as IBM's SNA and other real-time protocols that are designed around landline performance.

LEO Meets GEO

Motorola's Celestri plans an initial LEO constellation of 63 satellites coupled with one GEO satellite over the United States. The LEO constellation and the GEO satellites will be able to communicate directly through a satellite-to-satellite network.

The hybrid configuration will enable Celestri to take advantage of LEO's shorter delays for interactive uses and GEO's power in the broadcast arena.

Space Security Unit

Once you get beyond the latency and bandwidth issues, there is another challenge: security. If data is being packaged up and broadcast into space, anybody with a scanner can just tune in. The air interface technologies that these systems use will make it more difficult for anyone to eavesdrop. Combined systems will use *Code Division Multiple Access* (CDMA), *Time Division Multiple Access* (TDMA), *Frequency Division Multiple Access* (FDMA), and a bunch of other xDMA protocols. On top of that, many of the networks will offer some internal security encryption system.

The Market for the Network

The total global telecommunications market is about $750 billion, and that's going to double in 10 years, chiefly due to expanded use of data communications. We'll use whatever service we have available to meet that demand: Fiber, ATM, *Synchronous Optical Network* (SONET), xDSL, Gigabit Ethernet, cable modems, satellites, and probably a few that haven't even been thought of yet.

All telecommunications systems will compete on availability, price, and speed. That means there are going to be two big winners: whoever gets its broadband service to consumers first, and whoever can offer the most bandwidth with at least reasonable latency. A sample of the techniques and the services being offered in the 2000 timeframe are shown in Table 24-3.

With the discussion of the various bands that these services will operate in, the bands are continually referred to. As a means of showing the bands and the frequency they fall into, Table 24-4 is a summary of some of the bands used in the satellite arena.

Table 24-3

Sampling of services planned

Service Offering	Astrolink	Teledesic	Celestri
Companies (parties) Involved	Lockheed	Bill Gates, Craig McCaw, Boeing	Motorola
Service Offerings	Data, video, rural telephony	Voice, data, video- conferencing	Voice, data, video-conferencing
Satellite Orbit	22,300	435	875 and 22,300
Satellite Band	Ka	Ka	Ka and also 40–50 GHz
Dish Size	33–47 inches	10 inches	24 inches
Bandwidth	Up to 9.6 Mbps	16 Kbps−64 Mbps or up to 2.048 Mbps	Up to 155 Mbps transmit and receive
Available (approx.)	Late 2000	2002	2002
Number Satellites	9 GEOs	288 LEOs	63 LEOs, up to 9 GEOs
Access Technologies	FDMA, TDMA	MF-TDMA, ATDMA	FDMA, TDMA
Intersatellite Communications	Yes	Yes	Yes

Table 24-4

Sample of the frequency bands

Band	Frequency Ranges Used
HF-band	1.8–30 MHz
VHF-band	50–146 MHz
P-band	0.230–1.000 GHz
UHF-band	0.430–1.300 GHz
L-band	1.530–2.700 GHz
FCC Digital Radio and PCS	2.310–2. 360 GHz
S-band	2.700–3.500 GHz
C-band	Downlink: 3.700–4.200 GHz
	Uplink: 5.925–6.425 GHz
X-band	Downlink: 7.250–7.745 GHz
	Uplink: 7.900–8.395 GHz
Ku-band (Europe)	Downlink: FSS: 10.700–11.700 GHz
	DBS: 11.700–12.500 GHz
	Telecom: 12.500–12.750 GHz
	Uplink: FSS and Telecom: 14.000–14.800 GHz;
	DBS: 17.300–18.100 GHz
Ku-band (America)	Downlink: FSS: 11.700–12.200 GHz
	DBS: 12.200–12.700 GHz
	Uplink: FSS: 14.000–14.500 GHz
	DBS: 17.300–17.800 GHz
Ka-band	Roughly 18–31 GHz

Table 24-5 shows a sample of the satellite technologies that use the various bands and the size of the terminal and dish in satellite technologies.

Table 24-5

Summary of bands and types of terminals used

System type	Frequency bands	Applications	Terminal type/size	Examples
Fixed satellite service	C and Ku	Video delivery, VSAT, news gathering, telephony	1-meter and larger fixed earth station	Hughes Galaxy, Intelsat
Direct broadcast satellite	Ku	Direct-to-home video/audio	0.3–0.6-meter fixed earth station	DirecTV, Echostar, USSB
Mobile satellite (GEO)	L and S	Voice and low-speed data to mobile terminals	Laptop computer/ antenna-mounted but mobile	Inmarsat, AMSC
Big LEO	L and S	Cellular telephony, data, paging	Cellular phone and pagers; fixed phone booth	Iridium, GlobalStar
Little LEO	P and below	Position location, tracking, messaging	6 Inches; omnidirectional	OrbComm
Broadband GEO	Ka and Ku	Internet access, voice, video, data	20-cm, fixed	Spaceway, Cyberstar, Astrolink
Broadband LEO	Ka and Ku	Internet access, voice, video, data, videoconferencing	Dual 20-cm tracking antennas, fixed	Teledesic, Skybridge, Celestri, Cyberstar

Satellite Characteristics

As already mentioned, satellite communications have three general characteristics that lead to interoperability problems with systems that have not been designed to accommodate them:

1. Delay (or latency)
2. Noise
3. Limited bandwidth

It is important to understand each of these in general terms and in more detail. Since the invention and installation of fiber optics, communications systems have been treated as ideal with very low latency, no noise, and nearly infinite bandwidth (the capacities are rapidly moving to Terabit bandwidth from current fiber systems). These characteristics still have room for doubt and development, but for satellite communications, the characteristics of fiber can make it very difficult to provide cost-effective interoperability with land-based systems. The latency problem is foremost among the three. Noise can be handled by the application of error control coding. Bandwidth efficiency is an important goal for satellite systems today, but will become increasingly important as the number of users increases as well as the data requirements.

Latency

We have seen that there is an inherent delay in the delivery of a message over a satellite link, due to the finite speed of light and the altitude of communications satellites. As we stated, there is approximately a 250-millisecond propagation delay in a GEO. These delays are for one ground station-to-satellite-to-ground station route (or "hop"). The round-trip propagation delay for a message and its reply would be a maximum of 500 milliseconds. The delay will be proportionately longer if the link includes multiple hops or if intersatellite links are used. As satellites become more complex and include on-board processing of signals, additional delay may be added.

Other orbits are possible including *Low Earth Orbit* (LEO) and *Medium Earth Orbit* (MEO). The advantage of GEO is that the satellite remains stationary over one point of the earth, allowing simple pointing of large antennas. The lower orbits require the use of constellations of satellites for constant coverage, and they are more likely to use intersatellite links, which result in variable path delay depending on routing through the network.

Noise

The strength of a radio signal falls in proportion to the square of the distance traveled. For a satellite link, this distance is very large and so the signal becomes very weak. This results in a low, signal-to-noise ratio. Typical bit error rates for a satellite link might be on the order of 10^{-7}. Noise becomes less of a problem as error control coding is used. Error performance equal to fiber is possible with proper error control coding.

Bandwidth

The radio spectrum is a finite resource, and there is only so much bandwidth available. Typical carrier frequencies for current point-to-point (fixed) satellite services 6/4 GHz (C band); 14/12 GHz (Ku band); and 30/20 GHz (Ka band). Traditional C and Ku-band transponder bandwidth is typically 36 MHz to accommodate one color television channel (or 1200 voice channels). One satellite may carry two dozen or more transponders.

New applications, such as personal communications services, may change this picture. Bandwidth is limited by nature, but the allocations for commercial communications are limited by international agreements so that the scarce resource can be used fairly. Applications that are not bandwidth efficient waste a valuable resource, especially in the case of satellites. Scaleable technologies will be more important in the future.

Advantages

Although satellites have certain disadvantages, they also have certain advantages over terrestrial communications systems. Satellites have a natural broadcast capability. Satellites are wireless, so they can reach geographically remote areas.

There are major changes occurring in the design of communications satellite systems. New bandwidth-on-demand services are being proposed for GEO. The Federal Communications Commission received seventeen filings for new Ka-band satellite systems last year.

TCP/IP Over Satellite

There has been a fair amount of press about TCP/IP, the networking protocol that the Internet and the *World Wide Web* (WWW) rely on, not working properly over satellite. The claims are that the TCP/IP reference implementation's 4K buffer size limits channel capacity and data throughput to only 64kbps. The argument is then used that the maximum buffer of 64K limits maximum throughput to 1Gbps. Therefore, the argument continues that GEO Ka-band satellite services are unsuitable for high-bandwidth applications, as the increased latency of a GEO connection decreases the available bandwidth.

TCP buffers are dimensioned as:

*bandwidth * delay* = buffer size

With a limited buffer size, a longer end-to-end delay decreases the space available to hold spare copies of unacknowledged data for retransmission. This limits the throughput on a lossless TCP connection.

However, this argument ignores the work done on larger buffer sizes for TCP in RFC1323, the "large windows" effort. Work to expand TCP beyond its original, 16-bit buffer space has been going on for several years. Moreover, the effort and result of "large windows" is supported by several versions of UNIX. The TCP buffer limit isn't as bad as it is made out to be. TCP copes with GEO delays quite nicely today. Individual, high-bandwidth GEO TCP links are possible with the right equipment and software. Military applications have been using TCP/IP in many GEO-based networks for years.

GEO links are suitable for seamless intermediate connections in a TCP circuit. There is nothing denying the use of many "small" TCP connections over a broadband link; GEO, fiber, and most broadband Internet connections contain a vast number of separate small pipes. The real issue with GEO vs. LEO is the acceptability of the physical delay for two-way, real-time applications such as telephony or videoconferencing. Even then, the physical delay of GEO is balanced somewhat by the increased switching times through a packet-based LEO network. Having multiple, narrow TCP pipes across satellites works well. Wide pipes with large buffer sizes can suffer from the higher *Bit Error Rate* (BER) of satellites.

For TCP, implementing Selective Acknowledgements (RFC2018) and fast recovery (RFC2001) also improves performance in the face of errors. There is much work in progress with the *"IP over Satellite"* and the *"TCP over Satellite"* working groups.

IP-over-satellite's primary advantage is that it can deliver large amounts of bandwidth to underdeveloped areas where fiber does not exist. IP-over-satellite is also subject to a number of adverse conditions that can significantly impact the effective throughput rates and network efficiency. Bit Error Rates, congestion, queue management, Window size, and buffer status can all have a serious impact on the overall IP-over-satellite performance curves.

Another problem is the asymmetry encountered in back channel bandwidth, which in many IP-over-satellite systems is only a fraction of the forward channel bandwidth. Because TCP emerged in advance of IP-over-satellite, TCP often gets mistakenly identified as the source of the problem. It is the operating environment and not TCP alone that is the problem. The design of TCP is not optimized for conditions encountered during satellite transmissions.

The standard GEO round trip hop of 44,600 miles creates problems with the inherent transmission-related delay, breaking down the flow of packets between the sender and receiver, especially at very high transmission speeds. TCP works well as a general-purpose protocol for a congested environment, such as the Internet. However, TCP perceives a satellite delay and bit error as congestion and inherently slows down.

Satellite and ATM

Work has also been done on ATM over Satellite. Standardization of ATM over Satellite is underway and several organizations are working in parallel. The ATM Forum's *Wireless-ATM* (WATM) group has done some work on ATM over satellite. The *Telecommunications Industries of America* (TIA) Communications and Interoperability group (TIA-TR34.1) works on interoperability specifications that facilitate ATM access and ATM network interconnect in both fixed and mobile satellite networks. TIA is collaborating with the ATM-Forum WATM group. The standard development for satellite ATM networks is provided in ATM Forum contribution 98–0828. Other groups working on ATM and Satellite transmission include the following:

- *NASA*   NASA is working on the interoperability of satellite networks with terrestrial networks. ATM over Satellite QoS results for Motion-JPEG is now publicly available.

- *European Space Research and Technology Center*   This group is also interested in ATM over Satellite research. They have carried out a broadband communications over satellite study.

- *COMSAT*   COMSAT is one of the major players in providing ATM over Satellite services.

- *Communications Research Center*   Communications Research Center Broadband Applications and Demonstration Laboratory is testing and demonstrating information highway applications by using ATM via satellite.

■ *RACE   Research and Development in Advanced Communications Technologies* in Europe has also funded various projects for ATM over Satellite. The results of the RACE research projects are published as *Common Functional Specifications* (CFS).

Charting the Rules for the Internet

Everyone knows about the Internet. Fewer people, however, understand how it works. Very few people know where the standards come from that shape how the Internet works. Those rules come from five groups, four of them loosely organized and consisting primarily of volunteers. For more details on the Internet, see Chapter 29 in this book.

The groups don't issue standards in the traditional sense, although their conclusions become the "regulations" by which the Internet operates worldwide. Instead, they agree on how things should be done through discussion. The result is a surprisingly effective, if sometimes convoluted, system for dealing with the fast-growing and often unpredictable nature of the Internet. Like the Internet itself, the groups that oversee it have made recent innovations that continue to evolve and restructure.

The basic rules governing the Internet come out of two task forces.

■ The *Internet Research Task Force* looks at long-term research problems.

■ The *Internet Engineering Task Force (IETF)* that concentrates on more immediate issues such as data transfer protocols and near-term architecture.

IETF is the oldest group dedicated to Internet issues, having held its first meeting in 1986. At the time, the nascent Internet was still primarily a federal research activity evolving out of ARPANET, a modest network developed by the U.S. *Defense Advanced Research Projects Agency*. The IETF's work is divided among eight functional areas related to such issues as network management, routing, security, applications, and so on.

The activities of IETF and the Internet Research Task Force are coordinated by the Internet Architecture Board. The board has no official role in operational matters and contributes in only a minor way to policy issues.

The Internet Architecture Board and two Internet task forces fall loosely under the umbrella of a fourth organization that is more formal called the Internet Society.

The Internet links thousands of networks and millions of users who communicate by using hundreds of different types of software. They are able to do so because of *Transmission Control Protocol / Internet Protocol* (TCP/IP), the general procedure for accurately exchanging packets of data adopted by ARPANET in the early 1980s. Other networks soon adopted TCP/IP as well, thus paving the way for the global Internet of today. TCP/IP has been updated several times. It continues to undergo modifications, including those designed to speed throughput, especially for long terrestrial and satellite links.

Tailoring IP Can Accelerate Throughput

Everyone who uses the Internet wants faster connections. The quest for speed has become a big marketing issue for terrestrial and satellite systems alike. However, no one system is inherently "best" for every application. Finding the most efficient way to connect is a matter of matching communication needs to the unique characteristics of each option. The hardware involved is an obvious factor in system speed. Nevertheless, the protocols used by computers to talk to each other ultimately control throughput. Internet protocols are constantly being revised, and faster versions are coming that promise quicker data delivery over links of all kinds. The work is coordinated primarily by the *Internet Engineering Task Force*. The IETF seeks continual improvement in the software that makes the Internet work, and it prefers solutions that benefit terrestrial as well as satellite network links.

The Internet works because all computers using it abide by the same rules, or protocols. The most fundamental of these rules is the *Internet Protocol,* (IP). Its primary function is to provide the "datagram" that carries the address to which a computer is sending a message.

Guaranteeing that messages are received accurately over the Internet is the primary job of the *Transmission Control Protocol* (TCP). TCP sends data in segments and waits for a confirmation message from the receiving computer before sending more. It reacts to traffic loads on the network, dynamically regulating the allowable rate of data transfer between the two computers in an attempt to maximize data flow without overwhelming relay points or the capacity of the receiving computer.

TCP facilitates sending large data files accurately over transmission routes that may be plagued with bit errors or other forms of poor link quality. It also reduces throughput when it doesn't receive acknowledgment as quickly as it expects. TCP assumes that any delays are due to traffic congestion within the network. It responds by cutting back on the transmission rate; it then slowly speeds up again as long as no further delays are detected. The idea behind its "slow start" mechanism is to match the output of the sending computer to the maximum throughput allowed by the network at any moment. For the most part, TCP works just fine. However, pushing the ultimate throughput capabilities of TCP has revealed shortcomings, especially on high-capacity links with high latency. The problem is that high latency slows TCP's slow-start mechanism dramatically, causing the protocol to restrict the amount of data that can be "in flight" between the sending and receiving computers and preventing the link from being fully utilized.

The problem affects any high-delay path, whether terrestrial or by satellite. For high bandwidth links, the result is a sharp drop in efficiency. For example, a message controlled by the most commonly used form of TCP can move through a cross-continent, OC−3 fiber link at only about 1.1 megabits per second, even though the line can handle 155 Mbps of throughput.

A packet of data can propagate across the United States and back through a dedicated copper wire or fiber in about 60 milliseconds. It takes eight to nine times as long—500 milliseconds for the same packet to make the round trip via satellite.

Individual *Low Earth Orbit* (LEO) spacecraft promise very short propagation delays—as little as 12 milliseconds from ground to satellite and back —because they fly closer to the ground. However, a single LEO spacecraft can't "see" from coast to coast, so several such birds are required to relay a packet across the United States. LEO birds also are in motion relative to the ground, so their packet latency varies continually, a characteristic that forces current forms of TCP to reassess the allowable throughput rate constantly. The net result is cross-country transmission delays for LEO systems akin to those of long terrestrial links.

Signal latency isn't the only factor in selecting an appropriate transmission medium, however. Each alternative has its own attractions and challenges. More importantly, continuing adjustments to TCP promise to improve performance for any high-latency network link.

Low Earth Orbit Satellites (LEOs)

Introduction

Quite a bit of discussion regarding wireless communications has already been covered in earlier chapters. However, a combination of two separate services and technologies are merging as new services for broadband communications. These two services include long haul communications and the use of personal communications services. The technologies include the use of satellite and the cellular concepts combined. Worldwide communications services can be achieved by these two combined services, therefore some diligent effort should be made to understand just what is happening in this arena. This truly brings home the concept of communications from anywhere to anywhere. The thought of being out in the middle of a lake and receiving a call, or rafting down a river and making a call, boggles the mind. This is especially true when thinking about some of the more rural areas in the world where no telephone service infrastructure exists today. Yet, in a matter of a few years these remote locations, on mountaintops, in forests, in valleys, or on the sea will all be reachable within a moment's notice. The infrastructure of a wired world will not easily lend itself to this need, due to timing and cost issues. Therefore, the use of a wireless transmission system is the obvious answer.

However, the use of cellular and personal communications devices still leaves a lot to be desired. First, the deployment of these services is always going to be in the major metropolitan areas, where the use and financial payback will be achieved. Thus, in the remote areas it will be decades, if not longer, until the deployment ever works its way close to the remote areas. Enter the ability to see the world from above the skyline! The industry decided to attempt servicing remote areas from a satellite capacity. This is not a new concept; the use of satellite transmission systems has been around for over thirty years. However, the application for an on-demand, dial-up satellite service is new. This will have to be a lucrative business venture, because the costs are still quite high. Look at the Iridium project that took over five years to build and launch, only to meet with lax reception from the marketplace. This drove Iridium Inc., into bankruptcy in its first year of operation.

There are still approximately one dozen suppliers competing for space segment and frequency allocation to offer voice, broadband data, paging, and determination services. In each case, the organizations have selected various approaches on how to launch their service offerings and the use of orbital capabilities to provide the service. In general, this discussion will cover the most widely discussed today called *Low Earth Orbit* (LEO). However, some added discussion on the use of *Mid-Earth Orbit* (MEO) and *Geosynchronous Orbit* (GEO) satellite-based communications, offerings, and applications must be addressed. As mentioned, there are over a dozen organizations applying for the rights to use various orbital slots and frequencies to provide service, in the various orbits as shown in Table 25-1. The licenses requested are being discussed at great length around the world as the future interoperable service.

The table reflects the fast pace that is being created since the obvious train of thought is that the first company in the business, regardless of the orbit used, will gain the market share. Unfortunately, this did quite not happen with the Iridium network, which opened for service in 1998–1999. This is an expensive situation if some other carrier gets there first and takes the market by force. Of course, there are still those who feel there is no need and that these systems will go bankrupt within the first few years.

Low-Earth Orbit

In December 1990, Motorola, filed an application with the FCC for the purposes of constructing, launching, and operating a low earth orbit, global mobile satellite system known as Iridium. This was the hot button that sparked the world into a frenzy. Iridium was a concept of launching a series

Table 25-1

A summary of the number of competitors and the various orbits being sought

Orbit	Number Of Competitors	Status
Low Earth Orbit (LEO)	8	Pending licenses granted based on very specific areas of coverage
Mid-Earth (MEO)	4	Experimental licenses Orbit granted for specific areas of coverage
Geosynchronous Orbit (GEO)	4	Licenses have already been issued for some, are experimental with others.

of 66 satellites[1] around the world to provide global coverage for a mobile communications service operating in the 1.610–1.6265 GHz frequency bands. The concept was to use a portable or mobile transceiver with low profile antennas to reach a constellation of 66 satellites. Each of the satellites would be interconnected to one another through a radio communications system as they traversed the globe at 413 nautical miles above the earth in multiple polar orbits[2]. This would provide a continuous line of sight coverage area from any point on the globe to virtually any other point on the globe, using a spot beam from the radio communications services on-board each of the satellites. The use of this spot beam concept, which had been discussed for years in the satellite industry, allowed for high frequency reuse capacities that had not been achieved before. Iridium wanted to provide the services outlined in Table 25-2. Motorola also suggested that an interconnection arrangement would be set up with all providers around the world through an arrangement with the local *Post Telephone and Telegraph* organizations (PTTs). The concept was sound, and the approach would have provided for the coverage that was lacking in the past to remote areas. In the table, the two columns are used as exclusive of each other. The services can be provided in any of the coverage areas regardless of which service is selected.

Each of the satellites was relatively small in terms of others that had been used. The electronics inside each were very sophisticated, and the use of gateway controllers on earth provided the command and control service for the administration of the overall network. With the Iridium network, cellular communications was expanded by providing coverage in areas where cell sites would not have been practical. This is summarized in Table

Table 25-2	**Type Of Service**	**Coverage Areas**
Services and coverage through the Iridium Network	Voice Communications	Air
	Data Communications	Land
	Paging	Water
	Radio Determination Services	

[1] Originally the Iridium proposal was for 77 satellites, but Motorola amended this number after the World Administrative Radio Council meeting in the spring of 1992.

[2] The original concept was to use seven polar orbits with eleven satellites in each. This would provide worldwide coverage, much similar to an orange slice concept.

25-3, which shows a feature comparison of the cellular concept versus the Iridium concept.

The areas of coverage in the cellular and PCS networks primarily served populated geographic areas. There were areas that were not financially or operationally prudent to install services. Consequently, very rural areas, mountain terrain, and the like did not receive coverage at all. Under the LEO concept, all areas not previously served could be easily accommodated. Iridium even provided coverage in the extreme rural areas, as well as in the recent Kosovo, Yugoslavia crisis. The services addressed by these satellites included the obvious, such as two-way voice but also other services that were not as evident. Table 25-4 is a summary of the actual services that were provided. This includes a minor description of the service and a potential target market. The list is straightforward in a single line description.

Table 25-3 Summary of cell sites vs. Iridium Cells	**Cellular Networks**	**Iridium Network**
	Sites are fixed	Sites are the moving targets
	Users move from site to site	User stays put, sites move the user from satellite to satellite
	Areas of coverage are 3–5 miles across	Areas of coverage are 185–1100± miles across
	Coverage sporadic, not totally ubiquitous	Worldwide coverage

Table 25-4 A summary of the initial features available on LEO Networks		
	Radio Determination Services (RDSS)	Will allow for the location of vehicle fleets, aircraft, marine vehicles, etc. RDSS will also be an integral locator service for all voice communications devices.
	Voice Communications (VC)	Will allow on demand, dial-up digital voice communications from anywhere in the world.
	Paging (P)	A one-way paging service. The paging service includes an alphanumeric display for up to two lines, but will expand to short messaging services (SMS).
	Facsimile (Fax)	A two-way facsimile service.
	Data Communications (DC)	An add-on device will allow the transmission of two-way data. This capability will also allow for two-way messaging (e-mail) service across the network.

In Figure 25-1, the concept of the LEO arrangement is shown. In this particular case, the satellites are traversing the earth's surface at a height of 400+ nautical miles above the earth, in a polar orbit. In the polar orbit, the satellite moves around the earth's poles and pass over any specific point along its path very quickly. The satellites move at approximately 7,400 meters ± per second in different orbits. Therefore, as one target site moves out of view, a new one comes into view at approximately the same time. A handoff will take place between the individual satellites (using the K_a band).

In Figure 25-2, the ground telemetry and control services are represented, called gateway feeder links. These also use spectrum in the K_a band. Iridium used approximately 16.5 MHz of bandwidth in the L band. The L band is also used from the handset to the satellite, whereas the K_a band is used from satellite to satellite communications, as shown in Figure 25-3. The use of this L band allows low-powered handsets to communicate within the 413 nautical mile distances with the satellites.

The market for the handsets and the add-on devices was supposed to be very lucrative. As one might expect, the handsets operated in a dual-mode

Figure 25-1
The LEO concept

420 N
miles

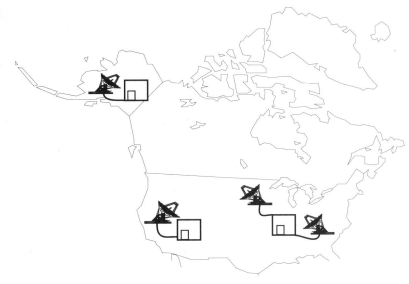

Figure 25-2
Ground station
telemetry and control

Figure 25-3
Satellite to satellite
communications is
handled on a K_a
band.

capability. Table 25-5 shows a summary of the spectrum and frequency bands requested by Iridium. This table shows the immediate request for bandwidth and the projected additions that were projected over time. This may well change with the financial problems now faced at Iridium. The issue here is the ability to create such a network, not to precisely pinpoint how it will look in five years.

Table 25-5

A summary of the bands and the bandwidth for Iridium

Band Requested	Immediate Bandwidth Needed	Future Additional Bandwidth
L-Band for set	16.5 MHz (operating in the 1.610–1.6265 GHz to cell communications range)	Up to 100 MHz in the L-band for the future
Ka Band for Gateway Feeder Links	200 MHz (100 MHz in the 18.8–20.2 GHz range for downlink; and 100 MHz in the 27.5–30.0 GHz range for uplink)	Possibly remain at the 200 MHz, or additional 100–200 MHz for the future
Ka Band for inter-satellite communications	200 MHz (all in the 22.55–23.55 GHz range for the intercommunications connection)	Possibly remain at the 200 MHz or additional 100–200 MHz for the future

So What Happened?

With the market ripe for a global communications system, and the lead in getting the service up and running, how did Iridium fail? One can only speculate, but the cost of launching Iridium was over $5 billion (US) and the project was nearly a year late. Moreover, during launch, several factors plagued Iridium (lost satellites, explosions, and failed launches), putting significant financial strain on the operating budget. When Iridium became available, the initial costs were touted as being $5–7 (US) per minute for an international call. In 1992 when the model was created, this price may have been realistic. However, in the time it took to get off the ground, the costs for dial-up voice communications plummeted. No one was willing to pay the prices asked by Iridium. A better price may have been one I predicted in 1995 stating that they will have to charge from $1–3 (US) per minute. That may have been one of the total downfalls for the Iridium network. Moreover, the handset price was too high, ranging upwards of $3,000 each. This does not mean that they cannot come back and reestablish themselves as the leader in the LEO industry, but there are many providers right on their heels (such as GlobalStar and Teledesic). These companies and their networks are aggressively chasing behind Iridium and plan to offer the broadband communications services initially when they open their doors. Iridium felt that they could do this later, possibly because of their lead in establishing the market niche.

The Benefits of These Service Offerings

Motorola established a list of benefits from the deployment of the Iridium network services, which at first glance may look biased toward their services. Upon further evaluation, these benefits can be derived from any network of this type. Therefore, generically these are addressed and kept in the context of any LEO network. The benefits lean toward the end user as shown. These are summarized as follows:

- Ubiquitous services: With continuous and global coverage, any-to-any connections can occur. As users travel either domestically or abroad, the service travels with them. It will eliminate the need for special access arrangements and special numbers that must be dialed. Users should never be out of range from their network. Remote areas with limited demand and finances now have the capability to connect anywhere in the world.

- Spectral efficiency: As already mentioned, the frequency reuse patterns for the bandwidth allocation will be significant. No other satellite system has achieved these reuse ratios. Iridium was first to claim this capability of efficiency. RDSS portion of the Iridium network is contained in the same spectral arrangement, freeing up 16.5 MHz of spectrum. This is a quantum leap in the efficient use of the spectrum.

- Public benefits due to flexible design: The digital technology deployed allows the total connection for all voice and data services on a seven day by twenty-four hour basis. This allows the flexibility of service provisioning. The low earth orbit overcomes some of the limitations of the higher transport systems, such as the delay in the round trip transmission. Because the satellites are low, the user set needs a lower power output device. This orbit has been selected to be the most flexible.

- The potential to save lives: It's common for the news media to publish stories of people stranded in remote areas with no life support systems who died because of their inability to communicate. The press today is filled with stories of cellular and PCS users notifying authorities of casualties. If only people in remote areas had a means of notifying authorities and/rescue parties, their lives could be spared in the event they get into a life-threatening situation.

■ Capabilities of the vendor: Motorola states that they are uniquely qualified to provide these types of services due to their background in the development and sales of other ancillary equipment that works in the wireless world. Specifically, they have been one of the major developers in the production, research, and development of private mobile-radio services.

■ LEO deployment promotes international communications: The LEO networks deliver modern digital-transmission services to remote areas of the world. The FCC and the United States government are attempting to use telecommunications as a strategic and economic tool to foster development in these areas. Their goals are to

1. Promote the free flow of information worldwide

2. Promote the development of innovative, efficient, and cost-effective international communications services that meet the needs of users in support of commerce and trade development

3. Continuous development and evolution of a communications service and network that can meet the needs of all nations, and specifically those of developing nations.

The above goals can be met with a mobile communications network such as that proposed in the LEO networks. An alternative is the *Global Services Mobile* (GSM) standard that is emerging throughout the international arena.

Deployment and Spacing of Satellites

Motorola's concept of seven different orbits on a polar orbital path would lend itself to the potential of collisions in mid-air if the spacing were not correct. Therefore, the spacing design had to take into account that the satellites were all traversing the same end-points where the paths would all cross. One can imagine that the two poles of the earth are the midpoints in the orbit. The Iridium network originally consisted of 77 satellites, however, a modification to the plans dropped the number of craft to 66. To assign or determine the best approach, Motorola took into account six critical criteria in determining the best approach. These are the following

1. The need to provide a single global coverage over the entire earth's surface for availability at all times. This means that each subscriber must have at least one satellite in view at all times to provide coverage.

2. Some portion of each orbit must be available to allow for low power outputs, which accommodates the recharging of the communications power subsystems. This minimizes the size requirements of the craft.

3. The relative spacing of the satellites and the line-of-sight relationships had to allow the on-board systems to control cross system linkage.

4. Costs for the entire constellation was a concern in the selection of the orbit and spacing requirements. Minimizing costs was a portion of the decision-making process.

5. The angle of incidence from the end user to the spacecraft as measured from the horizon to the line-of-sight communications process allowed link margins to accommodate the low-powered user device. The slant angle was selected at 10° to meet these criteria.

6. The final criteria was the operational latitude of the spacecraft. Systems operating over 600 nautical miles were affected more by radiation, which drives the cost up for the systems. Altitudes lower than 200 nautical miles required much more fuel and control over the positioning of the craft, thereby driving up the cost. At 413 nautical miles, the fuel consumption and command and control over the craft position coupled with the better radiation performance, the cost and coverage ratios provided a better solution.

The satellites were spaced at 32.7 degrees apart, traveling in the same basic direction and moving at approximately 16,700 miles per hour from north to south and 900 miles per hour westward over the equator. Given this path, each satellite was designed to circle the earth approximately every 100 minutes. At the equator, a single device would provide coverage. However, as the craft moved toward the poles, overlap occurred, increasing the levels of coverage above and below the equator. The expected life cycle of the satellites was around five years.

The basic building blocks of the Iridium system provided the model for other networks ready to launch (such as Globalstar systems). In each case, the systems used proven technology for radio transmission in well-established frequency bands. The basic system was composed of the following:

1. The space segment comprised of the constantly moving constellation of satellites in a low earth orbit

2. A gateway segment comprised of earth station facilities around the world

3. A centralized system control facility

4. The launch segment to place the craft in the appropriate orbit

5. A subscriber unit to provide the services to the end user

The Space Segment

The space segment consisted of small satellites, operating in the low earth orbit, which were all networked together as a switched digital communications system. In Figure 25-4, a sample outline of the beam coverage with Iridium is shown. Each satellite used up to 37 separate spot beams to form the cells on the earth's surface. Multiple relatively small beams allowed

Figure 25-4
The spot beam
pattern from Iridium

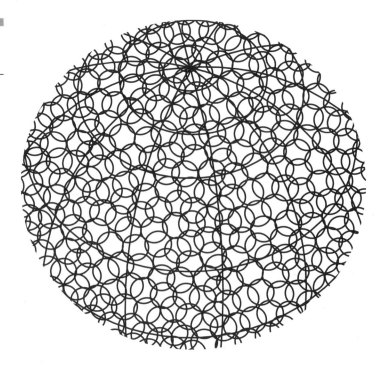

the use of higher satellite antenna gain and reduced RF power output. The system was designed to operate with up to 250 independent gateways, although in its initial deployment only between 5–20 gateways. Figure 25-5 is an overall view of the network communications through the gateways and the handsets.

The Cell Patterns

The satellites had the capability of projecting 37 spot beams on the earth. The spot beams formed a series of overlapping, hexagonal patterns that would be continuous. The center spot beam was surrounded by three outer rings of equally sized beams. The rings worked outward from the center beam in rings of 6, 12, and 18 spot beams. This pattern of a 37-cell pattern is shown in Figure 25-6, using the center beam as the starting point where the pattern can be seen. A spot beam was 372 nautical miles in diameter, and when combined they covered a circular area of approximately 2,200 nautical mile diameter. The average time a satellite was visible to a subscriber was approximately nine minutes. A seven-cell frequency reuse was formed to produce the actual pattern shown in Figure 25-7.

Figure 25-5
The overall system through the gateways

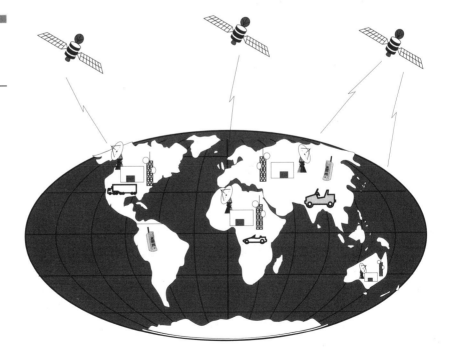

Figure 25-6
The 37-cell patterns

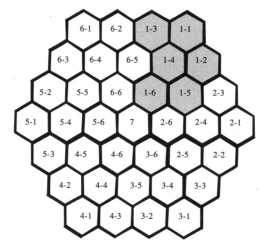

Figure 25-7
A seven-cell
frequency reuse
pattern

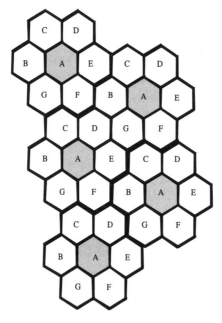

Traffic Carrying Capacity

A 14 slot TDMA format is used allowing each Iridium cell to be assigned an average of two time slots. The average traffic capacity with a utilization of 16.5 MHz is 174 full duplex voice channels per cell.

Modulation Techniques

The modulation process and multiple access capabilities of Iridium were modeled after the conventional terrestrial cellular networks, and particularly after the international GSM standard. Combining frequency and time division multiple access, the system also used a data or vocoder voice and digital modulation technique (i.e. QPSK, MSK, and so on). Each subscriber unit operated in a burst mode transmission by using a single carrier. The bursts were controlled to appear at the precise time necessary to be properly integrated into the TDMA frame.

The Gateway Segment

The gateway segment controlled user access to the network and provided interconnection to the *Public Switched Telephone Networks* (PSTN). Two initial gateways in the United States divided the eastern and western parts of the country. Two added sites will be located in eastern and western Canada to provide coverage throughout Canada and Alaska. Each of the gateways had the necessary terminal and switching equipment to support the operations.

The Earth Terminal

The key to the earth terminal function was the use of up-and-down link capabilities with three separate RF front-ends supporting continuous service and high up-time availability. One of the RF front-ends was used as an active up/downlink with one satellite. A second RF front-end was used to establish communications with the next active satellite. The third provided a backup capability in case of an equipment failure and provided for diversity against any of the atmospheric conditions that can plague a typical satellite communications system (i.e. sunspots, equinoxes, etc.), causing a degradation of service. As the satellites were the moving targets and the gateways were fixed, the antennas followed and tracked the nearest two satellites to them. The communications active channels were handed off from the current satellite to the next active one coming into view, while the first disappeared from view. This handoff was designed to be transparent from the user's perspective.

The Switching Equipment

Each gateway housed the necessary switching equipment to interface between the communications payload in the Ka band and the voice/data channels from the PSTN. The switching systems performed the following functions:

- Transferred common channel signaling information from the PSTN to the RF portion of the Iridium network
- Transferred line and address signaling information from the PSTN to establish circuit switched calls
- Supplied in-band tones and announcements to PSTN users calling onto the Iridium network with necessary progress tones and conditions
- Digitally switched PCM signals between channels derived from the terminal channels to the PSTN and provided channels to support the necessary in-band signaling capability for call control and progress.

Interconnecting to the PSTN

Voice connections were designed to be fully compatible with applicable ANSI T1 standards for the United States and the CCITT G and Q Recommendations (International T1/E1 standards) for digital transmission systems using either SS7 and R1 signaling. The data channel specifications were compatible with the OSI standards and with CCITT V and X series recommendations.

The System Control Portion

The *system control segment* (SCS) of the network provided the necessary control over the entire satellite constellation. The SCS managed and controlled all the system elements, providing service on both the short- and long-term picture. The functionality of the SCS was divided into either active control of the satellites and control over the communications assets within the satellites. The SCS monitored and managed the overall network control sequencing, traffic load balancing, and peak conditions, as necessary.

Other Competitors to Iridium

As already mentioned earlier, several other suppliers have all the rights and licenses to launch similar communications services, using constellations of Low Earth Orbits, either Mid Earth Orbits or Geostationay Orbits. This is the personal communications service of the future. The companies (at least seven others in the United States) are all vying for the licenses to use the L band communications bandwidth to provide global communications. These include some of the other big names in the business such as Qualcomm, Loral, TRW, Microsoft, and McCaw Communications. Each of the companies has a different approach to use, but they all share a common goal. Get there soon and offer high-speed digital communications with Internet access, e-mail, voice over, IP, and demand video services.

LORAL-QUALCOMM

Loral-Qualcomm's (located in Palo Alto, California) approach is to use approximately 48 satellites in a big LEO arrangement similar to Iridium, but with less craft. Their network handle is Globalstar systems. Qualcomm expects to use its expertise in the specialized mobile radio business and a *Code Division Multiple Access* (CDMA) arrangement that they pioneered. This service is currently two years behind schedule, but in the process of launch with expected availability in 2000. Loral-Qualcomm will draw from its experience in the business by deploying their CDMA technology and attempt to reuse the frequency spectrum as much as possible. Many of the initial service offerings from this company will include fleet tracking, locator services, and ultimately the extension of voice and data communications. The uniqueness of the Loral-Qualcomm services is that the system is designed to fully complement the cellular and PCS industries by providing single service coverage from a single telephone to a single number, through an intelligent network. Loral-Qualcomm did not plan to use a dual mode operating set, but a single set that can interface to any network service without having the user make a decision. Globalstar systems constellation is shown in Figure 25-8. These folks stand to be a big contender in the LEO business.

Figure 25-8
Globalstar systems
constellation

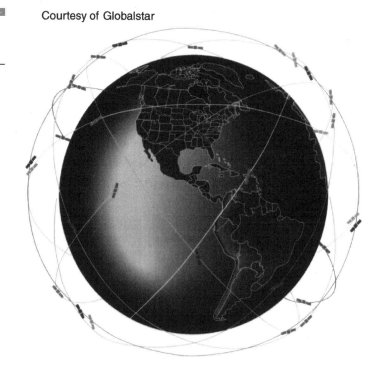

26

The T Carrier Systems (T1/T2 and T3)

Introduction

Meeting the high demands for voice, data, and multimedia communications, the larger corporations in the world are moving to the high-speed digital communications of the T Carrier system. This includes many of the copper and optical fiber-based systems. The hottest installation method today is still the T1. Although T1 has been around since 1958 when it was first created, and then rolled out into the carrier communities in 1960, one would think this technique became old. Yet, in 1998 many of the long distance carriers announced that they were experiencing a shortage of T1 ports, and installation times moved from 6 days *after receipt of order* (ARO) to approximately 120–150 days ARO. This supports the statement that T1 is still the hottest method of connecting some of the services discussed in other chapters, such as:

- Frame Relay
- ATM
- Point-to-point data lines
- Internet access
- Signaling systems
- Computer-to-Telephony Integration
- ISDN
- xDSL (HDSL, CDSL)

Why has this become such a popular method? The implementation of digital high-speed communications can be directly linked to the cost and convenience factors. T1 became a very price competitive access method for nearly all the communications services needed.

Yet, there has been other activity in the form of T3 or DS3 services to support the higher demands for the various multimedia and video applications. In both cases, T1 and T3 are now very popular in the commercial industry, as a means of providing the connectivity desired to meet the continuing communications development within organizations. The pricing models now make these services more readily affordable. Moreover, the enhancements to the digital circuits make this a tried and true technology. For example, today's tariffs make both T1 and T3 very price efficient and attractive. Even if a customer does not have enough volume to justify the use of a full T1 or T3, the prices allow the consumer to use portions of the services while holding excess capacity for new applications, without causing undue budgetary hardships. A T3, for example, costs the same as between 5–8 T1s terminated at the same location. This is beyond a financial decision and moves us to think of these services as application supportive.

Once a consumer or organization makes the "leap" into the digital circuit, the availability of bandwidth to host other services that were heretofore unavailable becomes paramount. Newer graphics and video services can be turned on, as needed, if the bandwidth is no longer the constraint.

This chapter will discuss the way the multiplexing and access of the digital services is facilitated. There are many other excellent publications that go deeper into the technologies. The reader can choose to do further study if so desired.

The Difference Between T-x and DS-x

In many cases, the terms become problematic because we tend to intersperse them. Therefore, when discussing the different services, it becomes imperative to discuss the difference between a T-x and a DS-x.

T-x (such as T1 and T3) refers to the services acquired from a carrier or local provider by physical layers one and two. The T carrier is a physical set of wires, repeaters, connectors, plugs and jacks, etc. When referring to a T1, we are actually describing the physical layer interface to the provider networks. A T1, for example, is a four-wire circuit (unless one of the HDSL or SDSL techniques is used with only one pair of wires). The four-wire circuit is installed between the local provider and the customer's premises, as shown in Figure 26-1. In this example, the provider will install the physical wires, or use four wires from a 50 pair bundle on the local loop. From the Central Office, the four wires will be cleaned up to remove splices, bridge taps, and load coils from the wires. At approximately 3,000 feet from the egress point at the CO, the provider will install a digital repeater (or regenerator, as it is also called). Thereafter, every 5,000–6,000 feet, another repeater is required until the last leg of the circuit. At approximately 3,000 feet from the customer's entrance point, the last repeater is installed. The provider then terminates the circuit at the demarcation point in a jack. The customer uses a plug to connect the CPE to the circuit. These are all mechanical and electrical devices allowing the installation of the T carrier.

Figure 26-1

A typical T1 carrier
installation

After the circuit is installed and the carrier is terminated, the customer then generates traffic (voice, data, video, etc.) across the circuit. This traffic is carried on the carrier at a digital rate, or what is called the *Digital Signal* level 1 (DS-1). DS-1 operates at a Digital Signaling rate of 1.544 Mbps. The traffic is therefore called a DS-1. (Note in many cases this may be annotated as DS1.)

One can therefore see that there is a difference between the physical carrier and the signaling rate of the traffic carried on the circuit.

DS-1 Framing Review

DS-1 operates in a framed format (some cases exist where a customer may request a nonframed format, but this is rare). The framer establishes the frame in a device called a *Channel Service Unit* (CSU). A frame consists of 193 bits of information created in a 125 μseconds. In a DS-1 there are 24 time slots associated with the signal. Each time slot carries 8 bits of user information. When the CSU sets up the frame and has 24 slots of 8 bits each, it then adds one overhead bit for framing delimitation. This creates the 193 bits referred to above.

24 channels   @ 8 bits per channel + 1 framing bit = 1 frame
24   × 8   + 1   = 193 bits/frame

Continuing the math, the 193 bit frames are generated 8,000 times per second:

193 bits/frame ×   8000 frames/sec.   = 1.544 Mbps

This framing and formatting is shown in Figure 26-2.

Figure 26-2
A frame of
information occurs
8,000 times/second

The frames of information are serially transmitted across a four-wire circuit, and operate full duplex. This gives the end user a 1.544 Mbps in each direction simultaneously.

Pulse Coded Modulation (PCM)

Before going any further into the DS-1, it may be appropriate to review the modulation technique used to create the digital signal. When DS-1 was first created, it was designed around converting analog voice communications into digital voice communications. To do that, voice characteristics were analyzed. What the developers learned was that voice operates in a telephony world in a band-limited channel operation. The normal voice will produce both amplitude and a frequency change ranging from 100 to approximately 5,000 times a second. These amplitude and frequency shifts address normal voices. However, the telephone companies decided, long ago, that carrying true voice would be too expensive and not provide any real added value to the conversation. They then determined that the normal conversation from a human actually carries the bulk of the information when the frequency and amplitude shifts vary between 100 and 3,300 times per second. Armed with this information, the developer determined that reasonable and understandable voice could be handled when carried across a band-limited channel operating at 3,000 cycles of information per second (what was termed as a 3 kHz channel). Taking all the electromagnetic spectrum available to them, the developers then channelized the frequency spectrum (in *Radio Frequencies* [RF]) and in electrical spectrum available on the cabling systems they had) to smaller capacities. The norm was set at 4 kHz channels. This is the foundation of the voice telephone network.

From there, the providers of the infrastructure (the telephone companies) placed bandpass filters on the facilities to limit the amount of electrical information that could pass across their wires (or any other communications facility). Using a standard 4 kHz channel, they limited the bandpass to no more than 3 kHz (see Figure 26-3).

Following upon this thought, the developers then wanted to convert the analog communications to a digital format. Here they studied the way the user modulates voice conversation. Now they had something to work with. The voice modulated at a normal rate operates in the 3 kHz range, but now must be converted to a digital signal consisting of 1s and 0s. Using the Nyquist theorem from 1934, the developers used a formula to convert the continuously changing amplitude and frequency shifts to a discreet set of values represented by the 1s and 0s. Using a three-step process, as shown in Table 30-1, they determined that they could carry digital voice. The table represents the three steps followed to do this conversion.

Once the values were determined, the developers used another process. Setting the values in place, they had to determine where along the wave the sample fell. They created a value system to show the 256 levels by using the table shown in Table 30-2. First, the digital PCM signal is representative of both positive and negative values. To reflect where along the wave the sample fell, the 8th bit in each sample is used as a sign (+/-) value. The other seven bits therefore represent the actual sample value. There are 128 points on the positive side of the wave and 128 points on the negative side of the wave. In this table, we look at the major stepping points. Two different formats are shown in the table, the PCM used in North America (and Japan) is Mu-Law, whereas the rest of the world uses an A-Law method. These are different as shown.

Figure 26-3

Band-limited channel limits the amount of information carried to 3 kHz

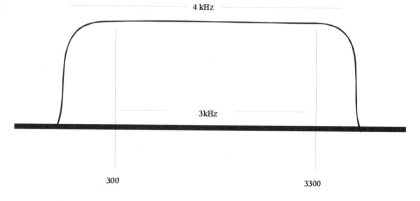

Table 26-1

The three-step process used to create the digital signal

Process	Result
1. Sample the analog wave at twice the highest range of frequencies that can be carried across the line. Using a 4 kHz channel capacity, the highest range of frequencies allowable is 4,000.	A sampling rate of 8,000 times per second: **4,000 × 2 = 8,000**
2. Quantify the values using a logical pattern of 1s and 0s to represent the height of the signal at any point in time. This deals with the amplitude shifts only.	Using an 8-bit sequence, the result is a total of 256 combinations of amplitude that can be represented. Although there may be more amplitude heights, (values) the 256 quantities were determined to be sufficient. **2^8 = 256 possible combinations**
3. After the values are determined from the samples, the final step is to encode the signal into a digital format and transmit information in its digital format onto the wires.	A sample value of 5 on the positive side of the wave will then be represented as an 8-bit data stream. **Binary 5 = 00000101**

Table 26-2

Summary of values for PCM in Mu-Law and A-Law formats

Coded Numerical Value	Bit Number		Comments
	Mu-Law 12345678	A-Law 12345678	The left-most bit (bit number 1) is transmitted first. It is the most significant bit. This bit is used as the sign bit; it is a 1 for positive values, and it is a 0 for negative values.
+127	10000000	11111111	
+ 96	10011111	11100000	
+ 64	10111111	11000000	
+ 32	11011111	10100000	
0	11111111	10000000	
0	01111111	00000000	
- 32	01011111	00100000	Note that 0 has two different values. Bits 2–8 are inverted between A-Law PCM and Mu-Law PCM. In A-Law all even bits are inverted prior to transmission.
- 64	00111111	01000000	
- 96	00011111	01100000	
- 126	00000001	01111110	
- 127	00000000	01111111	

The E-1 Pattern

Because the A-Law was shown in the table for comparative purposes, it is also wise to introduce the concept of the E-1. In North America, the DS-1 is the way we use our digital services. However, most of the world adopted a different standard called the E series. The E-1 is the European equivalent of a DS-1. There are many differences between the two that make them incompatible. The first is the companding method of creating the digital values as shown in the previous table. Above and beyond the companding values (using A-Law), there are differences between the formats used. An E-1 is a digital transmission rate that operates at 2.048 Mbps compared to the DS-1, which operates at 1.544 Mbps. Moreover, the E-1 carries 32 channels of 64 Kbps, whereas the DS-1 only carries 24 channels of 64 Kbps. Other multiplexing forms create higher speed E series just as we do with the DS signals. The following table summarizes the DS-n series compared to the E-n series. See Table 30-3 below for the channel speeds of the digital transmission systems.

Table 26-3

Comparison of North American and European digital services

DS-n	Channels	Speed	E-n	Channels	Speed
DS-0	1	64 Kbps	E-0	1	64 Kbps
DS-1	24	1.544 Mbps	E-1	32	2.048 Mbps
DS-2	96	6.312 Mbps	E-2	128	8.448 Mbps
DS-3	672	44.736 Mbps	E-3	512	34.368 Mbps

The Framing Protocols: D4 Framing

Before transmitting the frames across the circuit, the CSU device uses protocols that describe other activities. One protocol is a framing convention to preserve timing, signaling, and check for errors. The older version of this protocol is called D4 framing. D4 is comprised of 12 samples from each of the 24 inputs. The samples are held for a short time in a buffer by the CSU.

When all 12 passes are completed, the CSU then places the framing bit on each of the frames and transmits the information across the link.

When preparing the 12 frames, however, the CSU will also do a few other things. In the 6th and 12th frames, the CSU will steal (rob) one bit from each of the 24 inputs and use this bit for signaling. This means that the CSU steals the least significant bit from the sample and reassigns it. The unfortunate result of this framing convention is that it relegates the end user to only getting 7 data bits and 1 signaling bit every sixth frame. The 7 bits are trusted, whereas the eighth bit is suspect for use as data (because it is signaling). This results in a data carrying capacity of 56 Kbps (7 bits times 8,000 frames/sec). This is shown in Figure 26-4, using the robbed bit signaling approach.

Contrasting the E-1 and DS-1 Frame

Once again differences exist in the format of the framing patterns for the E-1 and the DS-1. Whereas the DS-1 is shown with 24 channels operating at 64 Kbps, plus a framing bit (1 bit added to the 24 byte samples) to create a frame of 193 bits, the E-1 uses 32 channels of 64 Kbps. However, the first channel of an E-1 frame (time slot 0) is used for synchronization, alarms

Figure 26-4
Robbed bit signaling in the D4 framing format

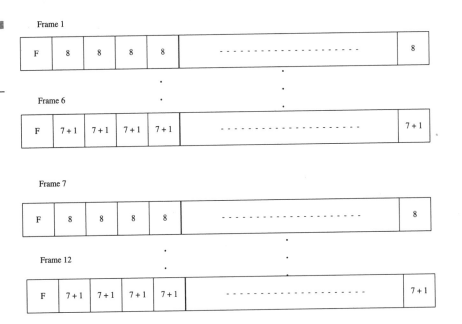

and international carrier use combined. This means that only 31 channels are available to carry voice or data traffic. This frame format is shown in Figure 26-5. The use of the pattern shown in this frame format uses Timeslot 0 (TS0) for framing and synchronization. Then TS16 is used for signaling for the various channels in the payload, whereas the DS-1 uses robbed bit signaling as discussed previously or common channel signaling on TS24. These differences are significant enough to prevent an even mapping from a DS-1 to an E-1.

Extended Superframe Format (ESF)

As a means of overcoming the limitations of the D4 framing protocols, the industry developed a newer protocol called ESF. This is an extension of the 12-frame format in order to double the size in the buffers at the CSU. When

Figure 26-5
An E-1 frame format

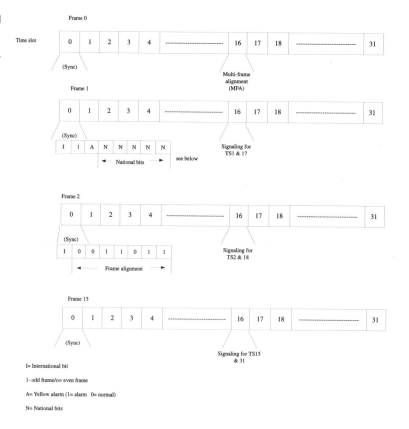

I= International bit

1- odd frame/o= even frame

A= Yellow alarm (1= alarm 0= normal)

N= National bits

the CSU has 24 frames of information, the CSU will use the framing bit for more than just a locator bit. It will define where the signaling is handled, and use this framing bit for other things such as the following: (Note: Remember there are 8,000 framing bits per second)

- 2 K bits are used for framing
- 2 K bits are used for error checking (CRC−6)
- 4 K bits are used for a maintenance and diagnostic capability to troubleshoot the circuit and improve availability.

The framing format of the ESF is shown in Figure 26-6.

After the framing bits had been addressed, another choice was used to handle the T1 circuit. As described previously, the robbed bit signaling gets in the way of data transmission. It is not a problem for voice communications, but definitely impacts the data side of the business. To solve this problem, a technique called common channel signaling was introduced. The use of channel number 24 was assigned for strictly signaling (call setup and teardown) for the other 23 channels. By using a dedicated out-of-band signaling channel, the other 23 channels can carry all 64 Kbps for data, yielding a higher throughput per channel, but with a penalty of losing one channel from the T1. The format of the common channel signaling arrangement is shown in Figure 26-7. This is a choice a user has to make. If all the user wants is voice communications, the robbed bit signaling will suffice. If, however, the user wants to transmit data, then the common channel signaling will be a potential benefit to the data transmission.

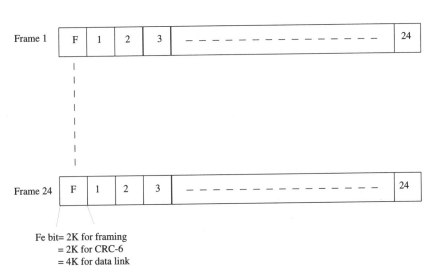

Figure 26-6
Extended Superframe Format improves uptime and error checking

Figure 26-7
Common channel
signaling uses
channel number 24

	TS1	2	3	4		23	Comm Channel
F	8	8	8	8	-------------------------------	8	SIG

Other Restrictions

Another problem surfaced when T1 was introduced for data transmission. In order to synchronize the carrier services, T1 uses a byte synchronization plan. In every eight bits there must be at least one pulse (represented as a digital one) for the transmitters to derive their timing on the line. If voice is the application, there are minimal problems with this constraint. As long as a party on the conversation is talking, there are sufficient numbers of ones pulses transmitted to keep the line synchronized. Yet, when data is transmitted, there may be strings of zeros transmitted continuously. This string of zeros will cause the transmitters to drift and lose timing.

To solve this problem, the Bell system introduced a "ones density requirement," which states that in every eight bits transmitted there must be at least one digital one. Further, no more than 15 consecutive zeros may be transmitted in a row; otherwise the timing will be lost. The solution to meeting this rule is handled in the CSU. If the CSU receives 8 zeros in a row, it will strip off the least significant bit and substitute a one in its place. This will meet the requirements of the ones density.

However, when the zero is stripped off and a one substituted in its place (called pulse stuffing), there is no way to know when this is done or when it is not. Consequently, this solution meets the timing and synchronization demands of the network, but leaves the user with the risk that the 8th bit is wrong. This leads the user to only trusting the first 7 bits, but ignoring the 8th bit. Ultimately, the result is 7 usable bits instead of 8 or a 56 Kbps throughput on the line instead of 64 Kbps. (7 bits times 8,000 samples = 56 Kbps). This results in a lot of wasted bandwidth and a limited throughput for data transmission. The pulse stuffing mechanism is shown in Figure 26-8.

B8ZS

It did not take long before the user began to demand better efficiency of the T1 line. Therefore, a newer technique is used called *Bipolar* (or binary) *8 Zero Substitution*. With B8ZS, the CSU is responsible for substituting a

Figure 26-8

Pulse stuffing allows
the timing to be
preserved

Data before pulse stuffing

| 0 | 0 | 0 | 0 | 0 | 0 | 0 | 0 | 0 | 0 | 0 | 0 | 0 | 0 | 0 |

Data after pulse stuffing

| 0 | 0 | 0 | 0 | 0 | 0 | 0 | 1 | 0 | 0 | 0 | 0 | 0 | 0 | 0 | 1 |

long string of zeros, eight at a time, with a fictitious 8 bits to meet the demands of the line synchronization, while at the same time allowing the receiving device to recognize that 8 zeros were originally sent. The B8ZS inserts a bit pattern of 0001 1011, easily satisfying the demands of the synchronization plan. By inserting these bits, the 4th and the 7th bits will be set as violations to the bipolar rule (*alternate mark inversion* [AMI]). Receiving a bit pattern of 0001 1011 with a bipolar violation in positions 4 and 7 are recognized as a flag that 8 zeros were intended. Therefore, the receiving CSU will strip off the fictitious word and reinsert all zeros to the receiving device. This allows for clear-channel 64 Kbps data transmission. The B8ZS is shown in Figure 26-9.

With all these problems and solutions, one can only wonder why everyone uses the T1. The primary reason is the 1.544 Mbps throughput (or slightly less based on the restrictions) for voice and data transmission. The second reason is the cost/benefits ratio associated with bundling the services onto a single, four-wire circuit. Third, is the benefit of all digital circuitry to the door. It is because of these three reasons that T1 is the most widely used digital transmission service in North America.

T-2 Transmission (or DS-2)

For all its work, the T1 was a beginning step to the North American Digital Hierarchy. From there, other multiplexing steps were instituted to get higher speed communications services. A second means of achieving the high-speed communications is to multiplex four DS-1s into a DS-2. The DS-2 (physically it is a T-2) is a data channel carrying 96 channels of 64 Kbps each, plus overhead yielding a nominal 6.312 Mbps data transmission speed. To create the DS-2, four DS-1 signals are combined. This creates a

Figure 26-9

B8ZS inserts an 8-bit pattern that is recognizable by the receiver as substituted

DS-2 frame as shown in Figure 26-10. The DS-2 frame is often referred to as the DS-2 M frame and consists of four subframes. Each of the subframes are labeled: M1 though M4. A subframe consists of six blocks of information; each block contains 49 bits.

The first bit in each block is an overhead bit. A DS-2 frame has 24 overhead bits. The remaining 48 bits in each block are DS-1 information bits. Carrying this out then there is:

48 DS1 bits/block X 6 blocks/subframe X 4 subframes/ DS-2 frame = 1,152 DS-1 information bits

The four subframes do not represent each of the separate DS-1 signals; instead a bit interleaving of the four DS-1 signals forms the DS-2 frame.

The overhead bits precede the data bits in each of the blocks. The data bits are interleaved, where 0i designates the time slot devoted to DS-1 input i. After 48 information bits, 12 from each of the DS-1s, a new DS-2 overhead bit is inserted. The total number of DS-1 bits transmitted per second in a DS-2 frame is therefore:

DS-1 rate $\times$ **4 DS-1 signals/DS2** **= information bits**
1,544 Mbps $\times$ **4 DS-1 signals** **= 6.176 Mbps**

The rate chosen for the DS-2 is 6.312 Mbps. This allows extra overhead for bit stuffing to synchronize the signals.

Figure 26-10
A DS-2 M frame

1.) 6 blocks per subframe = 49 × 6294 bits bit interleaved
2.) DS2 M-Frame = 4 sub frames = 4 × 294 = 1176 bits

DS-2 Bit Stuffing

The four DS-1 signals are synchronous to themselves, but asynchronous to each other. This means they may be operating at different rates. To synchronize the signal, the multiplexers use a bit stuffing technique (this was referred to as pulse stuffing in the DS-1 discussion). Bit stuffing is used to adjust the incoming rates because they all differ.

Framing Bits for the DS-2

Framing bits (F-bits) form the frame alignment much the same as the DS-1 framing bit. There are a total of 8 framing bits in a DS-2 frame (two in each subframe). F bits are located in the first bit position in blocks 3 and 6 of each subframe. The frame alignment pattern is "01," which repeats every subframe. The framing bits are shown in Figure 26-11. Other overhead bits are also shown in the framing pattern. These include the following:

- *M-bits (multiframe)* There are four M-bits per DS-2 frame. These are always located in the first bit position in each subframe.

- *C-bits* The C-bits are used to control bit stuffing. Three C-bits per subframe are allowed, also shown in Figure 26-11.

Figure 26-11
The DS-2 overhead
bits shown

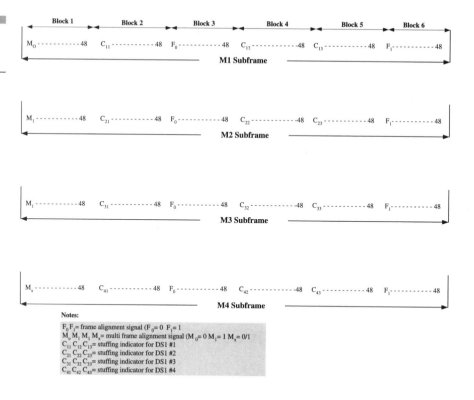

What this all boils down to is the information bits, overhead bits, and other inserted stuff bits all equate to a DS-2 framed format that produces the 6.312 Mbps data rate. The DS-2 rate is a provider service, primarily used in the intermediate steps for achieving a DS-3 (called also a T-3) and for use in delivering a DS-2 to a remote terminal over a subscriber loop carrier with 96 channels (called an SLC-96). Very few customers ever had the availability to use a DS-2. But one must not view this as an unnecessary multiplexing rate because of the original method of multiplexing from a DS-1 to a DS-3; the intermediate rates were needed to control and format the signals. This is a piece of history we should all remember.

One caveat bears mention. With the emergence of the ADSL services, the local providers were trying to deliver a download rate from the network to the consumer for a 6 Mbps video rate, 2 Mbps for data and POTS. The upload rates were lower, but the download rates approximated the DS-2 rate. (It is not classified as a DS-2; the reference here is for analysis to the speeds and aggregates we are now seeing from the suppliers).

DS-3 Service (T-3)

The next step in the evolution of digital transmission services is the T3 or DS-3. This is a high-speed, communication signal used by the larger organizations and the public carriers (local and long distance carriers). Because of its pricing and delivery, this was never practical except for the largest of corporations and providers. The delivery on a copper-based plant used coaxial cable to reach and support the 44.736 Mbps rate of speed. Newer versions deliver the DS-3 on fiber optics or microwave radio, but these are strictly media used to carry the signal. Getting the signal is what is important.

The DS-3 Frame Format

A two-step process is used when multiplexing to a DS-3. First, four DS-1s are multiplexed together to form a DS-2. Second, seven DS-2s are multiplexed together to form the DS-3. This is referred to as the M13 format and protocol. Literally, the same method is used to multiplex the seven DS-2s into a DS-3 as was used to multiplex four DS-1s into a DS-2. In Figure 26-12, the framing format is shown, which uses the M-13 Asynchronous Protocol. The DS-3 frame consists of seven subframes, labeled as 1 through 7. Each subframe consists of eight blocks of data and each block contains 85 bits.

Figure 26-12
A DS-3 frame

|←—85—→|←—85—→|←—85—→|←—85—→|←—85—→|←—85—→|←—85—→|←—85—→|

7 M Subframes

OH+84 bits	OH+84 bits	OH+84 bits	OH+84 bits	OH+84 bits	OH+84 bits	OH+84 bits	OH+84 bits
OH+84 bits	OH+84 bits	OH+84 bits	OH+84 bits	OH+84 bits	OH+84 bits	OH+84 bits	OH+84 bits
OH+84 bits	OH+84 bits	OH+84 bits	OH+84 bits	OH+84 bits	OH+84 bits	OH+84 bits	OH+84 bits
OH+84 bits	OH+84 bits	OH+84 bits	OH+84 bits	OH+84 bits	OH+84 bits	OH+84 bits	OH+84 bits
OH+84 bits	OH+84 bits	OH+84 bits	OH+84 bits	OH+84 bits	OH+84 bits	OH+84 bits	OH+84 bits
OH+84 bits	OH+84 bits	OH+84 bits	OH+84 bits	OH+84 bits	OH+84 bits	OH+84 bits	OH+84 bits
OH+84 bits	OH+84 bits	OH+84 bits	OH+84 bits	OH+84 bits	OH+84 bits	OH+84 bits	OH+84 bits

8 blocks/M Subframe × 85 bits	=	680 bits per M Subframe
NOTES: 7 Subframes × 8 blocks of 85 bits	=	4760 bits per M Frame
1 O/H bit per block × 8 blocks x 7 rows	=	56 O/H bits
Data bits = 5460 - 56	=	4704 per M Frame

The first bit in each block is an overhead bit, which make 56 overhead bits in the DS-3 frame. The remaining 84 bits in a block are the DS-2 information bits. Doing the math on this, the total DS-2 bits in a DS-3 frame is as follows:

84 DS-2 bits/block $\times$ **8 block/subframe** $\times$ **7 subframes/DS-3**

84 $\times$ **8** $\times$ **7 = 4704 bits**

The seven subframes do not represent each of the separate DS-2 signals. Rather, the DS-3 frame is formed by bit interleaving the seven DS-2 signals. The interleaving process is the same as that used for the DS-1s to the DS-2. After 84 information bits, 12 from each DS-2, an overhead bit is inserted. To get to the nominal number of DS-2 information bits in a DS-3 frame in a one-second period, we see the math as follows:

DS-2 rate $\times$ **7 DS-2** = **Number of bits carried**

6.312 Mbps $\times$ **7** = **44.184 Mbps**

The overall rate of the DS-3 is 44.736 Mbps, allowing extra overhead for bit stuffing and the overhead bits used in the framing format.

DS-3 Bit Stuffing

The seven DS-2 signals are asynchronous to each other (because they may not have been formed in the same multiplexer). They may be operating at different rates. Bit stuffing is used to adjust the different incoming signals. This is the same reasoning behind the bit stuffing in each of the preceding discussions. Stuffing bits are used to justify the signals within the DS-3 frame, and they provide enough consistency for the multiplexers to derive the channels from within the DS-3 signal. In the 8th block of each subframe, an added stuffing slot is reserved to do this justification. The last block therefore may contain the following two choices:

- 1 Overhead bit plus 84 information bits
- 1 Overhead bit, plus 1 Stuffing bit plus 83 information bits

The stuffing bits reserved in the 8th block will occur in each of the 7 subframes randomly. However, a stuff bit will occur in 7 of 18 frames, but not always in the same place or the same pattern. This helps to make the transmitters asynchronous because one never knows how many bits will be used for overhead in each of the seven subframes.

The DS-3 Overhead Bits

As already stated, there are overhead bits allocated within the DS-3 frame for the purposes of aligning the frames, providing indication of stuffing and other low-level functions. These are labeled as follows:

- *F-bits*   The framing bits form the frame alignment for the DS-3 signal. The F bits are located in the first bit slot in blocks 2, 4, 6, and 8 of each subframe. The frame alignment pattern is "1001" and repeated in every subframe.

- *M-bits*   The multiframing bits form the multiframe alignment signal. Three M-bits are used in each DS-3 frame. The M-bits are located in the first bit slot in block 1 of subframes 5, 6, and 7. The DS-3 equipment uses the M-bit pattern "010" to locate the seven subframes.

- *C-bits*   These are used to control the bit stuffing. There are three C-bits per subframe, thus 21 C-bits per DS-3 frame. These are numbered a C_{ij}, with i designating the subframe and j designating the position of the C-bit in the subframe.

- *X-bits*   Two X-bits are designated for each DS-3 frame. These are located in the first bit slot of block 1 in subframes 1 and 2. The X-bits must be identical in any DS-3 frame (either 00 or 11). The source may not change the X-bits more than once per second. These bits are used for service messages as a low-level, transmitter-to-receiver form.

- *P-bits*   The P-bits (or parity bits) carry parity information. There are two P-bits per DS-3 frame. The P-bits are located in the first bit slot of the first block in subframes 3 and 4. A bit-interleave parity check computes parity over the 4,704 information bits contained in the DS-3 frame. The state of the parity bits is always identical. The two bits are set to a "1" if the previous frame contained an odd number of ones, or the two bits are set to "0" if the previous frame had an even number of ones.

SONET and SDH

Introduction

Synchronous Optical Network (SONET) is a standard developed by the *Exchange Carriers Standards Association* (ECSA) for ANSI. This standard defines an optical telecommunications transport for U.S. Telecommunications. SONET standards provide an extensive set of operational parameters for optical transmission systems throughout the industry. The North American industry uses the SONET specifications, whereas the rest of the world uses a close "cousin" defined as *Synchronous Digital Hierarchy* (SDH). Between the two sets of standards, the industry attempted to define the roles of transport for the telecommunications providers using optical fibers as the transport medium.

SONET provides more, though. It defines a means to increase throughput and bandwidth through a set of multiplexing parameters. These roles provide certain advantages to the industry, such as the following:

- Reduced equipment requirements in the carriers' network
- Enhanced network reliability and availability
- Conditions to define the overhead necessary to facilitate managing the network better
- Definitions of the multiplexing functions and formats to carry the lower level digital signals (such as DS-1, DS-2, and DS-3)
- Generic standards encouraging interoperability between different vendors' products
- A flexible means of addressing current as well as future applications and bandwidth usage

SONET defines the *Optical Carrier* (OC) levels and the electrical equivalent rates in the *Synchronous Transport Signals* (STS) for the fiber-based transmission hierarchy.

Background Leading to SONET Development

Prior to the development of SONET, the initial fiber-based systems used in the PSTN were all highly proprietary. The proprietary nature of the products included the following:

SONET and SDH

1. Equipment

2. Line coding

3. Maintenance

4. Provisioning

5. Multiplexing

6. Administration

The carriers were frustrated with the proprietary products because of interoperability problems, sole-source vendor solutions (which held the carriers hostage to one vendor), and cost issues. These carriers approached the standards committees and demanded that a set of operational standards be developed that would allow them to mix and match products from various vendors. In 1984 a task force was established to develop such a standard. The resultant standard became SONET.

Synchronizing the Digital Signals

SONET involves the synchronization of the digital signals arriving at the equipment. Keeping in mind that the signals may be introduced in one of three ways, it is important to attempt to get everything on a common set of clocking mechanisms. In digital transmission, the normal way of synchronizing traffic is to draw a common clocking reference. In the hierarchy of clocking, systems use a stratum clocking architecture. The stratum references in North America come in a four-level architecture. These are shown in Table 27-1.

In a set of synchronous signals, the digital transitions in the signals occur at the same rate. There may be a phase difference between the transitions in the two signals, but this would be in specified ranges and limits. The phase differences can be the result of delay in systems, jitter across the link, or other transmission impairments. In a synchronous environment, all the clocks are traceable back to a common reference clock (the Primary Reference Clock).

If two signals are almost the same, they are said to be plesiochronous. Their transitions are close (or almost the same) and variations are contained within strict limits. The clocking between the two different sources, although accurate, may be operating at a different rate.

Table 27-1

Summary of
clocking systems

Stratum Reference	Location Used	Accuracy
1	Primary Reference drawn from a GPS or the National Reference Atomic clock	$\pm$ 1 pulse in 10^{11}
2	Toll Offices (Class 1–4)	$\pm$ 1.6 pulses in 10^8
3	End Offices (Class 5)	$\pm$ 4.6 pulses in 10^6
4	Customer Equipment (Multiplexer, Channel Bank, and so on)	$\pm$ 32 pulses in 10^6

Finally, if two signals are randomly generated and do not occur at the same rate, they are said to be asynchronous. The difference between two clocks is much greater, possibly running from a free running clock source.

Any one of these signals, synchronous, plesiochronous, or asynchronous, may arrive at a SONET multiplexer to be formulated and transmitted across the network. SONET defines the means of synchronizing the traffic for transmission.

The SONET Signal

SONET defines a technique to carry many signals from different sources and at different capacities through a synchronous, optical hierarchy. The flexibility and robustness of SONET are some of its strongest selling points. Additionally, in the past many of the high-speed, multiplexing arrangements (DS-2 and DS-3) used bit interleaving to multiplex the data streams through the multiplexers. SONET uses a byte-interleaved multiplexing format. This is a strong point because it keeps the basic DS-0 intact throughout the network, making it easier to perform diagnostics and troubleshooting. Byte interleaving simplifies the process and provides better end-to-end management.

The base level of a SONET signal is called the *Synchronous Transport Signal Level1* (STS-1), operating at 51.84 Mbps. The first step in using the SONET architecture is to create the STS-1. Other levels exist in multiples of the STS-n to create a full family of transmission speeds. The SONET hierarchy is shown in Table 27-2.

Table 27-2

Summary of
electrical and
optical rates for
SONET

The SONET hierarchy			
Electrical Signal	Optical Value	Speed	Capacity
STS-1	OC-1	51.84 Mbps	28 DS-1 or 1 DS-3
STS-3	OC-3	155.520 Mbps	84 DS-1 or 3 DS-3
STS-12	OC-12	622.08 Mbps	336 DS-1 or 12 DS-3
STS-24	OC-24	1.244 Gbps	672 DS-1 or 24 DS-3
STS-48	OC-48	2.488 Gbps	1344 DS-1 or 48 DS-3
STS-192	OC-192	9.95Gbps	5376 DS-1 or 192 DS-3

Other rates exist, but these are the most popularly implemented

Why Bother Synchronizing?

In the past, transmission systems have been primarily asynchronous. Each terminal device in the network was independently timed. In a digital synchronous transmission system, clocking is all-important. Clocking uses a series of pulses to keep the bit rate constant and to help recover the ones and zeros from the data stream. Because these past clocks were independently timed, large variations occurred in the clock rate, making it extremely difficult (if not impossible) to extract and recover the data. A DS-1 operates at 1.544 Mbps $\pm$ 150 pps, whereas a DS-3 operates at 44.736 Mbps $\pm$1789 pps. These differences mean that one DS-1 may be transmitting at up to 300 pps different than the other (assuming that DS-1 is at −150 pps and the second one is at +150 pps). The differences can make it difficult to derive the actual data across a common receiver.

Back in the section on the T-carriers (Chapter 26), we discussed the asynchronous method of multiplexing a DS-3. In that section we saw that four DS-1s were bit interleaved together to form a DS-2, and that bit stuffing occurred. From there, seven DS-2s were bit interleaved together to form the DS-3, but there were several possible steps where bit stuffing occurred at the multiplexing point. The stuff bits were random occurring in seven of 18 frames, causing confusion in delivering and demultiplexing the signal. Moreover, when a problem occurs on a DS-3 using the M13 Asynchronous Multiplexing technique, the entire DS-3 must be demultiplexed to find the problem. This is inefficient.

Therefore, the method of synchronously multiplexing in a SONET architecture provides for better efficiency and problem resolution. Using SONET, the average frequency of all the clocks will be the same. Every clock can be traced back to a common reference, which is highly reliable and stable. Bit stuffing can be eliminated in the preparation of the STS-1 signal; therefore, the lower-speed signals are more readily accessible. The benefits outweigh the possible overhead associated with the SONET multiplexing scheme. In SONET, the hierarchy of clocking follows the master-slave clocking architecture. Higher level (stratum 1) clocks will feed the timing across the network to lower level devices. Any jitter, phase shifts, or drifting by the clocks can be accommodated through the use of pointers in the SONET overhead. The internal clock in a SONET multiplexer may also draw its timing from a *Building Integrated Timing System* (BITS) used by switching systems and other devices. This terminal will then serve as the master for other SONET nodes downstream, providing timing on its outgoing signal. The receiving SONET equipment will act in a slave mode (loop timing) with their internal clocks timed by the incoming signal. The standard specifies that SONET equipment must be able to derive its timing at a stratum level 3 or above.

The SONET Frame

SONET also defines a frame format in which to produce the basic rate of 51.84 Mbps (the STS-1). Each of the additions to the multiplexing rates is a multiple of the STS-1. The basic format consists of a frame that is 80 bytes (columns) wide and 9 bytes high (rows). The basic STS-1 signal is then applied into this 810-byte frame. The frame is shown in Figure 27-1. The frame will occur 8,000 times per second. If we calculate the math on this we have the following:

810 bytes $\times$ **8 bits/byte** $\times$ **8000 frames/sec** = **51.84 Mbps**

Overhead

From the 810-byte frame, overhead is allowed in several ways to handle the *Operations, Administration, Maintenance, and Provisioning* services (*OAM&P*).

The first part of the overhead is defined as the transport overhead. The transport overhead uses the first three columns and all nine rows. This creates 27 bytes of transport overhead. As shown in Figure 27-2, the transport overhead is divided into two pieces. The first three columns and the first

Figure 27-1
The SONET frame

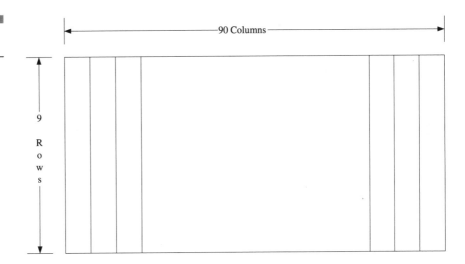

Figure 27-2
The transport overhead is divided into section and line overhead.

three rows (9 bytes) are used for section overhead. The remaining six rows in the first three columns (18 bytes) are used for line overhead.

The remaining 87 columns and 9 rows (783 bytes) are designated as the *Synchronous Payload Envelope* (SPE). Inside the SPE, an additional one column, nine rows high (9 bytes) is set aside for Path Overhead. This is shown in Figure 27-3. After the Path Overhead is set aside, the resultant payload is 774 bytes. In these 774 bytes, the services are then mapped into the frame. The STS-1 payload can carry the following:

- 1 DS-3

- 7 DS-2s

- 21 E-1s

- 28 DS-1s

Combinations of the above payloads are also allowable. Two columns are reserved as fixed stuff columns; these are columns 30 and 59. The remaining 756 bytes carry the actual payload.

Figure 27-3
The SPE shown

SPE: 87x9 = 783 bytes

Inside the STS-1 Frame

The SPE can begin anywhere inside the STS-1 envelope. Normally, it begins in one STS-1 frame and ends there. However, it may begin in one STS-1 frame and end in another. The STS payload pointer contained inside the transport overhead points to the beginning byte of an SPE. The possibilities then are that the frame can carry a locked payload or a floating payload. Floating, in this regard, refers to the payload floating between two frames. The overhead associated with SONET is designed to let the receiver know where to look for its payload and extract the information at a starting point. This floating frame is shown in Figure 27-4.

Figure 27-4

A floating payload inside two frames

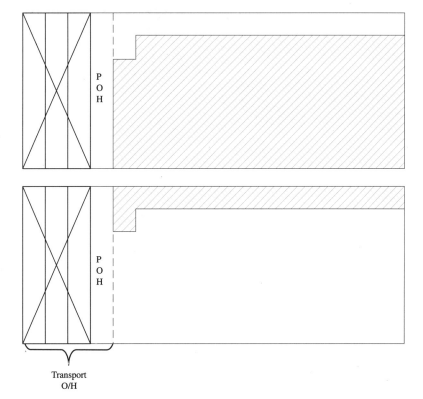

P
O
H

P
O
H

Transport
O/H

SONET Overhead

Sonet provides significant overhead, allowing the simpler multiplexing techniques and improved OAM&P operations. In the next discussion the overhead is examined to better understand why so much is dedicated to the overhead functions.

In Figure 27-5, the architecture of the SONET link is shown as defined by ECSA and ANSI. In this reference the link architecture is broken down into three parts:

- *Section* The section is defined as the portion of the link between two repeater functions, or between a repeater and line terminating equipment. Sufficient overhead is allowed to detect and troubleshoot errors on the link between these two points (STE).

- *Line* The line overhead provides sufficient information to detect and troubleshoot problems between two pieces of *Line Terminating Equipment* (LTE).

- *Path* The Path Overhead provides sufficient overhead to detect and troubleshoot problems between the end-to-end path terminating *Pieces of Equipment* (PTE).

Figure 27-5
SONET link
architecture

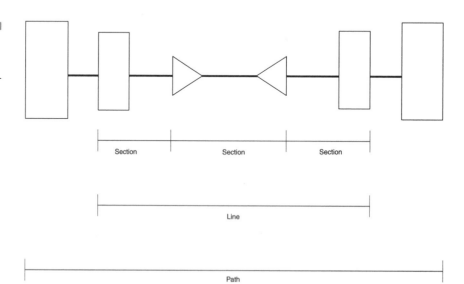

Section   Section   Section

Line

Path

Section Overhead

The section overhead is shown in Table 27-3 and Figure 27-6. This overhead contains 9 bytes of information that is accessed, generated, and processed by the section terminating equipment. It supports functions such as the following:

- Performance monitoring
- A local orderwire
- Data communications channels for OAM&P messages
- Framing information

Table 27-3

The individual bytes are defined in the designation and their use.

Byte	Description
A1-A2	Framing bytes: These two bytes provide an indication of the start of an STS-1 frame
J0/Z0	Section Trace (J0) or Section Growth (Z0): this byte in each of the STS-1s in an STS-n was originally defined as the STS-1 ID (C-1) byte. It has been redefined as either the section trace byte (in the first STS-1 of STS-n) or as a growth byte (used in the 2nd through nth STS-1s)
B1	Section *Bit Interleaved Parity* code (BIP-8) byte: This is a parity code (using even parity) used to check for transmission errors over a section of the link (a repeater). The value is calculated over all bits of the previous STS-n frame after scrambling and then placed in the B1 byte of the STS-1 before scrambling. This byte only defines for STS-1 number 1 of an STS-n signal.
E1	Orderwire: This is a 64Kbps channel for a voice communications orderwire between two repeater functions or a repeater and a line terminating equipment. It allows for two technicians to talk to each other while they troubleshoot problems.
F1	Section User Byte: This byte is set aside for users' choice. It terminates at all section terminating equipment within a line. It can be read or written at each section terminating equipment on the line. This may be used for proprietary user maintenance and diagnostic systems, or for SNMP functions on a user network.
D1–3	Section *Data Communications Channel* (DCC) bytes: This creates a 192 Kbps data channel for diagnostics and testing for OAM&P. It uses a message-based channel for remote locations to control alarms, monitoring, and other maintenance functions.

Figure 27-6
The section overhead

A_1	A_2	J_0/Z_0	
B_1	E_1	F_1	
D_1	D_2	D_3	

Line Overhead

In Table 27-4 we see the line overhead, which occupies rows 4 through 9 in the first three columns. The line overhead used between two pieces of line terminating equipment provides for the OAM&P functions. This is more extensive than the section overhead. Figure 27-7 also shows its position in the overhead of the STS-1 frame.

Table 27-4

The individual bytes of the line overhead

Byte	Description
H1-H2	STS Payload Pointers: These two bytes indicate the offset bytes between the pointer and the first byte of the STS SPE. The pointer bytes are used in all STS-1s within an STS-n to align the transport overhead in the STS-n and perform frequency justification. These bytes can also be used to indicate when concatenation is used or to indicate path *Alarm Indications Status* (AIS)

Byte	Description
H3	Pointer Action byte: This byte is allocated for frequency justification. The H3 byte is used in all STS-1s within an STS-n to carry the extra SPE byte to either increment or decrement the pointer.
B2	Line *Bit Interleaves Parity* code (BIP-8) byte: This parity code byte is used to determine if a transmission error has occurred over the line. It uses even parity and is calculated over all bits of the line overhead and STS-1 SPE of the previous STS-1 frame before scrambling. The value is placed in the B2 byte of the line overhead before scrambling.
K1-K2	*Automatic Protection Switching* (APS) channel bytes: These two bytes are used for protection signaling between two line terminating devices for bidirectional automatic protection switching and for detecting *Alarm Indication Signal on the Line* (AIS-L) and *Remote Defect Indication* (RDI) signals.
D4-D12	Line *Data Communications Channel* (DCC) bytes: This creates a 576 Kbps channel capacity for message signaling from a central location for OAM&P. These bytes are available for generic, internal, or external messages, or vendor-specific messages. The use of a protocol analyzer is needed to access the DCC.
S1/Z1	Synchronization Status Byte (S1): The S1 byte is used for synchronization status of network elements. It is located in the 1st STS-1 of an STS-n. Bits 5–8 convey the synchronization information.
	The Growth (Z1) byte is located in the second through nth STS-1s of an STS-n and allowed for growth.
M0-M1	STS-1 REI-L (M0) byte is defined for STS-1 in an OC-1 or for the STS-1 electrical signal. Bits 5–8 report on the *Line Remote Error Indication* function (REI-L, which used to be called the FEBE).
	STS-n REI-L (M1) byte is located in the third STS-1 (in order of appearance in the byte interleaved STS-n electrical or optical signal) in an STS-n (where $n \geq 3$) and is used for the REI-L function.
Z2	Growth (Z2) byte: This byte is located in the 1st and 2nd STS-1s of an STS-3, and the 1st, 2nd and 4th through nth STS-1s of an STS-n (where $12 \leq n \leq 48$). These bytes are reserved for future growth.
E2	Orderwire byte: This orderwire byte provides a 64-Kbps channel between line equipment for an express orderwire. It is a voice channel used by technicians, and is ignored as it passes through the repeaters.

Figure 27-7
The line overhead of
a SONET frame

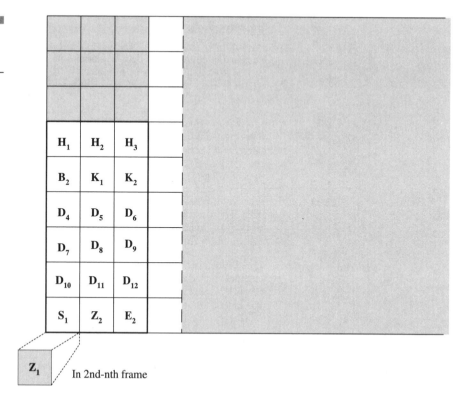

Path Overhead

The final piece of overhead is the Path Overhead contained inside the SPE. Path Overhead contains 9 bytes of information starting at the first byte of the STS SPE. The Path Overhead provides for communication between two path-terminating pieces of equipment (PTE). The PTE is where the actual multiplexing and demultiplexing function takes place as the services are mapped into the SONET frame. The functions supported by the POH are as follows:

- Performance monitoring of the STS SPE
- Signal label the individually mapped payloads
- Path status
- Path trace

The Path Overhead is shown in Table 27-5 and in Figure 27-8 to correlate them in the overall function of the STS frame.

	Byte	Description
Table 27-5	J1	STS Path Trace byte: This user programmable byte repetitively transmits a 64 byte, or a 16-byte (E.164) format string. This is used so the receiving end of the path to verify it is still connected to the sending end.
The Path Overhead defined	B3	STS Path *Bit Interleaved Parity* code (path BIP-8) byte: This byte is even parity to check if an error has occurred across the entire path.
	C2	STS Path Signal Label byte: This byte is used to identify the content of the STS SPE, including the status of the mapped payloads.
	G1	Path Status Byte: This byte is used to convey path terminating status and performance characteristics back to the originating end. Bits 1–4 are used for ATA Path REI-P (was referred to as the FEBE) Bits 5, 6 and 7 are for an STS Path RDI (RDI-P) signal and bit 8 is not defined.
	F2	Path User Channel Byte: The byte is used for user communications between end-to-end elements. This may be a proprietary network management system or an SNMP system to determine the status and alarms of the far end on the path.
	H4	VT Multiframe Indicator byte: This byte provides a general multiframe alignment indicator for payload containers. Currently, it is only used for VT payloads.
	Z3-Z5	Growth bytes: These bytes are undefined at this time.

Figure 27-8

Path Overhead in the STS frame

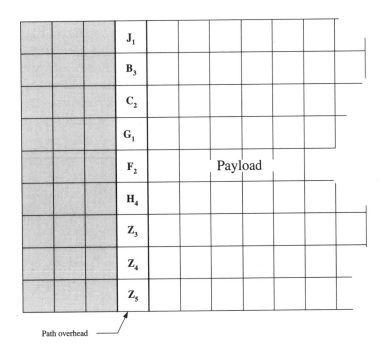

Path overhead

Virtual Tributaries

SONET is more than the STS frame; it also defines the sus-STS levels of payload that can be carried. The STS-1 payload can be subdivided into virtual tributaries, synchronous signals to transport lower-speed signals. The normal sizes defined by SONET are shown below in Table 27-6.

SONET Multiplexing Functions

The primary principles of the SONET multiplexers are as follows:

■ Mapping the tributaries into the STS-n frame

■ Aligning the information by using the pointer information to determine where the first byte of the tributary is located

■ Multiplexing lower order signals are adapted to higher order signals

■ Stuffing of bits necessary to handle the various lower-speed, asynchronous channels and filling up the spare bits to keep everything in alignment

The SONET equipment provides these functions. SONET can carry very large payloads, as we have seen in the hierarchy of the data speeds. Up to now, we have seen the primary mapping and layout of an STS-1 (OC-1). SONET equipment can add the value necessary to protect investments by either lower-rate multiplexing or higher-rate multiplexing. If one looks at an OC-3, for example, the multiplexer will produce a larger STS frame. In this case, as shown in Figure 27-9, the frame is three times larger or 270 bytes (columns) wide and nine rows high for a total of 2,430 bytes. Note

Table 27-6

The values of the virtual tributaries defined for SONET

VT	Bit Rate	Equivalent DS-n level	Required Bytes (rows × columns)
1.5	1.728 Mbps	DS-1	27 (9 × 3)
2	2.304 Mbps	E-1	36 (9 × 4)
3	3.456 Mbps	DS-1C	54 (9 × 6)
6	6.912 Mbps	DS-2	108 (9 × 12)

Figure 27-9

The STS-3 (OC-3) frame

from this figure the overhead for the STS-n is located in the beginning of the frame, whereas the *Path Overhead* (POH) is located at the start of each payload.

ADD-Drop Multiplexing: A SONET Benefit

Another major benefit of the SONET specification is the ability to perform add-drop multiplexing. Even though network elements are compatible at the OC-N level, they may still differ from vendor to vendor. SONET doesn't attempt to restrict vendors to all providing a single product, nor does it require that they produce one of every type out there. One vendor may offer an add-drop multiplexer with access to the DS-1 level only, whereas another may offer access to DS-1 and DS-3 rates. The benefit of an add-drop multiplexer on a Wide Area Network is to drop (demultiplex) only the portions of the optical stream required for a location and let the rest pass through without the demultiplexing process. It would be extremely inefficient to have to demultiplex an entire OC-12, for example, only to drop out one DS-1. The ability to extract only what is necessary helps to prevent errors, loss of data, and other delays inherent with older technologies. The add-drop multiplexer makes this attractive for carriers to use in rural areas where they may bundle many lower-speed communications channels onto a single OC-1 or OC-3 to carry the information back to the central metropolitan areas. Moreover, beyond just dropping a digital signal out of a higher-speed OC-n, the carrier can fill in what has been vacated. (For example, if a DS-1 is dropped off along the optical path, a new DS-1 can be multiplexed back into the OC-3 in its place.) This allows the carriers considerable flexibility. In Figure 27-10 an add-drop multiplexer is shown. Here portions of the bandwidth can be dropped off and additional new signals can be added in place of the data stream dropped out of the higher-speed signal. A single stage add-drop multiplexing function can multiplex various inputs into an OC-n signal. At an add-drop site, only those signals that need to be accessed are dropped and inserted. The remaining traffic continues through the network switching system without requiring special processing.

SONET Topologies

Several different topologies can be used in a SONET network by using the various multiplexers. These include the normal topologies most networks have been accustomed to over the years. They include the following:

- Point-to-point
- Point-to-multipoint

Figure 27-10
Add-drop
multiplexing with
SONET

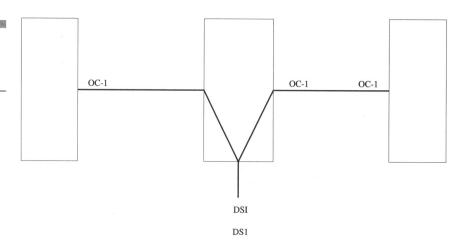

- Hub and spoke
- Ring

It is the variations that allow the flexibility of SONET in the Wide Area Networks built by the carriers, but now becoming the method of choice at many large organizations.

In each of the topologies, larger organizations are finding the benefits of installing highly reliable, interoperable equipment at the private network interfaces and access to the public networks.

Point-to-Point

The SONET multiplexer, the entry level PTE for an organization (or the equipment installed by the LEC at the customer's premises to access the network), acts as a concentrator device for the multiple lower-speed communications channels, such as DS-1 and DS-3. In its simplest form, two devices are connected with an optical fiber (with any repeaters as necessary) as a point-to-point circuit. As the entry-level point into a SONET architecture, the inputs and outputs are identical. In this environment, the network can act as a stand-alone environment and not have to interface with the public switched networks. See Figure 27-11 for the point-to-point multiplexing arrangement.

Point-to-Multipoint

The next step is to consider the point-to-multipoint arrangement. This will use a form of add-drop multiplexing to drop circuits off along the way. In a large corporate network spanning the country (or any subset), a single, high-speed link may be employed. The SONET *Add-Drop Multiplexer* (ADM) is

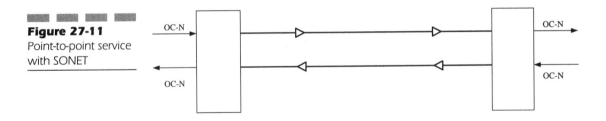

Figure 27-11
Point-to-point service with SONET

used for the task at hand, dropping circuits out without demultiplexing the entire high-speed signal. In Figure 27-12, the ADM is installed between two far-end locations so that signals can be added or dropped off as necessary. This is a better solution than renting three different circuits between points A-B, A-C, and B-C, which adds to the complexity and cost. By using a circuit from A-B-C with ADMs, the service can usually be more efficiently accommodated.

Hub and Spoke

The hub and spoke method (sometimes referred to as the star network) allows some added flexibility in the event of unpredicted growth or constant changes in the architecture of the network. SONET multiplexers can be hubbed into a Digital Cross-Connect where it is concentrated and then forwarded on to the next node (Figure 27-13). This is used in many larger orga-

Figure 27-12
ADMs installed along the way

Figure 27-13
Hub and spoke in a SONET multiplexer network

nizations where regional offices are located and district or branch offices are tied into the network through the hub. Once again the flexibility is there if a major change occurs in the network architecture or in the event of major campaigns in the organization. Hubs will act as the cross-connect points to link the various echelons in the network together. These may be developed in a blocking or nonblocking manner. Typically, some blocking may be allowed.

Ring

Next, the glory of SONET! In a ring architecture, where SONET automatic protection switching is employed, the best of all worlds comes to fruition. In this topology, ADMs are used throughout the network, and a series of point-to-point links are installed between adjoining neighbors. The bidirectional capability places the most robustness into the network; however, unidirectional services can also be installed. The primary advantage of the ring architecture is survivability in the event of a cable cut or a failure in a network node. The multiplexers have sufficient intelligence to reroute or reverse direction in the event of a failure. If more than one fiber link is installed, the systems could use alternate paths, but they must recover in milliseconds (which APS on SONET is designed to do). Figure 27-14 shows the ring topology with dual fibers run (bidirectional service) between the ADMs.

Figure 27-14

Ring architecture of SONET multiplexers

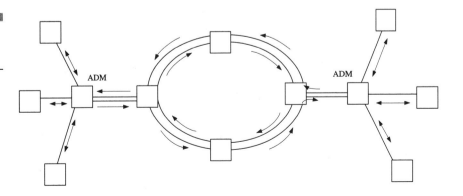

Evolution of SONET in the Rest of the World

Following the work by ANSI in developing the SONET standards, the rest of the world began to move in tandem. The ITU-TS decided to define a synchronous standard that would address internetworking between the ITU and ANSI transmission hierarchies. In 1989, the ITU published the *Synchronous Digital Hierarchy* (SDH) standards. SDH is a world standard, whereas SONET is a North American standard. However, SONET is considered a subset of the SDH standards. Prior to the publication of the SDH standard, different rates and speeds were available in the North American community (which the United States, Canada, and Japan) compared to the rest of the world. The ITU was looking for a means of accommodating the differences in a standardized package so that both standards could coexist. SDH marks one of the first steps in making that happen. However, although SDH sets a rate of multiplexing and the derived speeds (and the means of achieving those speeds), some differences between SONET and SDH exist. The packaging of the tributaries is different.

In Table 27-7, a comparison of the SONET and SDH architectures is shown. SDH defines the *Synchronous Transport Module* (STM) level N, whereas SONET defines the STS-n and the OC-n.

Table 27-7

Comparison of SONET and SDH rates

SONET Signal	Rate	SDH Signal	SONET Payload	SDH Payload
OC-1	51.84 Mbps	STM-0	28 DS-1 or 1 DS-3	21 E-1
OC-3	155.52 Mbps	STM-1	82 DS-1 or 3 DS-3	63 E-1 or 1 E-4
OC-12	622.08 Mbps	STM-4	336 DS-1 or 12 DS-3	252 E-1 or 4 E-4
OC-48	2.488 Gbps	STM-16	1344 DS-1 or 48 DS-3	1008 E-1 or 16 E-4
OC-192	9.953 Gbps	STM-64	5376 DS-1 or 192 DS-3	4032 E-1 or 64 E-4

A. The OC-n also has a compatible STS-n rate.

B. Even though the rates are the same, the packaging of the frames for SDH and SONET is different.

Synchronous Digital Hierarchy

Ever since the standards bodies approved the recommendations for the SDH (and SONET), the services have been effectively used to improve and revolutionize the industry. Significant cost efficiencies and performance improvements have been shown. SDH provides a means for the rest of the world to use the capabilities of fiber-based transport systems and multiplexing architectures to improve upon the older *Plesiochronous Digital Hierarchy* (PDH), which was inefficient and expensive.

Digital networks continue to expand in complexity and penetration within the carriers' networks, now moving closer to the consumer's door. High-speed communications prior to the formulation of SDH in 1990 operated at speeds of up to 139.364 Mbps. However, the carriers implemented coaxial and radio-based systems operating at 140 Mbps– 565 Mbps. The networks were severely constrained due to the high cost of the transmission medium (coaxial cable, especially). The multiplexing rates used plesiochronous rates, which led to the European PDH.

After the development of fiber and the enhancements of integrated circuitry, the newer transmission speeds and complex networking architectures became realistic. In Europe the evolution and deployment of ISDN also led to the proliferation of the B-ISDN standards, which allow a simple multiplexing technique. In the United States, the Bell breakup prompted the local carriers to look for interoperability and improvements in network management because of the proliferation of the number of carriers providing long distance services.

The ITU-TS agreed that something had to be done to improve and standardize the multiplexing and the interoperability, while at the same time taking advantage of the higher capacity of optical fiber. Older bit interleaving of multiplexers should be replaced by byte interleaving to afford better network management. The new standard appeared as SONET in the North Americas, drafted by Bellcore. Later, this same standard developed into the SDH/SONET standard, as approved by the ITU. Although SONET and SDH were initially drafted in support of a fiber, radio-based system, supporting the same multiplexing rates became available.

28

Wave Division Multiplexing

Introduction

In the past few years, several major improvements have been made in the use of optical fiber communications. In earlier discussions regarding the use of SONET (see Chapter 27), the issue of bandwidth surfaced. Not only does the issue of bandwidth keep coming up, but also the problem of just how quickly we consume all the bandwidth that is made available. No matter how much or how fast we improve our spectrum availability, newer applications crop up that literally *eat up* all the capacity available to us.

To solve this problem, the use of frequency division multiplexing with our light-based systems became a topic of research. What ensued was the ability to introduce various wavelengths (frequencies) of light on the same fiber cable, and the resultant increase in possible throughput. We have seen increases that approximate between 16–30 times the original capacity of a single fiber. These capacities are now being pushed to the limit with variations being promised of up to 128 times the capacity of the existing fiber technologies. What this means is that the old days of having to replace the in-place fibers with new technology have been replaced with newer technology that uses the in-place fiber, necessitating only the change in electronics on the line. We can expect to see these advances provide virtually unlimited bandwidth without en masse changes in the infrastructure. The future holds the promise of reduced costs, increased bandwidth utilization, and ease of implementation that meet the demands of our higher speed communications.

Wave Division Multiplexing

Ten years ago, the implementation of the OC−48 SONET specification had the industry believing that limitless bandwidth was available. One can just imagine that a mere decade ago the 2.5 ± Gbps capacities of the optical fiber networks was innovative and exceeded our wildest imaginations about how we would ever fill these communications channels. Yet, the industry began to recognize the changes in consumption patterns. In the demand for multimedia communications, video WAN started to erode even the highest capacities available. To solve this problem, researchers began to experiment with the use of more than one light beam on the same cable. Light operates in the frequency spectrum similar to the older cable TV systems, employing *Frequency Division Multiplexing* (FDM), as shown in Figure 28-1.

Figure 28-1

Frequency Division Multiplexing has been used by cable TV operators to carry more information on their coaxial cables.

By using different radio frequencies on a cable TV system, the carriers were able to expand the number of TV channels available to them on the same coaxial systems. Why not do the same things with the various frequencies of light? Thus, a new era was born in the use of wavelength division multiplexing. By adding a tightly separated wavelength of light on the same cable, more capacity could be obtained.

In normal fiber transmission, the use of two standard wavelengths was originally deployed as shown in Figure 28-2. First, a red band of light was used to transmit the signal from the originating end. A blue band was used at the opposite end. Therefore, the same cable could carry both send-and-receive traffic by using different color bands. The original OC−48 transmission (operating at 2.5± Gbps) was exciting using a single wavelength of light and driving the signal over 20–30 miles of fiber. However, the carriers soon recognized that skyrocketing demand for the bandwidth would outpace the throughput of this single wavelength transmission.

Depending on the usage of various types, the life cycle of the OC−48 was anticipated to be approximately two years. Clearly, the rapid increase in Internet access, cellular communications, high-speed data, and the multimedia improvements were on a collision course with the capacity of this OC−48 architecture. Choosing to move to a higher capacity multiplexing technique in *Time Division Multiplexing* (TDM) was a definite consideration. This led to the higher rate multiplexing at the OC−192 level (9.953 Gbps ±), a four-fold increase in the speed and throughput of the SONET networks. Yet, even though the increases were achieved to the OC−192 level, the initial implementation was with one wavelength of light.

At the same time, OC−48, using two wavelengths produced a 5 Gbps throughput on the same fibers, proving that the technology could work. Shortly after the introduction of the OC−192, strides were taken to introduce OC−48 running four wavelengths (10 Gbps), or a single OC−192 using one wavelength.

Shortly after 10 Gbps were demonstrated, the designers began to experiment with a 20 Gbps capacity, using eight wavelengths of OC−48 or two

Figure 28-2

The colors of light separate the bands carried on the cable

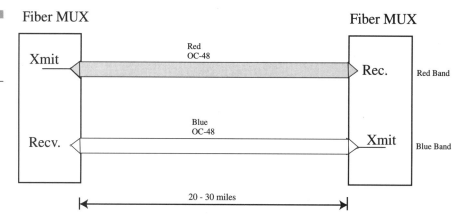

wavelengths at the OC−192 rate. Now the stage was set to push the envelope as far as possible! It was only a matter of a few years before the developers began introducing quantum leaps in their multiplexing ability. Now with multiples of OC−48 and OC−192, the capabilities of fiber-based transmission exceeded their developers' wildest imagination. Capacities for the DWDM service were now ranging from 160 Gbps to as much as 400 Gbps. This is the first of many steps we can expect to see in the near future. Some rumors indicate 128 wavelengths of OC−192 being possible shortly after the turn of the century. That is 1.2 Tbps!

Table 28-1 summarizes the capacities of the DWDM today and what the future may hold. This table shows the level of multiplexing, as well as the possible throughput we can expect for the future. This table also shows the number of wavelengths on the single mode fiber.

Fiber Optics Summarized

Before proceeding too far from the beginning of dense wave division multiplexing, it may be appropriate to look at the fiber optic world and its relationship to the demand of the industry. Moreover, the types of fiber in use determine the capacity being delivered today, so a quick review of the variations of fiber may help to define the overall benefits of the multiplexing schemes on the glass.

Fiber has been used in North America since the 1960s. The carriers recognized the limitation of the copper cable they had been installing for

Table 28-1

A summary of current capacity of the DWDM services

DWDM level	Number λ	Number OC–48	Number OC–192	Total Throughput
1	1	1	–	2.5 Gbps
	2	2		5.0 Gbps
2	4	4		10 Gbps
	1		1	
3	8	8		20 Gbps
	2		2	
4	4 Hybrid	1	2	25 Gbps
5	16	16		40 Gbps
	4		4	
6	8 Hybrid	4	4	50 Gbps
7	32 Hybrid	32	8	80 Gbps
8	16	12	4	100 Gbps
9	16	–	16	160 Gbps
10	32 Hybrid	16	16	200 Gbps
11	32	–	32	320 Gbps

years. First, the copper was very expensive. Moreover, it was also highly susceptible to noise and interference (both *Radio Frequency Interference* [RFI] and *Electromagnetic Interference* [EMI]). The carriers tried to get more capacity on the wires to make the cable plant more efficient and cost-effective. However, modulation techniques left the carriers wanting more. Most of the cabling systems were installed by using some form of analog transmission system. In the late 1950s and into the 1960s, the Bell System developed a digital modulation technique, which could take advantage of the limited bandwidth and capacities of the copper cables.

Radio-based systems were also in heavy use on the carriers' networks. The limited frequencies, radio interference, and line-of-sight demands of radio all led to frustration. As a result, the use of fiber optics became very attractive to the carriers. Fiber is made up of glass, which is an inert material that is unaffected by electromagnetic and radio frequency interference.

The thickness of the glass is used to determine the throughput and the light capacities in a carrier-based network. Two primary types of fiber were developed over the years to support the communications demands of the network. These two types are multi mode fiber and single mode fiber. Multi mode fiber is thicker and therefore limited in its bandwidth. Single mode is thinner and purer so it has greater capacities.

Multi Mode Fiber

When fiber was first being used in the communications area, the technology was still developing. Multi mode fiber was used as a means of carrying communications signals on the glass more reliably. Multi mode, as the name implies, has multiple paths where the light can reach the end of the fiber. Single mode fiber, on the other hand, has only one path that the light can travel to get to the other end. Using a large piece (chunk) of pure glass, the developers extruded the fiber into much thinner glass strands.

The multi mode fiber is the thicker glass by today's standards. The multi mode was developed in two different types, step index and graded index, both operating differently. Figure 28-3 is a representation of the multi mode fiber, using a step index mode. In this case, the glass is very thick at the center core (approximately 120–400 microns). The thickness of the glass is crucial to the passage of the light and the path used to get from one end to another. The step index is the thickest form of fiber; using the core of 120–400 microns thick, the light is both refracted and reflected inside the encased fiber. Because of the density of the glass, the center core refracts the light in different angles. At the same time, there is an outer cladding on the glass used to reflect the light into the center of the glass. This combination of refraction and reflection of the light, along with the density of the glass causes the light to take different paths (or bounces) to the end of the cable. Different light beams inserted into the fiber will take different lengths of path and therefore different timing periods to get to the end of the cable.

Figure 28-3
Step index fiber optics

120 - 400
Microns

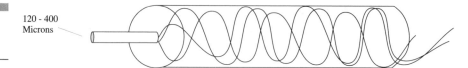

Graded Index Fiber Optics

A second form of multi mode fiber was the use of a graded index of the glass, as opposed to the step index. Essentially, the grading of the refraction levels of the light was changed to compensate for the amount of refraction and the density of the glass. In the step index, modal dispersion (spreading out of the light) was in part responsible for a reduction of the throughput on the fiber. By using a grading of the density of the glass from the center core out, the capacity of the fiber was increased. Simply stated, the more dense the glass was, the more refractive the surface of the glass was; or the more refraction taking place, the longer the path. By having a step index, the path of the outer part of the glass was longer than the path in the center of the glass. This meant that the light arrived in different times because the length of the path was longer. Grading the center core to be a higher level of refraction and the outer parts of the glass to be thinner (and less refractive), could use the characteristics of the glass to get approximately the same length of a wave on the cable and therefore increase the speed of throughput. The graded index is shown in Figure 28-4. The better the grading of the index, the more throughput we can expect. Currently, the two forms of graded index fiber use either a 62.5-micron or a 50-micron center conductor with a 125 micron outer cladding on the glass.

Single Mode Fiber

As fiber became more popular and research was stepped up, a newer form of glass was developed. If the glass could be made so thin and so pure in the center, the light would have no choice but to follow the same path every time. A single path (or single mode) between the two ends allowed the developers to speed up the input because there was no concern about varying lengths of the path as shown in Figure 28-5. Using a single mode fiber today, the thickness is approximately 8.3–10 microns thick at the center. There is still an outer cladding on the outer edges to reflect the light back into the center of

Figure 28-4
A graded index fiber

62.5
or
50 microns

Figure 28-5
Single mode fiber is
so pure and so thin
that the light has
only one path from
end to end.

the glass. The outer cladding is still approximately 125 microns. The single mode fiber is where most of the activity is being generated today. Many of the telephone carriers deployed a multi mode fiber in their networks when it was first introduced. However, over the years the multi mode has given way to single mode fiber throughout the public carrier networks.

Benefits of Fiber Over Other Forms of Media

The benefits of fiber over other forms of media are shown in Table 28-2. However, it is important to note that the fiber has been used extensively in the long distance telecommunications networks and the local telephone company networks. It is more recently that a single mode fiber has worked its way into the end user networks (LANs and CANs). With single mode fiber, the speeds are constantly being upped, while the error performances are continually being improved. This table concentrates on the bit error rates and the speed of the cabling systems.

Wave Division Multiplexing

Now, back to the wave division-multiplexing concept. There is no mistake that the demands and needs for telecommunications services are growing.

	Advantage	Description
Table 28-2 Fiber based advantages over other media	Lower errors	BER approximates 10^{-15} for fiber, whereas copper will be in the 10^{-4} to 10^{-8} range
	Attractive Cost per foot	Cost per foot on fiber is now approximately $.20 compared $.13 for copper
	Performance	Immune from RFI and EMI without extra cost of shielding on copper
	Ease of Installation	Ease of installation due to lower weight and thickness
	Distances	Greater distance with less repeaters. Now can achieve 30–200 miles without repeaters. Copper and radio limited to less than 30 miles
	Bandwidth Improvements	Fiber nearing 1 Tbps, copper 100 Mbps and coax 1 Gbps
	Capable of carrying analog and digital	Using TDM and WDM, the fiber is both digital and frequency multiplexed increasing capacity

The combination of voice, data, and multimedia applications that are constantly putting pressure on the infrastructure add to the growing problem. Where normal digital transmission systems use *Time Division Multiplex*ing (TDM) and cable TV analog technologies use *Frequency Division Multiplexing* (FDM), *Wave Division Multiplexing* (WDM) is a combination of the two schemes combined. The use of TDM in a multiplexer breaks the bandwidth down to specific timeslots such as those found in SONET based networks. However, by using the combination of Frequency (different wavelengths) and Time (timeslots), the fiber can be used to carry orders of magnitude more than traditional fiber-based systems. Figure 28-6 shows the combination of the two multiplexing arrangements in a single format. Furthermore, there are several different wavelengths and colors of light that can be used to produce far more capacity.

No one can mistake the fact that telecommunications capacity needs are rapidly growing. DWDM was developed to produce even better results on a single fiber pair than the original techniques deployed on the backbone networks. Expanding capacity with DWDM can produce some significant improvements for today's technology. An example of this is the fact that some DWDM multiplexers can produce up to 240 Gbps on each pair of fibers. This equates to a call carrying capacity of up to 3.1 million simultaneous calls on a single pair of fiber. Now many of the carriers have installed

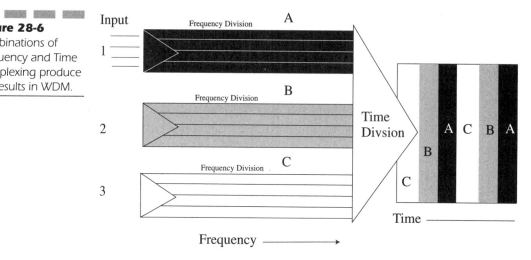

at least 12 pairs of fiber in their backbone networks. Performing the math yields a total of 37.2 million calls simultaneously on the fiber, using current technology before more fibers are required. This is obviously very attractive to the carriers who would prefer to get more service on their existing infrastructure rather than installing more glass in the ground. Wave Division Multiplexing is particularly useful when congestion begins to build up on the existing carrier-based networks. Surely the equipment is expensive, but the cost of adding multiplexing equipment or changing it out, is far less expensive than digging up the ground to lay more fibers. The electronics for the DWDM multiplexers are rapidly becoming very cost efficient.

Using different light wavelengths, DWDM simultaneously transmits densely packed data streams on a single fiber. By combining DWDM with special amplifiers and filtering equipment on the links, the carriers can achieve unprecedented throughput on their existing single mode fiber. Current technology supports approximately 16 different lengths on a single fiber (OC–192 is 10 Gbps, using 16 wavelength produces up to 160 Gbps in each direction). As stated earlier in the beginning of this chapter, 128 wavelengths are targeted for early in the new millennium. Table 28-3 is a summary of some of the benefits of the DWDM usage.

Why DWDM?

Clearly, the use of the fiber technologies has taken over the industry. However, the more information that can be generated over a single fiber, the

Table 28-3	**Industry demand**	**What causes the demand**
Summary of demand to justify the use of DWDM	Need for bandwidth	Internet access, *Cellular and Personal Communications* (PCS) and data/voice integration.
	Need for reliable communications	In order to guarantee the reliability those customers are demanding, the carriers have been committing alternate routes and spare fiber capacity to back up existing infrastructure.
	More capacity	In order to get the *service level agreements* (SLA) and maintain the network in a fashion expected by the consumer, the carriers have to install backup circuits and fibers in their highest density routes.
	Higher performance on the network	Network dependency has become the norm. All forms of traffic must run on the existing infrastructure, and the carriers must provision capacity to meet the ever-changing demands of the network.

better the carriers like it. To be specific, before digging up the ground to lay more fibers, it is far less expensive to use the multiple wavelengths of light on a single fiber. Thus, as the carriers seek to gain more utilization on the existing fiber, the incentive will be for the manufacturers to continue to proliferate this technology.

The major manufacturers have obviously put all their resources into the development of the dense wave division multiplexing techniques to meet the rising demand from the carriers and the end users alike. Using multiple wavelengths allows the carriers to achieve the 320 Gbps capabilities today, with terabit speeds being discussed before the new millennium. Using a few different techniques, the major suppliers have developed keen interest in the ability to utilize DWDM. *Multiwavelength Optical Repeaters* (MOR) help in deploying this technology out to the carrier networks. A MOR (or MOR plus) is an amplifier used to support at least 16 wavelengths on the same fiber in each direction. Using 16 wavelengths (16 λ) operating at the OC$-$192 rate yields a total of 160 Gbps in each direction or 320 Gbps in one direction. In Figure 28-7, a typical multiplexer is used to generate a bi-directional flow of information through the fiber multiplexer. This figure shows the major components of the systems used by Nortel, but can be the same as used by others (for example, Lucent Technologies, Siemens, or Alcatel). In either case, the major point here is that the use of a DWDM coupler aggregates the capacity of the fiber by using two different light bands (red and blue) and multiple wavelengths to achieve the results we have grown to expect.

Figure 28-7
A typical DWDM
multiplexer

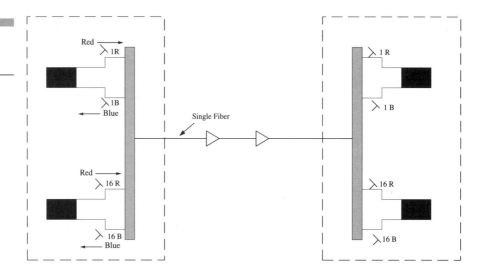

By using the fiber-based example for the increased capacity of the fiber, the DWDM method appeals to many of the providers and carriers alike. The use of fiber-based multiplexing also adds some other enhancements that were not traditionally available in the past. Because the signal never terminates in the optical layer, the interfaces can be bit-rate and format independent, allowing the service providers the opportunity to integrate DWDM easily with their existing equipment and infrastructure, while still gaining the access to the untapped resources and capacities in the existing fiber.

DWDM combines multiple optical signals so they can be amplified as a group and transported over a single fiber, increasing the capacity. Each signal carried on the DWDM architecture can operate at a different rate (for example, OC−3, OC−12, OC−48 and so on) and in a different format (for example, SONET, ATM Cells, data, etc.). Using a mix and match approach allows the carriers to achieve different rates of speed and aggregated throughput, depending on the multiplexing equipment used. Future DWDM developments are touted as being able to carry up to 40 Lambda of OC−192 or a total of 400 Gbps. As a comparison, 400 Gbps will be 90,000 volumes of the encyclopedia in one second!

DWDM requires that the transmission lasers have a very tight wavelength tolerance so that the signals do not interfere with each other. After all, the systems carry 16 or 32 different wavelengths on a single fiber. It is imperative that the tolerances be held tightly to prevent mass destruction of the data signals being generated on the fibers. The *International Telecommunications Union* (ITU) has specified 100 GHz standard spacing between the wavelengths, and most of the vendors are now manufacturing lasers at the standards-based wavelengths for the combination into DWDM.

The most advanced today, using OC−192 lasers, has a tolerance of 0.1 nm, which is small enough to preclude any interference between two adjacent wavelengths. Designing OC−192 DWDM systems is ostensibly more difficult to design than the older OC−48 systems. The OC−192 systems being deployed in the networks also use both unidirectional and bi-directional systems in support of the fiber and DWDM multiplexers. However, the future may change things very quickly. Where OC−192 offers a 10 Gbps operation, the future offers an OC−768, which will create a fourfold increase in the capacities in the network. At the OC−768 rate, a multiplexing scheme will produce 40 Gbps on a single fiber rate (SONET rate), which then can use the DWDM capabilities to create a throughput of up to four times the 320 Gbps rate currently available. That equals terabit speeds (1.28 Tbps)! However, the use of spacing on the various lambdas will actually allow up to 400 Gbps and offers future speeds of up to 1.68 Gbps. What OC−768 will also bring to the table is a capability to concatenate the OC−192 levels. Today's current technology can use a concatenated OC−48, but stops there in the bonding of higher speeds. In the future, as ISPs and traditional telephone carriers need to expand their capacities, the need to draw a 10 Gbps concatenated throughput will appear. Today, this is not a big concern. However, in three to five years, this will become a commonplace occurrence. Table 28-4 is a summary of some of the other things that we are expecting to occur over the next five to ten years in the area of fiber transport systems. This table takes advantage of the work being done in many of the labs and research facilities around the world. The ability to push the envelope is what the communications industry is all about.

These rates are all projected to be available in the next decade. However, as the technology seems to be doubling every six to twelve months, the rates shown above could happen as soon as 2005! The issue is not how long it will take, but the ability to drive better and faster data and other communications needs on the same fibers. The future holds a lot of excitement with these speeds.

Table 28-4

A look at what DWDM and fiber rates will bring over the next decade

Technology and capacities	Current vs. Future technology
DWDM at OC−192 and 40 λ	Current technology capable of carrying 40 different wavelengths (using ITU 100 GHz spacing) at 10 Gbps each or a total of 400 Gbps. Currently, the industry has achieved a total of 320 Gbps with the 400 Gbps rate in the very near future.
DWDM at OC−192 and 80 λ	The current spacing of 100 GHz, as specified by the ITU, is under attack. The near future holds the promise of doubling the number of wavelengths by using a 50 GHz spacing, allowing up to twice as many wavelengths on the same fiber (80) each operating at 10 Gbps.
DWDM at OC−768 and 40 λ	This is a turn-of-the-century technology with up to 40 Gbps per wavelength and 40 wavelengths or a total of approximately 1.6 Tbps.
DWDM at OC−768 and 80 λ	By the year 2002, we can probably expect to see the OC−768 plus the use of 80 wavelengths or a 3.2 Tbps throughput on the fiber.
DWDM at OC−192 and 160 λ	By the year 2005 we can expect to see that the wavelengths' decreased spacing and tighter tolerances on the fiber lasers will yield a total of 160 different wavelengths at 10 Gbps each or back to the 1.6 Tbps rate
DWDM at OC−768 and 160 λ	This is likely a 2008–2009 technology that will again double the capacity and yield a total of 160 different wavelengths at the OC−768 rate (40 Gbps) or netting 6.4 Tbps on a single fiber.

The Internet

Introduction

We are accustomed to succinct definitions of systems and equipment. The Internet is a little difficult to describe in 25 words or less. The Internet is really a collection of privately owned networks. By private, we mean that it is not owned by the government but by private and public companies. Standards are set by the *Internet Architecture Board* (IAB) and the *Internet Engineering Task Force* (IETF). The IAB members are elected by the Internet Society. The IETF membership is voluntary.

A Brief History

The Internet started as *Advanced Research Projects Agency Network* (ARPANET). Because universities were working on research for the military, a means was desired to be able to share research papers. Some say it started before ARPANET by informal agreements between universities to dial each other up and exchange e-mail. At the time, there were no formal protocols such as IP. These informal dial-up connections were using acoustic-coupled modems at 1200b/s. Actually, there was no real network—only these agreements to forward mail.

ARPANET started with the first packet switching technology. The first packet software was called an *Intermediate Message Processor* (IMP) and was built under contract by Bolt, Bernak, and Newman. The goal was to remotely log into computers in Berkeley from UCLA. The initial experiment didn't work, but it provided the basis for continued improvement, which lead to ARPANET. The Internet began then as an interconnection of IMPs that could relay traffic between hosts. (A host was a mainframe or a mini-computer. Today, any device can interface to the network and run the networking protocols.)

The military changed the name of the project to *Defense Advanced Research Projects Agency* (DARPA). Under DARPA, the Internet continued to grow and expand. Outgrowths of this work were WMICS, WINS, MILNET, and eventually DDN. The military, therefore, had no need any longer for DARPANET and handed it over to the *National Science Foundation* (NSF) and the Internet became NSFNET. The NSF was funding most of the research anyway, and the universities were using some of that funding to support their Internet habit. The goal of the NSF was to provide shared access to supercomputers. Each university was naturally lobbying for its own supercomputer. Providing remote access via the Internet would be a lot less expensive for the NSF. The NSF funded the Internet from the mid-1980s until 1992 when benefits previously reserved for the academic community became apparent to all. What was happening was that students, who had had access to the Internet while on campus, were cut off upon graduation. They were therefore very creative in figuring out ways to hang onto their Internet accounts.

This created a set of "haves" and "have-nots." The pressure built for the commercialization of the Internet. Continuing the funding became the pressing question. Under the NSF, the Internet had grown to include virtually all-major schools. Since funding was from the government, access was easy. If your school wanted access to the Internet, you simply called up your nearest associated school who had access to the Internet and asked for a port on their router. You then leased to communications lines of the appropriate bandwidth. You sent some money to the next school to help them pay for the additional router capacity and communications capacity that they would need to have to support your additional traffic going through their system.

Notice that we said that the Internet was a loose collection of many networks, and here we just added another network. (By the early 1990s, the potential of the Internet was well recognized, and everyone was clambering for access.)

The initial purpose of the Internet was to exchange data and make research papers available to the entire academic community. Universities would therefore have open hosts connected to the Internet. This meant that anyone could essentially log onto the host as an anonymous user, search through the directories on the host server, and download whatever files were wanted. It's interesting to compare this free and easy access with today's world of networks protected by firewalls. Actually, it wasn't that free and easy because in the UNIX operating system, files and directories have permissions as to who can read and write to them.

The mechanism for doing this is a utility called *File Transfer Protocol* (FTP). In order to FTP, your system (host) had to have the FTP client and the target system had to have the FTP server ready and willing to process your request. If one is familiar with UNIX and the directory structure of the target host, this isn't too difficult. For the average techno-peasant, however, it is beyond grasp. Surfing or browsing, as we know it today, was not easy. Only the patient, persistent, and knowledgable browsers were able to find what they were looking for.

The Internet is not a *Bulletin Board Service* (BBS). A BBS is essentially a host into which anybody can dial and collect and receive information or messages. These services abounded in the late 1980s and early 1990s. The

Internet is not a timesharing service either, although some hosts provide that service. Timesharing became popular in the 1970s when computers were too expensive to own for smaller companies so they would do remote computing, using dumb computers using services such as Tymnet and *CompuServe Information Service* (CIS). CIS grew into an information resource provider and created agreements with Telenet and Tymnet to provide worldwide access while they developed their own worldwide network. The original CIS access was via dumb ASCII terminal at low data rates. They had a character-oriented information interface that was no friendlier than the Internet. Given the period, it offered a powerful service.

In the early 1990s, AOL saw the benefits of the Internet and began to offer an information service via a *Graphical User Interface* (GUI) that ;AOL;GUIwas very easy to use. Although AOL was a proprietary network, it rapidly added various information services in addition to the ever-present e-mail. It is likely that its phenomenal growth is due in significant part to its ease of setup and use. Several information services came and went. We like to maintain that AOL and CIS are not Internet service providers because their primary business is providing interactive information services. All their computers are in one room. Rather than being connected to a network of computers, you are connected (for all practical purposes) to one huge computer that offers many different services. By comparison, the Internet consists of millions of computers connected to millions of networks, each accessible from any other network. It is not obvious to the casual observer that there is much difference today because the browser interface cleverly hides the complexities of connecting to multiple different computers and both systems seemingly have endless amounts of information available. Today, both AOL and CIS permit seamless movement between their own information services and those available on the Internet. Selecting a link to information on the Internet causes the automatic launching of a browser.

AOL was the most successful of the information services, eventually buying CIS. This was primarily due to marketing, low cost, and ease of use. (CIS was a little late introducing a GUI and when they did, it was obscure, nonintuitive, and hard to use.)

AOL's membership snowballed because the more users there were, the more advertisers were willing to pay for advertisement and offer services via AOL. The more services they offered, the more attractive they became to more subscribers. The best part was that AOL could now charge more for its banner ads because they would be viewed by that many more folks when they signed on.

Early Internet Services

The Internet was, as indicated, difficult to use in the beginning, requiring knowledge of telnet and the UNIX command structure. (Not to be confused with the Telenet packet switching network) Telnet is a UNIX utility that lets you remotely log into another host. Once you logged in you had free reign of that system with two significant limitations:

1. You had to have an account on the target system (in the early days most systems would let you log in as anonymous).

2. You could only access the directories and files for which general user permission was granted.

This permission system was so clever that a user could be granted permission to see a directory and even run programs in it, but not be allowed to modify its content.

As indicated, when you found the file or program for which you were looking, you launched the FTP utility and downloaded it to your system. Searching for and finding things was difficult at best.

The next step in improving utility of the information on the Internet was the development of Archie, which was designed at McGill University in Canada. Archie essentially indexed FTP sites. The target machine had to run the Archie server, while your machine ran the Archie client. The good part was that the Archie server was accessible via an Archie client, e-mail, or telnet. Archie was a great catalog on the Internet. Archie returned a list in UNIX language that gave you host and file names to use when you set up the FTP session. Archie was a step in the right direction of making the Internet easier to use, but you still needed to know how to use telnet and FTP.

GOPHER

Gopher was also a client-server system. It provided a menu of choices, and it allowed you to set up bookmarks of locations you liked. Like previous Internet accesses, it was designed around the user having a dumb ASCII terminal. Thus, it provided menus and minimalist keystrokes to make a selection. That was the good part. Unfortunately, the single keystrokes were not always intuitive. The intent was that you would launch the Gopher client from your own host and communicate with the Gopher server on the remote host. It was possible to telnet to a remote host and then launch Gopher, but this was painfully slow and created excessive network traffic. Remember that the interconnection of hosts was often with 56 kb/s lines.

Veronica

There was some speculation as to the source of this name, but most people think it was named after Archie's girlfriend. Others insist it stood for a very easy, rodent-oriented net-wide index to computer archives, but this is unlikely. Veronica is an extension to Gopher that facilitated searches and returned a list of *hits* (hits in this case were file names that matched the name in the original search request). Veronica was also a client-server arrangement, where the Veronica server keeps the database being searched.

By selecting one of the returned hits, you could get more information about it or transfer it back to your computer. Veronica therefore, automated the FTP process, but was still dumb terminal- and menu-oriented.

WAIS

Wide Area Information Service (WAIS) was another database search engine that allowed you to enter keywords and search a database for entries. Although the number of databases was large, finding this was still not easy because you were using Gopher to search WAIS.

WWW

The World Wide Web has essentially replaced all of these older search engine capabilities. The early WWW often resorted to Gopher or WAIS to actually do the transfer.

The two developments that made the WWW useful were browsers, hypertext and hyperlinks. Hypertext was a way of encoding formatting information, including fonts, in a document while using plain ASCII characters. Hyperlinks were essentially addresses imbedded in the hypertext Web page. By selecting the hyperlink, you were taken directly to a new page.

Browsers

Browsers basically automated the searching on keyword function that we had done via menus using Gopher. Browsers today are very large and complex programs. Look on your hard drive, and you will find that the Netscape application is about 1.3 megabytes. With each revision, it gets larger and more complex as new features are added. Microsoft's browser is buried in the operating system, and it is therefore difficult to tell how big it is.

Marc Andreesen was instrumental in creating one of the first browsers called Mosaic. It became the foundation of Netscape's Navigator. Netscape's success can be attributed to the fact that it followed the AOL model of giving away the browser software. Netscape even gave away their source code to its browser on the theory that thousands of heads are better that a few, and eventually it will result in a better product. This same philosophy made Linux such a strong operating system for PC platforms.

Browsers are also part of the client-server world. The money lies in getting the service providers to buy your suite of server software that provides the data to the browsers. This suite of software may provide many other features and capabilities such as calendar, proxy, LDAP, and mail servers as well.

Hypertext

In the UNIX world (where this entire Internet started), there were only two kinds of files:

1. ASCII text-based files that were all the documentation, source code, and configuration files

2. Binary files that were the executable program files

Unfortunately, no one considered fancy formatting, multiple fonts, graphics, and tables. ASCII text was boring in the modern world of animated color. The question became, how can we add this capability?

Individual vendors' products, such as Word (registered trademark of Microsoft), utilize proprietary code sets in which the font and size (for example) are imbedded. Other products, such as WordPerfect (registered trademark of Corel), chose to imbed special tag characters, indicating the beginning and end of special font and size groups of characters. Unfortunately, with special (non-ASCII) characters imbedded in the text, these were no longer text documents, but binary documents that could only be operated on by proprietary vendor-specific programs that were not universally available or free.

How then could we keep the documents on the Internet open, free, standardized, and comprised strictly of ASCII characters so that anyone could read them? How could we extend the capability without making the previous version obsolete? A major problem with a specific vendor's product was that when the new one came out, the older versions couldn't read the newer versions' formats. This was a major inconvenience, designed to force the users to upgrade to the latest version.

The solution was to go with the tag approach. Rather than using special characters as tags, we simply used a sequence of ASCII characters, which meant something specific to the browser and did not impair the ability of a dumb ASCII terminal to read them. For example the tag <title>Important Subject</title> will cause the browser to display that line as a title. For the curious, you may view the source of a Web page and see all these tags. All you have to do is go to the View menu and select Source. You are then presented with all the original ASCII information. Although a little difficult to read because of all the tags, all the text is there as are all the references to other Web pages (HREF) and all the font and formatting information. While we are on the subject, you might try converting one of your text documents to Rich Text format. Here the formatting stuff is all in the beginning of the document-just another (standardized) way of sending formatting information in plain ASCII text format.

If you are using an older browser, it simply can't properly display the text within the new tag, but the text is still there, and you can read it. Fortunately, new versions of the browser are readily available and free for the downloading.

Hypertext then allows the standard ASCII text characters to define special formatting which the browser can display.

Hyperlink

A hyperlink was simply a pointer to another Web page. It was the complete address to find the specified Web page. The link visible on the browser-presented Web page might say something innocuous like "more info." If you viewed the source and searched for the HREF or the "more info," following the href= will be the actual path to that page. Selecting the hyperlink caused a lot of background processing. The browser took the *Universal Resource*

Locator (URL) and fabricated a query to that location just as though you had filled in that value manually in your browser window. It then set up a connection, downloaded the desired page, and terminated the connection.

URL

The *Universal Resource Locator* (URL) was simply the Internet address of the Web page. URLs were displayed in this form www.tcic.com. This URL took you to TCIC's main Web page. Selecting a hyperlink from that page took you to a *Universal Resource Identifier* (URI), which pointed to the files in the directory tree of the host server. Each slash (/) in the name identified a directory level on the server. In some cases where a document management system was employed to build and provide Web pages, these slashes were actually logical divisions within the resource and had nothing to do with actual directories. Today, the trend is to use dynamically built Web pages. They can be better customized (see the discussion on cookies) to the user's needs, you don't have to store thousands of different Web pages, and the processors are fast enough to create them quickly.

DNS

One of the most interesting and important parts of the Internet is the *Directory Name Service* or the *Domain Name Service*. In short, the DNS system permits human readable names to be quickly translated into IP addresses that are needed to route the packets across the network.

As is described in the addressing section, the IP address is structured by network and then by host. It is read from left to right by the router in trying to find the proper destination network.

The human readable addresses are hierarchical from right to left. For example, take the address Bud@TCIC.com. First, we know that it is somewhere in the .com domain. If I am George@biguniversity.edu, I need to find Bud's address. It is very likely that the university has no idea of what the address might be. The local e-mail system therefore makes a query of the unnamed "." domain. There are several of these servers around the Internet, which are updated daily by the InterNIC as names are registered.

There is some controversy over the InterNIC's absolute control over domain names. (Someone has to be in charge to prevent chaos.) To register a domain name, one (or one's ISP) must contact the Internic and pay an annual fee. (Prior to the commercialization of the Internet, this function was performed by *Stanford Research Institute* (SRI). There are proposals to open other domains (for example, .biz) and let other entities administer the names within that domain.

Each country has a domain and administers these domain names themselves. All public entities such as cities, counties, and states are under the .us domain. Some of the country domains are uk for united kingdom, cz for Czech, .de for Germany, .au for Australia, etc.

Each domain then has its own DNS server so when George is trying to send e-mail to Bud, his e-mail server asks the "." domain for Bud's address. The "." domain replies that it can only provide the address of the .com DNS. We then ask the .com DNS that replies with the DNS address of TCIC. Since TCIC is really under an ISP, what we really get is the address of the DNS server at that ISP. Finally, we get Bud's real address. Now, this address is put into the e-mail packets, and they are sent on their way. In addition, we, the users, never knew about all the fooling around that went into finding the address in the first place.

JAVA

The early Internet was strictly ASCII text-based. Then came the inclusion of *Graphical Interchange Format* (GIF) files. As indicated previously, these were simply references in the hypertext document to a location that contained the graphic file that was displayed. Next, came the desire to automate or animate Web pages.

Sun Microsystems invented a clever (and, to some, controversial) language called Java, which is a registered trademark of Sun Microsystems.

The basic problem was that to animate a page the local machine had to execute some software. This opened the door to viruses. Sun's clever idea was to have the browser execute the program, rather than the host hardware. The good part was that you were somewhat protected from malicious programs. The bad part was that the browser was interpreting the Java language on the fly, and this interpretation was slow. (Faster machines help a lot here.) The original idea was that the browser could prevent the Java script from doing any damage (like wiping out the hard drive). Unfortunately, the more powerful we needed the Java to become, the more capability it needed on our host machine. Microsoft, naturally, has its own approach to page automation called Active-X. The good part is that it runs as machine code. The bad part is that it runs as machine code and can therefore contain viruses. The user is given the opportunity to download or not download the Active-X code. If you trust the source, go ahead and run it. If you are not sure, cancel the download and do without the animations.

Surfing the Web

When you select a hyperlink, the browser creates a packet requesting a Web page and sends it to the specified URL. Your browser actually sets up a connection to the server. The server replies with the requested file (Web page) and the browser displays the page. Your browser now stores this page in its cache memory so that after you have followed several other links, you can easily get back by selecting the back button. The button retrieves the page out of cache rather than having to fetch it from the source (which as you have experienced, could take a while). You should periodically empty your cache. First, depending on the settings of your browser it may never use the cache again. Second, if the browser does always check cache first, regardless of age, when you sign onto the site the next time a month later, the page you see is the old one from cache. Actually, this is an exaggeration because the browser has a setting for how old a page can get before a fresh copy is fetched. You can just throw away the cache folder (it is safer to discard the contents); your browser will build a new one when you launch it. If you do a lot of surfing, this cache can take up a lot of disk space.

You can always force the browser to get a fresh copy of the page by selecting the Reload button. This causes the browser to ignore whatever was cached. The fact that each page can contain multiple references to other pages anywhere in the world is the reason it is called the World Wide Web. Links can take you anywhere including back where you started. There is no hierarchical structure to the Web.

Tracking Visitors

From a commercial point of view, we would like to know how many visitors or hits we have on our Web page. (Here hit is defined as someone accessing our page, for example, setting up a connection to us.) We can tally each connection and present that to potential advertisers as an indication of the popularity of our page. First, the number of connections or hits to our page is only approximately related to the number of viewers. Here is why: If you happen to set the home page to www.tcic.com, every time you launch your browser, you go there. It looks like a hit! The fact that you immediately go to some other book-marked page doesn't register. Second, depending on your browser settings, you could revisit a page multiple times, but your browser has that page cached. The Web page owner then has no way of knowing that you are frequently referring to his page during your online session. Third, if the client or user is surfing from behind a proxy, his local proxy server will provide your page whenever requested, saving network bandwidth, but the Web site owner again doesn't know that he has been hit once again.

Cookies

One of the more controversial aspects of the Web is the existence of cookies. Cookies are nothing but an encoded set of information that the Web server asks your browser to keep for it. The cookies simply contain information about you and the sites you visit. They may contain information, such as your credit card number, that you have entered while visiting a site. They may also contain links you selected from that Web site. If you visit a major outfitter.com and you look at hiking boots, the next time you log into that site, the first page may have backpacks and tents prominently displayed. The site read the cookies it left on your machine the last time you visited. It determined that you are an outdoor type (and decided it would be a high probability of sale for you on related outdoor equipment. However, you still have the ability to view all the other parts of the site because the index page always contains an index to other pages on the main page.

Cookies, then, are a convenient way for the vendor to keep information about site visitors without having to keep a huge database of all visitors whether they are casual or frequent visitors.

You can throw away the cookie file anytime you want. The browser will rebuild it when it needs it. You can also read your cookie file by opening it in a text reader. Depending on the browser, the cookie file normally contains ASCII text. The information is normally encoded as a set of numbers that are not meaningful to anyone but the originator of the cookie. Some Web pages won't work correctly unless you have the cookies enabled. You may leave "accept cookies" turned on and discard the file at the end of your session if you are paranoid about the information that might be stored therein.

Search Engines

Even if we have links to follow, there is no good way to find a specific set of information. We still need a database catalog that we can search that lists sites that might contain the information we want. The search engines (such as Yahoo! and Alta Vista) provide a database to search very similar to the way Gopher and Veronica tools did in the past.

Enterprising individuals also developed "Web crawlers" that would follow hyperlinks based on a key word and fetch all the associated pages. You could start these crawlers, go to bed, and wake up with a full disk. Today's databases are a combination of crawling and advertising. The business plan of the search engine provider is to offer advertising along with the database information. Companies that pay more get a better position on each page. In a few cases, the ordering of the hits on the page is a function of how much the information source paid to gets its listing put first. The goal of the game is to entice the Web surfers to your site. Once there, you can either sell them something or present them with advertising for which you have been paid. The more visitors to your site, the more you can charge for the so-called "banner advertisements." For folks looking for information, these banner ads are just background noise. Being enticed to look at one will often lead you on an interesting, but irrelevant, wild goose chase.

Standards

Within the Internet, standards are created and managed on a voluntary basis. The IETF makes recommendations to the IAB for inclusion as standards. Remember that the whole Internet started as a volunteer-based network, each building on what was done before. Anyone who determines a new feature is needed or a problem needs fixing creates a solution. That solution is implemented on one's own network, and when ready, it can be submitted to the IETF as a *Request for Comment* (RFC). It is then published on the relevant newsgroups for others to try it and comment on it. After it has survived this torture test, it is ready for formal adoption. Because the whole Internet is voluntary, it is up to the network administrator to decide whether to use that RFC or not. Failure to implement it, however, may mean compatibility problems with other networks.

This process is very practical and very different from that used by the formal international standards organizations, such as the *International*

Standards Organization (ISO) and the *International Telecommunications Union* (ITU). These have a formal membership and a proposal submission and review procedure that is designed to form a consensus.

Internet Operation

We have seen that the Internet is really a collection of independently owned and operated networks. The concept of network was previously dominated by the telephone model. It had a central switching exchange and all information ran through this exchange. The telegraph network also used centralized switching. Like the telephone switching system, it, too, had telegraph lines that connected the switching centers. The telegraph-switching center was called a message switch. These message switching centers were originally manually operated copy and relay centers. Later, paper tape with automatic punches and finally computers replaced the manual switching centers. It became clear that some mechanism was needed to distinguish between the message content and the control characters that were used by the computer to recognize the beginning and end of the message.

Message switching with its telegraphy history was message integrity-oriented. Each message was carefully acknowledged. Losing messages was not acceptable. This formed the basis for reliable message transfer.

About the time the Internet was taking shape, the X.25 network protocol was being developed. One might say that the X.25 protocol is the successor to message switching. It was essentially designed with the Telco model in mind. It was intended to make money by charging for packets, just as the message switching system was designed to charge for each message. The goal was to provide end-to-end message integrity.

A further impetus to the development of X.25 was the deregulation of the telephone equipment market brought about by the Carterphone decision of 1968. Heretofore, modems were only available from the Telco, and the data rates were slow. At that time, the Telcos controlled telephone calls between countries and all data (telex) messages went through what were known as international record carriers. Because they were the only carriers permitted to transport such international traffic, the cost was high. Modems unfortunately fell into the hands of the travelling public who would make dial-up telephone calls across international boundaries and transmit data at a fraction of the cost of the "record traffic." The Telcos tried to control this, but it was hopeless. Their next attempt was to offer a slightly lower cost *Public Data Network* (PDN), which was the X.25 network. Although it worked well, it really was still more costly (although more reliable) than using the *Public Switched Telephone Network* (PSTN). The X.25 network replaced the telex network to most parts of the world. When traveling, using the X.25 network to contact, for example, CompuServe is often the most cost-effective way to do it.

One of the large steps forward that X.25 provided was that it made the transport mechanism (packet switch) separate from the message content. The network was then optimized for packet transport, and was completely independent of the message or its content. The maintenance of end-to-end connectivity is a natural outgrowth of the Telco and message switching history.

The operation of X.25 is very different from the *Internet Protocol* (IP). It is important at some point to draw the distinction between the two protocols.

Without a full-blown discussion of the OSI Reference Model, let us start with the basic concepts of packet switching and the transport of packets across an X.25 network. Figure 29-1 is a picture of the Reference Model to be used as a reference as we discuss the operation of networking protocols. A few quick notes are in order:

1. The physical layer attaches to whatever network is at hand. There could be multiple physical connections in any given system.

2. The link layer handles communications across the physical link only between adjacent machines by putting data in frames.

3. The network layer must be consistent across the entire network, and it is where packet switching is handled.

Figure 29-1
The OSI model as the reference

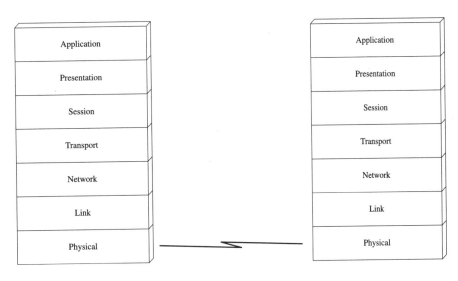

4. The Transport layer handles end-to-end transport and integrity of data. The communicating entities at transport layer have no idea (very theoretically) that the underlying layers exist or what their characteristics are.

X.25 is a *Connection Oriented Network Protocol* or service (CONS). The IP provides, by contrast, a *Connection-Less Network Service* (CLNS). This is the major difference, and as we shall see, it pervades the networking decisions made and the choices made for the functionality of the transport layer. The details of how the packets are handled within the packet switches are very similar in terms of how the queues are handled.

A connection is established by X.25 for the end user across the network. Although we said that transport is responsible for end-to-end connections, network layer can also choose to provide this capability. If our network layer is going to do this, then the transport layer need not be complex at all. You will note in the following discussion that the reference is to telephone terminology. The reasons for this are now clear.

An X.25 connection is established by the end system across the network. The calling party sends a setup packet containing the address of the source and destination systems. Figure 29-2 shows the basic network configuration. The packets travel on each leg of the journey in a link layer frame that

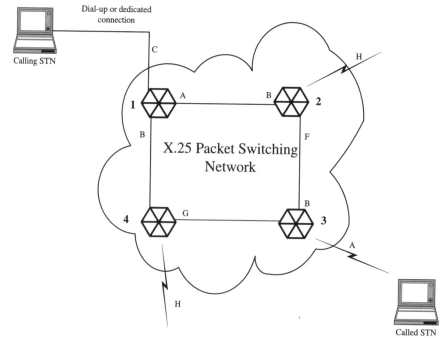

Figure 29-2

Basic configuration of X.25 networks

primarily provides error detection and correction by using the FCS on a link by link basis. The packet containing the addresses permits each packet switch along the way to make a routing decision about the best way to send the packet toward its destination.

The frame itself consists of a *Flag* (F) character at the beginning and end, as delimiters of the frame. The *address field* (A) is rarely used, but logically contains the physical address of the device on the network The control field tells what kind of frame it is (data bearing or control) and can be used to contain sequence and acknowledgement numbers. The *Frame Check Sequence* (FCS) is a very powerful, 16-bit checksum system that is able to detect 99.998% of all errors. Given the quality of the circuits of the day, such a powerful error detection scheme was necessary. The control field shown as C in Figure 29-3 carried the sequence and acknowledgement numbers.

Figure 29-3 shows the relationship of a packet being carried inside a frame. Note that the frame is terminated (FCS evaluated and acknowledgement sent on each hop). The packet is re-encapsulated in a new frame for its hop across the next link. Chapter 10 discusses Frame Relay, which is the successor to X.25.

In addition to the routing information, the first packet also contains the very important *Logical Channel Number* (LCN). You may think of it as a random number that is used by each packet switch instead of the address. There are several benefits to doing this. First, you don't need to put the complete address in each packet, and second, you achieve a relationship between packet switches that lasts for the duration of the call. (The importance of this will only become clear later). This relationship between packet switches is called a virtual circuit. You will find further discussions of virtual circuits in both Chapter 10 on Frame Relay and Chapter 11 on ATM.

Figure 29-4 shows the routing table (technically a switching table) in each packet switch shown earlier in Figure 29–2 that resulted from the call setup from caller to called system. This association of ports and LCN create the virtual circuit across the network. This has several major implications for overall network design.

1. All packets follow the same path between any source destination pair of end systems. (This makes maintaining sequencing of packets easy.)

Figure 29-3
The Frame carrying
the X.25 packet

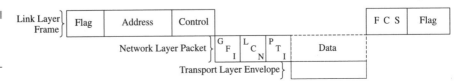

Figure 29-4
The routing table
example

Packet Switch #1				2				3			
Input		**Output**		**Input**		**Output**		**Input**		**Output**	
Port	ICN	Port	LCN	Port	LCN	Port	LCN	Port	LCN	Port	LCN
C	35	A	15	B	15	F	25	R	25	A	5

2. Acknowledgements can be done link by link, using frame acknowledgements and end-to-end by using network layer sequence numbers.

3. At setup time, we could determine the load on the packet switches and interconnecting links before choosing that path.

This would enable *Quality of Service* (QoS) and *Service Level Agreement* (SLA) adherence. It must be clearly stated here that X.25 doesn't provide these services. However the existence of a virtual circuit relationship between packet switches for the duration of the connection is a necessary condition to provide QoS and SLAs.

Connectionless Network Services

Now that we understand CONS, we can look at CLNS. The Internet Protocol is a CLNS. The apocryphal story as is oft repeated is that IP was designed during the height of the Cold War. The military (funded by ARPANET) wanted a bomb-proof network—one in which any of the lines or packet switches could be knocked out at any time and not seriously affect network operation. Therefore, the designers chose a dynamically routed CLNS. The philosophical differences between a CLNS and CONS are large, but each is a valid way to construct a network. The Internet has grown to become a high volume worldwide network. Even some of the original designers are amazed at its ability to continue to grow yet maintain its robustness. More than a few doubters have been predicting Internet meltdown. Congestion has occurred due to lack of bandwidth connecting routers. Adding more bandwidth has so far alleviated this condition. There was some discussion in the late 1990s that the demand for bandwidth would exceed the supply in the year 2001. The technology breakthrough of dense wave division multiplexing (discussed in Chapter 28) essentially solved the bandwidth problem. The solution isn't inexpensive per se, but it yields tremendous bandwidth for the price.

It is now easy to draw a comparison between X.25 and IP. Each IP packet (called a Datagram) is a stand-alone entity containing all the information needed to be routed from one end of the network to the other. The lack of any formal associations between packet switches means that (theoretically) packets can be routed instantly and dynamically without regard for the final destination. Figure 29-5 shows the relationship between frame and packet. Frames in the case of IP are only used for error checking. The *control field* (C) simply says that this frame contains data. Bad frames (for example, those failing the FCS check) are simply discarded. Error recovery will be left to the transport layer.

The contents of the IP packet header are shown in the more conventional representation in Figure 29-6. Note that each row is 32 bits (four bytes) wide. The first thing we notice about IP packets is the large amount of overhead in the packet header. We will not describe the use of all these fields.

Figure 29-5
The relationship between the frame and packet

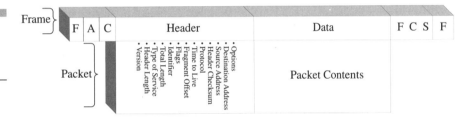

Figure 29-6
The IP header

0	8	16	24	32
Version	IHL	Type of Service	Total Length	
Identifier		Flags	Fragment	
Time to Live		Protocol	Header Checksum	
Source Address				
Destination Address				
Options and Padding				
Data				

The important fields for our current discussion include:

1. The Version field theoretically allows multiple versions of the protocol to exist in the network, thus permitting seamless upgrading of the protocol. IHL is the *Internet Header Length*, and is the length in bytes of the header.

2. TOS is *Type of Service* and was intended to permit priority handling of packets. It is generally not implemented, although some vendors provide proprietary implementations.

3. Total Length is the total length in bytes of the Datagram.

4. The next 32 bits including the identifier, flags, and fragment offset are rarely used.

5. TTL is the *Time to Live* indicator, and is essentially a hop counter. The originator sets it to a value a little larger than the maximum number of hops that it should take to get the packet to the destination. If the network can't deliver it in that many hops, then it is considered to be undeliverable and discarded. This prevents undeliverable packets from bouncing around the network forever.

6. Protocol is a pointer to the layer above network that is using the IP packet. According to OSI purity laws, that should be the transport layer. The designers of IP built in more flexibility so that other layers could use the packets.

7. Header Checksum is a simple checksum on just the header of the Datagram. This is not a *Cyclic Redundancy Check* (CRC) of the sort used by link layer to really protect the data.

8. Source and destination addresses are each 32-bits long. The format and use of the addresses is discussed in detail in the addressing section.

Options and Padding

IP provides some interesting options, one of which is forced routing. If forced routing is employed, the route to follow is specified in the options field. Padding is used to fill out the last field so that the data field starts on an even 32-bit boundary.

The source and destination fields are used by the routers to route the packet to the next router. The Internet uses a best effort, hop-by-hop routing system to keep the handling of the packets simple. Much is made of the fact that the Internet is a CLNS and therefore can route packets almost arbitrarily and those packets arrive out of order or not at all. While it is theoretically true that a router could choose a different path for the next packet with the same destination address as the previous packet, let us consider the practical realities. First, it the chosen route is good for the current packet, the chances are excellent that it is also good for subsequent packets. Second, the router normally has a limited number of entries in his routing table: the normal route, the secondary route, and the default gateway. Therefore, the number of different paths that the packet could theoretically take is large. In practice, it is relatively small. The practical result is that most packets follow the same path and generally arrive in order, but there are no guarantees.

TCP

For customers needing high-integrity, guaranteed, error-free delivery, a transport layer must be built on top of IP. *Transmission Control Protocol* provides the sequenced acknowledged reliable delivery mechanism. Figure 29-7 shows the format of the TCP packet called a segment. The important parts of the TCP segment for our discussion are the sequence number and acknowledgement numbers.

Although the IP packets carrying these TCP segments may arrive out of order, TCP can easily resequence them. The acknowledgement field provides the reliable delivery part.

Figure 29-7
The TCP header

Source Port	Destination Port
Sequence Number	(byte count)
Acknowledge Number	
Data Offset / Reserved / Flags / Window	(buffer space)
Checksum / Urgent Pointer	1st urgent byte
Options (+padding)	As needed
Data (variable)	

The Fields in the TCP Header

The fields in the TCP header are

1. *Port number.* The source and destination port numbers are used to identify the listening agent that is using TCP to transfer data. IP used the protocol field; TCP uses the port number. The port address, in conjunction with the IP address of the packet, is refereed to as the socket. Therefore, in order for a socket to exist one needs both the IP address and the port address. Port addresses below 1024 are assigned by (or reserved for assignment) by the IETF. When you surf the Web, your browser always uses TCP port 80. One can provide Web servers on other ports (for example, port 8080, but to use that port, the sender has to specify the port number after the host name because the default is port 80.

2. *Sequence number.* The sequence numbers are not packet numbers, but rather byte numbers such that if there were 100 bytes in a segment, the sequence number will increment by 100 each time. This helps the recipient reserve the proper amount of buffer space if it gets, lets say, the 10th segment first.

3. *Flags.* The flag section is used to set up and tear down connections, as well as to push the final segment and get acknowledgements for the segments.

4. *Window size.* Window size describes the flow control window. Rather than the more conventional mechanism of on and off flow control, TCP uses a credit system where credit (in bytes) is granted to the sender. When the sender has sent that many bytes, it must stop. The recipient keeps this updated to provide for a continuous flow of data.

5. *Checksum.* As on the IP packet, the checksum is not a CRC, but a simple checksum, and it is run on the entire TCP segment.

6. *Urgent Pointers.* The urgent pointer points to urgent data (if any). Theoretically, this handles higher priority data within a segment ahead of normal priority data.

7. *Options and Padding.* Here again, options can be added to TCP. These fields need only be added when the options are present. The field is always padded to even 32 bit boundaries.

TCP is an end-to-end protocol that makes sure that the data gets to the destination and in proper order. It accomplishes this by setting up a connection to the other end, much as X.25 set up a connection, except that it doesn't create a virtual circuit. Therefore, when you visit a Web site, TCP sets up a connection so that you reliably get a complete Web page. The page may consist of several segments and many packets, since each packet is about 576 bytes.

After the page has been delivered and acknowledged, the connection is torn down. When you select the next URL or hyperlink, a new connection must be made. Remember that there is no way to know whether the user will select another link on that same Web site or a hyperlink to some other site. Some browsers provide indications of this connection process on the bottom edge of the browser window.

UDP

In the interest of thoroughness in a discussion about Internet protocols, we need to mention User Datagram Protocol. It also is Transport Layer Protocol, but does not create connections. It is another "send and pray" protocol like IP. Who then, would want such a thing? Consider the problem of a database query. You don't need a connection, but you need an answer quickly. If the answer doesn't come back in a reasonable time, ask again.

Both the *Domain Name Service* (DNS) and *Simple Network Management Protocol* (SNMP) use UDP. These are both query/response protocols. Figure 29-8 shows the UDP header in all its simplicity.

The port addresses are the same as described for TCP. The length tells how long the segment is in bytes and the checksum is, as with TCP, a simple sum over the whole segment.

Figure 29-8
The UDP header

<--------------------------------32 Bits-------------------------------->

Source Port	Destination Port
Length	Check Sum
DATA	

IP Addressing

Now that we have a handle on the protocols that are going to be used, let's take a look at the structure and significance of addressing and then we will discuss how routing is done.

The addressing scheme is important for obvious reasons. Addressing is used by the routers to find the destination for the IP packets. One could imbue the routers with God-like knowledge of the network or the router could have a simple table to useto look up the next hop. If there is no entry in the table, then send it to the default router. The whole Internet works because of this default route mechanism. If any device doesn't know how to return a packet, it sends it to the default router, known as the default gateway.

Routers Versus Gateways

A point of clarification is needed about routers and gateways. In the Internet community, routers and gateways get a little confused. In the previous discussion, we use the term router almost exclusively. As pointed out, in the Internet community, it is referred to as the default gateway.

The Internet terminology started before terms were standardized. A gateway was a box that sat between two networks. Therefore, it was natural to refer to routers, especially ones connecting different networks as gateways.

Today, a router is a box that operates at network layer, reads IP addresses, and chooses the best hop it can find to the destination network. A gateway is a device that handles protocol conversion. A mail gateway is a great example. If you are using IBM's PROFS mail system, you have a mail gateway that reformats the messages and changes the characters in the message from IBM's proprietary *Extended Binary Coded Decimal Interchange Code* (EBCDIC) to standard ASCII characters.

When selecting the addressing scheme, the Internet designers chose their own path, rather than taking a page from the telephone folks.

The designers were thinking in terms of networks for universities, government, and the military. They anticipated that there would be a few very large networks for the military, government agencies, and very large universities; some medium-sized networks; and many little networks for smaller colleges.

Correspondingly, IP has three address formats, known as class A, B, and C addresses, respectively. A class D address format exists for multicasting to groups of destinations. Figure 29-9 shows the IP address formats.

You will note that in the IP packet format that the address is 32-bits long or four (4) bytes, which is represented in a dotted decimal format. Each byte is represented by a decimal number of the value of that byte. This makes it relatively easy for humans to read and use the addresses.

A four-byte address might look like this: 147.45.193.15.

We can tell immediately that this is a class B address.

The first bit of the address is 0 for Class A and 1 for class B. Class C addresses have the first two bits set to 1. Therefore,

Class A addresses are from 0 to 127.

Class B addresses are from 129 to 191.

Class C addresses are from 192 to 223.

Class D addresses for multipoint broadcast are from 224 to 239.

Class E addresses (reserved for special projects) are from 240 to 254.

You will note that certain addresses aren't valid or used. "All zeros" means this network. "All ones" is the broadcast address.

Class A addresses were reserved for the largest organizations. The format of the address has 8 bits reserved to define the network and 24 bits to define the host. Originally, a host was a computer that was smart enough to run IP and TCP. Today, every network-connected device contains a computer, and therefore, it has its own IP address. In fact, many devices have multiple IP addresses because they use different addresses for network management functions than the one they use for data transfer. Therefore, it is very easy to run out of IP addresses.

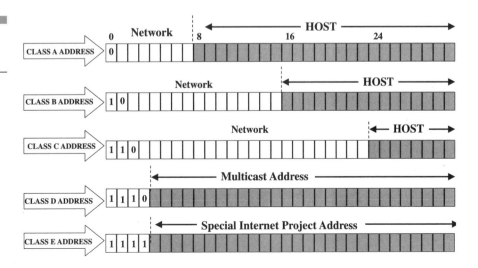

Figure 29-9
The Internet IP addressing scheme

A Class A address provides for 127 networks with over 16 million attached devices. A Class B address provides for 16,000 networks and 65,000 attached devices. A Class C address provides for about 2 million networks and 255 attached hosts on each small network.

Now that we understand the format of addresses, let's see how we can use them. The first point is that routers only route on the network address. The router doesn't care about the host address. The assumption here is that the job of the router is to find the destination network and that all the hosts behind that network address are all accessible thereafter. Imagine that the router is the interface to an Ethernet network where all the hosts are connected. While this is conceivable for a Class C address with 255 hosts, it is mind boggling when one considers a Class A address.

As Figure 29-10 roughly depicts, there are examples of the three classes of networks connected to the router. If a device on the 145.15.0.0 Class B network wants to send to a device on the 14.0.0.0 network, all it does is send it to the router that forwards it to the class A network. The router considers the job done when it finds the 14.0.0.0 network. Now, clearly, it is silly not to mention that it is unworkable to have 17 million devices on one Ethernet.

Figure 29-10
Different classes of IP addresses connected to a router

Subnetting

The solution to the problem is to subnet the 14.0.0.0 network into a bunch of smaller networks. We do not have to create subnets on even byte boundaries, but for the sake of discussion, we will break the Class A net into Class C like addresses. Figure 29-11 shows (without getting too carried away) what this might look like.

The first thing we note is that we have introduced another router because the only way to get between subnets is by using a router. So now, our first router (R−1) thinks it has found the 14.0.0.0 net when it finds a port on the 14.1.1.0 just as before. The second router (R−2) has to know that the 14.0.0.0 network has been subnetted into Class C nets, and that it isn't finished routing until it finds the Class C addressed network. The way the router and all hosts know this is by being given a subnet mask that contains all ones over the portion of the address that specifies the network.

Figure 29-11
A subnetted class A
address

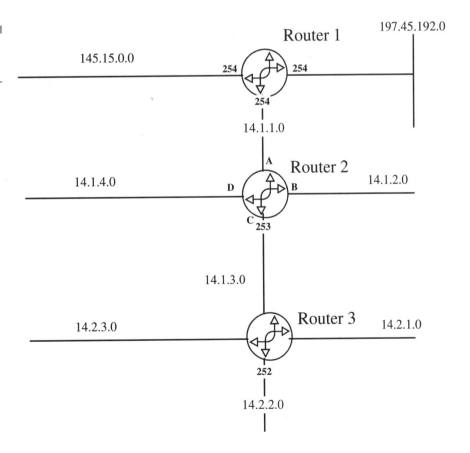

The subnet mask for hosts on the 14.0.0.0 network in Figure 29-10 would be 255.0.0.0. The subnet mask for hosts on any of the networks from routers 2 or 3 would be 255.255.255.0. Thus, these routers and hosts don't care if it is a Class A address; they route until they find the last network.

The good news is that we can now limit the size of our broadcast domains. Can you imagine what would happen on the 14.0.0.0 network (shown earlier in Figure 29-10) if we sent out a broadcast message looking for a particular station? Everyone, everywhere would get it. Since nearly all of the Class A and Class B addresses have been assigned, the only addresses available are Class C, and they are running out.

Network Address Translation

The solution to the problem is to use NAT. *Network Address Translation* (NAT) permits us to use any network addresses we want internally while looking to the outside world as though we were a Class C network. A few clever tricks make this work. First, we must have only one access point to the Internet.

If we use, for example, IBM's Class A address internally, you can imagine the havoc that would be created if one of our routers advertised that we were now IBM on the Internet.

To use NAT, our interface device to the Internet simply keeps a dynamic table that maps official Class C addresses to the address of our internal user's Class A address. An internal user, 14.1.4.25 for example, wants to surf the Internet. Our real address on the Internet (public address) is 245.123.237.0. The firewall then maps 14.1.4.25 to 245.123.237.1. The firewall will map this in both directions. There is a secondary side benefit to providing NAT. Hackers will never know our real internal network addresses or address structure. They won't be able to tell servers from terminals. It therefore makes it that much harder for them to hack into our network. With a properly configured firewall, it is almost impossible anyway.

There are some obvious limitations. Namely, that we could theoretically only have 254 folks surfing the Net at one time. If this is a serious limitation, another Class C address is requested from the ISP or the InterNIC.

Back to preventing havoc on the Internet. To prevent any possibility of the fact that we are using the 14.0.0.0 address from ever getting out and creating havoc, we can legally use the 10.0.0.0 address. This address is "illegal" on the Internet. Any Internet router discovering a 10.0.0.0 address in a packet will simply discard that packet. This effectively gives us a free Class A address that we can subnet any way we want. Our firewall still must do NAT because it is not polite to send garbage packets out on the Internet.

CompuServe and AOL use NAT between their internal address scheme and the real Internet.

DHCP, BOOTP, ARP, RARP

The next dynamic addressing scheme is that used by most ISPs. They don't have enough addresses to be able to give a real Internet address to every subscriber. So, what they do is dynamically assign one to you when you sign on and put it back in the pool of unused numbers when you sign off. This is called Dynamic Host Configuration Protocol. These all do essentially the same thing, although there are significant technical differences. When you sign on, you contact the terminal server at the ISP, who in your stead goes and gets an IP address for you. If you were to sign on from an Ethernet, you would send a broadcast message to all devices on you subnet asking the "keeper-of-unused-IP" addresses to assign you one. In a large network, there are two ways to handle addressing. One is to permanently assign every device (Host) an address. The other is to use DHCP. There are arguments for both approaches.

Address Resolution Protocol (ARP) is a little different. It is used by a router to find the physical address of a destination host when it has found the proper network. The router only has network layer information in the packet. It needs to have a physical (Ethernet) address in order to actually deliver the packet. It uses the ARP to broadcast to all devices on the destination looking for that device whose IP address matches the one for which it is looking. When the target host responds, the response contains not only the matching IP address but the physical, Ethernet address as well.

Reverse Address Resolution Protocol (RARP) works just the opposite. The host knows its ethernet address and sends out a broadcast to learn its IP address.

Routing

As we have indicated, IP routers route the datagrams to the next hop. Routing is accomplished by simply looking in the routing table to find the next hop. This, of course, begs the question of where this table does come from.

Table 29-1 shows the routing table for router R−2 in Figure 29-11 and Figure 29-12. We have taken the liberty of giving all the ports of the router the same final byte address, although it is connected to different networks. Using this convention makes it easier to troubleshoot the network since the router's address is always the same. If you are designing and building your own network, this is handy to remember.

In these diagrams, there is only a single path from any one location to another. This makes the routing table simple and static. We now introduce Figure 29-13. This is Figure 29-11 with two major changes. First, we are using the same diagram so we don't have to learn a whole new set of numbers, but now these are just addresses on the network and have nothing to do with NAT anymore. Second, we introduce router R−4.

Table 29-2, the routing table for router R−2, now has three additional entries marked by * because R−2 now has two ways to get to the 145, 197, and 14.2.3 network. Notice that the hop counts are larger in each case because the connection is not direct. Notice also that we could have put entries in for the 14.2.1.0 and 14.2.2.0 networks via R−1 as 3 hops away, but these entries as well as the one added to the 197 are bogus entries. If R−3 (or R−1 in the case of the 197 network) is down, there is no point in sending packets, although we think that there is a route.

Table 29-1

The routing table for route R-2

Destination	Next Hop	Hops	Port#
14.2.1.0	14.1.3.252	1	C
14.2.2.0	14.1.3.252	1	C
14.2.3.0	14.1.3.252	1	C
14.1.1.0	Attached	0	A
14.1.2.0	Attached	0	B
14.1.3.0	Attached	0	C
14.1.4.0	Attached	0	D
145.15.0.0	14.1.1.254	1	A
197.45.19.0	14.1.1.254	1	A
All others	14.1.1.254	1	A

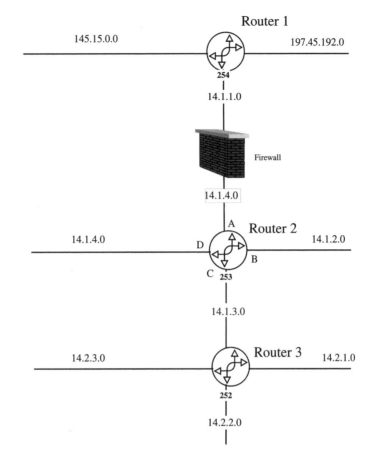

Figure 29-12
Class A subnetted
with a firewall to the
Internet

Table 29-2

The routing table
for route R-2.

Destination	Next Hop	Hops	Port#
14.2.1.0	14.1.3.252	1	C
14.2.2.0	14.1.3.252	1	C
14.2.3.0	14.1.3.252	1	C
14.2.3.0*	14.1.1.254	2	A
14.1.1.0	Attached	0	A
14.1.2.0	Attached	0	B
14.1.3.0	Attached	0	C
14.1.4.0	Attached	0	D
147.15.0.0	14.1.1.254	1	A
147.15.0.0*	14.1.3.254	2	C
197.45.19.0	14.1.1.254	1	A
197.45.19.0*	14.1.3.252	3	C
All others	14.1.1.254	1	A

Figure 29-13
A new router is
added to the
network.

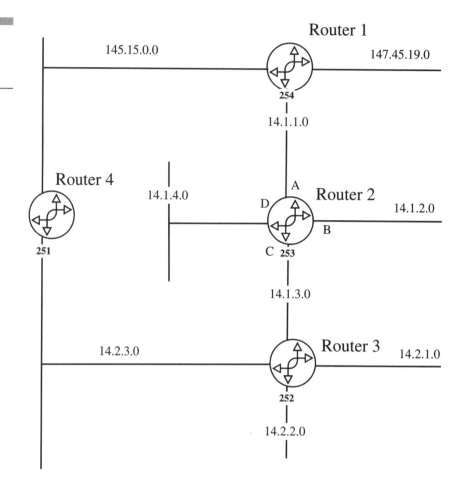

Dynamic Routing Tables

You have seen how much fun it is building routing tables by hand. Imagine doing this for 1,000 routers. What happens if there is a failure in the network? We would have to figure it out and send out new tables. For this reason, networks use dynamic routing, whereby they learn who their neighbors are and build their own routing table from that information. Two basic methods for doing this are used in the Internet.

Routing Information Protocol (RIP) was the original mechanism for sending routing table updates. Its operation is very simple. Each router packs up its routing table into packets and sends these packets to all of its neighbors. The neighboring routers then pick through the received table to look for better routes to destination networks than it already has. If it finds better routes, it puts them in its table. The process repeats itself. Clearly, if we do this once per day, then old or invalid routes will remain in the table for at least a day. The more dynamic we want our network to be, the more often we should do the update. Unfortunately, this consumes lots of network bandwidth when done too often. Every half a minute (30 seconds) to a few minutes seems to be a reasonable range. The irony here is that larger networks have a greater probability of something changing so they have to do it more often than smaller networks.

There are a couple of problems with RIP when the network gets big. The tables grow exponentially, and therefore it takes more and more router processing power to handle the table updates. The second problem is that good news travels fast, but bad news (unavailability of routes) travels slowly. RIP works fine in small networks.

To handle the problem of exponentially growing routing tables, we break the network into routing domains. These are called autonomous systems. The basic idea is that all the routers within a routing domain or autonomous system can exchange routing tables as usual. All routes to locations outside their domain are sent to the boundary router (frequently called a gateway router or simply gateway). The boundary router is also part of the local domain, and is part of the exterior domain. It uses a different protocol to exchange routing tables with its peers called *Boundary Gateway Protocol* (BGP). This keeps the volume of traffic between routing areas or autonomous systems manageable. Figure 29-14 depicts these routing domains. The Internet terminology (using "gateway" for router) is shown on the top of each figure, while the ISO standard terminology is placed below each figure.

The next approach to the routing problem is to use the *Open Shortest Path First* (OSPF) protocol. This protocol was developed and standardized by the *International Standards Community* (ISO). It differs from RIP in two significant ways.

First, only changes in network status are sent rather than the whole routing table. This significantly cuts down on network overhead.

Second, it creates hybrid metric to use in lieu of a simple hop count to the route selection process.

In RIP, the router cannot make a distinction between a T1 line and a 9600b/s line if they are both x hops to the destination. By adding this hybrid metric that includes a weighted average of hop count, data rate, error rate, delay, even maximum transmission unit size, a communications cost for that path can be calculated. True, it takes more router processing power to calculate all this, but it allows the router to make better route choices. OSPF isn't the only protocol that does this; each of the router vendors has invented a proprietary routing protocol.

Figure 29-14
Routing Domains

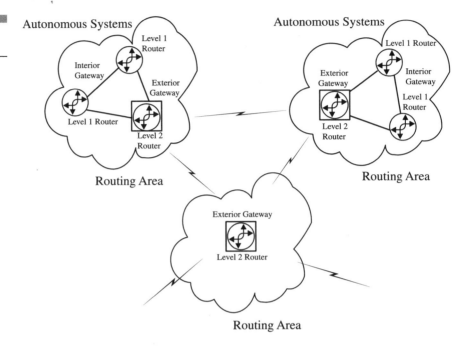

Although OSPF is more efficient in the total volume of update information sent there still comes a time in network size where the routing update information begins to affect the network bandwidth available. Therefore, ISO has defined a two-stage hierarchy of routing areas called level 1 and level 2 areas. These are the same as autonomous systems with a different name. Level 1 routers are the local routers, and they exchange local data. Level 2 routers are the gateway routers, and they are the interdomain routers.

Routing Versus Switching

There is a good deal of confusion over the issue of routing vs. switching. Now that we have discussed and given examples of each, let's see if we can make a clear distinction. This section began with a discussion of X.25 and packet switch. In X.25 we used the first packet to contain the address. That packet had to be routed to find the destination. In the process of doing the routing, a virtual circuit was set up by using the *Logical Channel Number* (LCN) that was now carried in every packet in lieu of the address. This is called a switching system because the table into which the packet switch must look to find out what to do with each packet is small. For example, only those currently active virtual circuits have entries. The packet switch can therefore process each packet very quickly. If the total number of virtual circuits was limited to some number like 1024, you could set aside a block of memory and use the LCN as an index into the table.

A router, on the other hand, must have a complete table of all possible routes including, of course, the default route. Clearly, you could make the table smaller by using the default route more often, but that just moves the problem to another router. As we've discussed, this table grows exponentially with the size of the network. The addresses are neither hierarchical nor structured nicely to facilitate table searches. *In addition,* we have the problem of keeping the tables updated. The big efficiencies that switching shows over routing is not so much in table searching, where there is an advantage, but in maintaining the routing tables. Most packet switch systems use static (manually built) routing tables or use OSPF. Some use a separate control network to manage routing tables. Therefore, the table update problem does not have the impact on a packet switching network as it does on a router-based network.

Real-time Applications

This subject is covered in detail in Chapter 30. However, two points need to be made. All of these packet switching systems suffer from highly variable interpacket delay. The connection-less systems suffer from greater variability or are rather less predictable.

This arises from two sources. First, the queuing and routing functions in the routers or packet switches, and second from the fact that the output port/path may be busy, requiring that the packet be queued until its turn arrives. The systems we have discussed have no priority schemes for handling one type of packet ahead of any other.

In order to handle real-time applications, such as voice-over IP, we must be able to control interpacket delay and ideally control absolute end-to-end delay. This is for all practical purposes impossible with the Internet as it stands since it is a CLNS and made up of many different providers.

The virtual, circuit-based networks have a better chance since there is a relationship between packet switches. To these, we need only add the capability to evaluate the internal congestion and that of its external path to the next packet switch. Then, during call setup, the packet switches could determine before setting up the virtual circuit whether it supports the delay or bandwidth requirements of the requestor.

MLPS

Multi-protocol Label Switching is an attempt to "Band-Aid" virtual circuits onto a CLNS IP network. The idea is to append a label (called also a tag, in which case it is refereed to as tag switching) to each packet. Since we already discussed X.25 and LCNs, this tag is essentially an LCN. Thus, the routers can create a virtual circuit across the network. We now have created an association between routers. The routers can make a determination with their peers as to whether they have the capability to support the *Quality of Service* (QoS) requested at virtual circuit setup time.

This additional capability leads to multiple priority queuing systems inside the routers and packet switches. Unfortunately, it can't address the in-and-out queuing times and the absolute transmission delay between packet handlers. One way to address this problem is to employ an ATM network (see Chapter 11).

Summary

One of the hottest vehicles used in the discussion of convergence between voice and data networks is the Internet. Only 10% of all the data in the world is running across the Internet. However, 90% of the data runs on corporate private and virtual private networks in 1999. This will probably be true for the next few years. However, one major difference exists, these corporate networks are modelled after the Internet creating Intranets and Extranets. Thus we must be very aware of what is happening in the world of data communications.

Moreover, the real time applications such as Voice over IP, streaming audio and streaming video applications are all placing added burden and traffic on the Internet. As the convergence continues, many changes will take place. These are the parts of the data network that draw the most attention.

Voice-over IP (VoIP)

Introduction

Lately, the hype in the industry is the convergence of the Voice and Data transmission systems. This is nothing new because we have been trying to integrate voice and data for years. However, a subtle difference exists with today's convergence scenarios:

- In the beginning, the convergence operation mandated that data be made to look like voice; then it could be transmitted on the same circuitry. Circuit-switched voice and data used the "age old" techniques of a network founded in voice techniques.

- Now, the convergence states that voice will look like data and the two can reside in packetized form on the same networks and circuitry. Circuit switching technologies are making way for the packet switching technologies.

These changes take advantage of the idle space in voice conversations, where it has been determined that during a conversation, only about 10–25 percent of the circuit time is actually utilized to carry the voice. The rest of the time, we are in idle condition by either of the following:

1. Listening to the other end
2. Thinking of a response to a question
3. Breathing between our words

In this idle capacity, the compression of voice stream can facilitate less circuit usage and encourage the use of a packetized form of voice. Data networking is more efficient because we have been using data packeting for years through packets, frames or cells.

The use of a packet switching transmission system enables us to interleave voice and data packets (video, too) where there is idle space. As long as a mechanism exists to recoup the information and reassemble it on the receiving end, it can be a more efficient use of bandwidth. It is just this bandwidth utilization and effective saving expectations that have driven the world into a frenzy over packetizing voice and interleaving it on a data network, especially the Internet.

A note on the cost of voice and data communications is probably in order here. (It is the opinion of this author.) *Currently the drive is to get free voice on a data network.* In the early years of data communications, data always was given a free ride on the voice networks. Telecommunications managers diligently fine-tuned their voice networks and allowed the data to run over the voice networks during the off-hours (after hours). This use of the circuitry was paid for through the dial-up voice communications.

This method of providing data over the voice networks crept into some of the business hours when real-time communications were needed. However, many times, the voice tie lines and leased lines (point-to-point circuits) usually had some reserve capacity. Therefore, the data was placed on the point-to-point circuits that were justified by voice. (History!). As the competition in the telecommunications market heated up, we saw the costs for a minute of long distance drop from $.50–.65 per minute in North America down to an average cost today of about $.05–.10 per minute. This significant drop in cost has been the result of competition and technological enhancements. At the same time, the data convergence took advantage of this falling cost factor. Meanwhile, the data was migrating from a point-to-point technique on a dial-up connection to a more robust packet form of transmission, using X.25 or IP packet switching techniques. The cost per bit of data transmission was dropping rapidly. Then the introduction of the Internet came to the commercial world. With the commercialization of the Internet in the early 1990s, the culture and the cost factors changed at an escalating pace. The cost of data transmission has been touted as being free on the Internet; the only cost being the access fees. However, the cost of data networking has been rapidly declining on private line (Intranets), drawing the attention from both a public and a private networking focus. This convergence has everyone looking for free voice by interleaving voice on the data networks.

I agree with the scenario and would not attempt to detract from that goal. However, the other side of the equation is that the circuit switched networks are continually driving the cost per minute for voice down. Closely aligned to this paradigm is the fact that the North American long distance companies are now offering specials in their service. For example, in 1999 many of the carriers offered $.05 per minute on nights and weekends, another carrier offered Fridays for free, and so on. Looking closely at their advertisements, we see a culture changing slowly, moving its way into the industry. More of the ads state "not quite free, but almost," of "free weekends, or free Fridays." It is this shift that makes one wonder if free voice on the data networks is as much a deal as we think. Alternatively, it may be the movement to draw the crowd in. I believe that voice will be free after the turn of the century with the strategy being "give them the voice, they will buy the data!" However, we have to wait and see how this plays out.

Voice over the Internet Protocol

The public telephone network and the circuit switching systems are usually taken for granted. Over the past few decades, they have grown to be accepted as almost 100 percent reliable. Manufacturers built in all the necessary stopgaps to prevent downtime and increase lifeline availability of telephony. It was even assumed that when the commercial power failed, the telephony business continued to operate. This did not happen accidentally, but through a very concerted development cycle jointly by the carriers and the manufacturers alike.

Access to a low-cost, high-quality worldwide network is considered essential in today's world. Anything that would jeopardize this access is treated with suspicion. A new paradigm is beginning to occur because more of our basic communications occur in digital form, transported via packet networks such as IP, ATM, and Frame Relay. Packet data networking has matured over the same period of time that the voice technologies were maturing. The old basic voice and basic data networks have been replaced with highly reliable networks that carry voice, data, video, and multimedia transmissions. Proprietary solutions manufactured by various providers have fallen to the side, opening the industry to a more open and standards-based environment.

In 1996 there was as much data traffic running on the networks as there was voice traffic. Admittedly, industry pundits are all still saying that 90 percent of the revenue in this industry is generated by voice applications. This may be an accounting problem, because on average 57 percent of all international calls originating in North America and going to Europe and Asia are actually carrying fax, not voice. Yet, they are considered dial-up voice communications transmissions because of the methodology used. Moreover, the voice market is growing at approximately 3–4 percent per year, whereas data is growing at approximately 30 percent per year. Since data traffic is growing much faster than telephone traffic, there has been considerable interest in transporting voice over data networks (as opposed to the more traditional data over voice networks). Support for voice communications using the *Internet Protocol* (IP), which is usually just called *Voice over IP* or VoIP, has become especially attractive given the low-cost, flat-rate pricing of the public Internet. In fact, toll quality telephony over IP has now become one of the important steps leading to the convergence of the voice, video, and data communications industries. The feasibility of carrying voice and call signaling messages over the Internet has already been demonstrated. Delivering high-quality commercial products, establishing public services, and convincing users to buy into the vision are all still in their infancy. The evolution of all networks begins this way, so there is no mystique in it.

IP telephony will also have to change, somewhat. We will expect it to deliver interpersonal communications that end users are already accustomed to using. These added capabilities will include (but not be limited to) the following:

- *Calling Line ID* (CLID)
- Three-way calling
- Call transfer
- Voice mail
- Voice-to-text conversions

Users are very comfortable with the services and capabilities delivered by the telephone companies on the standard dial-up telephone set, using the touch-tone pad. IP telephony will have to match these services and ease-of-use functions in order to be successful.

IP telephony will not replace the circuit switched telephone networks overnight; this will be a co-existence for the near future. In 2003 analysts expect that the amount of IP telephony will amount to 3% of all voice traffic domestically and approximately 10–15% of international traffic. This amounts to 50 billion minutes of traffic, so it is consequential. One must be prepared for both alternatives to carrying voice in the next decade. Thus, the differences between the two opposing network strategies will be ironed out, and the world may shift into a packet switched voice network over the next decade.

Quality of Service

One of the arguments against IP based telephony today is the lack of *Quality of Service* (QoS). The manufacturers and developers will have to overcome the objections by producing transmission systems that will assure a quality of service for lifeline voice communications. Mission critical applications in the corporate world will also demand the ability to have specified grade of service available. Applications that will be critical include some of the technologies we discussed in Chapters 6 and 8. The CTI applications with call centers being Web-enabled, interactive voice recognition, response, and other speech activated technologies will demand a quality of service to facilitate the use of these systems. Each will demand the grade and quality of service expected in the telephone industry.

Another critical application for IP telephony will be the results of quality of voice transmission. Noisy lines, delays in voice delivery, and clicking and chipping all tend to frustrate users on a voice network. Packet data networks carrying voice services today may produce the same results. Therefore, overcoming these pitfalls is essential to the success and acceptance of Voice over IP telephony applications. Merely installing more capacity (bandwidth) is not a solution to the problem, but it is a temporary fix. Instead, developers must concentrate on delivering several solution sets to the industry such as those shown in Table 30-1.

The QoS requirements for IP Telephony can therefore be summarized as shown in Table 30-2, which considers the layered approach that vendors will be aggressively pursuing. IP Telephony datagrams entering the network will be treated with a priority to deliver the QoS expected by the end user. The routers and switches in the network will assign a high priority marking on each datagram carrying voice, and treat these datagrams specially. Queues throughout the network will be established with variable treatments to handle the voice datagrams first, followed by the data datagrams.

Table 30-1

Different approaches to QoS

Strategy	Description
Integrated Services Architectures (Int-Serv)	Int-Serv includes the specifications to reserve network resources in support of a specific application. Using the RSVP protocol, the application or user can request and allocate sufficient bandwidth to support the short- or long-term connection. This is a partial solution because Int-Serv does not scale well, because each networking device (routers and switches) must maintain and manage the information for each flow established across their path.
Differentiated Services (Diff-Serv)	Easier to use than Int-Serv, Diff-Serv uses a different mechanism to handle the flow across the network. Instead of trying to manage individual flows and per-flow signaling needs, Diff-Serv uses DS bits in the header to recognize the flow and the need for QoS on a particular datagram-by-datagram basis. This is more scalable than Int-Serv, and does not rely solely on RSVP to control flows.
802.1p Prioritization	The IEEE standard specifies a priority scheme for the Layer 2 switching in a switched-LAN. When a packet leaves a subnetwork or a domain, the 802.1p priority can be mapped to Diff-Serv to satisfy the layer 2 switching demands across the network.

Table 30-2

QoS Requirements
for IP telephony

Layer Addressed	Technique	Variable
1	Physical Port	Variations of port definitions or the prioritization of port interfaces based on application
2	IEEE 802.1p Bits	Dedicated paths or ports for high bandwidth applications, but very expensive to maintain
3	IP addressing	RSVP protocol (Int-Serv) DS bits in the IP header (Diff-Serv)

Given the differences, why should we even consider the use of IP telephony? If the IP telephony world is only going to account for 3 percent of domestic traffic by 2003, is it worth all the hassles? The answer is a mixed bag, but overall the efficiencies of VoIP will outweigh the need to develop better control mechanisms to satisfy the telephony industry.

Applications for VoIP

Voice communications will remain a basic staple for the industry. The public switched telephone networks will not be replaced, or dramatically changed for that matter, in the near term. The immediate need then for VoIP providers is to deliver equal services, similar to the PSTN, at a lower cost of operation and to offer a suitable competitive product as an alternative. This alternative will increase the competition in the industry and force stodgy providers into a new era of meeting the consumer's demands. The first and foremost yardstick that everyone is using is the pricing model. Although switched voice is now down to $.05– .10 per minute, just what is the equilibrium price that should be used? When we consider the IECs' costs per minute is less than $.00125, the $.10 per minute charge they levy on corporate or residential consumers appears excessive. These pricing models are also changing as new technology is implemented in the SONET and DWDM architecture as discussed in previous chapters. In Figure 30-1, we see one of the basic scenarios of how the new providers may challenge the existing infrastructure providers of telephony service. This assumes that a telephone to telephone access method is used through the networks. This, by the way, is the preferred method because it uses the tools and

Figure 30-1
Telephone to
telephone through
the IP network

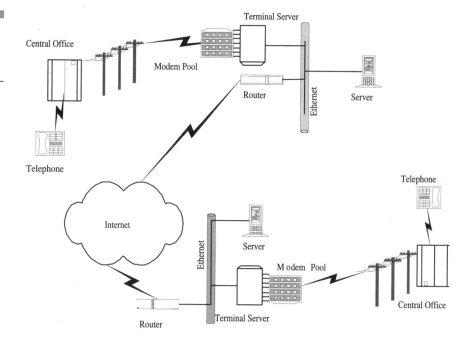

techniques to which everyone is accustomed. The basic telephone set has
become a staple in the business by having an ergonomically designed device
that is easy to use. This is the best way to handle the voice over the Internet or IP for a call. The telephone sets are ergonomically designed to have
the mouthpiece close to the human's mouth and the ear piece that helps to
screen out a lot of the ambient noise in the room. Moreover, the sets are not
actually doing any of the compression or conversions; these are handled by
the devices in the network such as gateways and gatekeepers, specifically
designed for this task.

VoIP can be used for just about any requirement for a telephony call. It
can be used for fax and for video applications as well. From a single point-
to-point call to a multiparty conference call, and a Web-enabled call center
application, the manufacturers have to devise tools and techniques to support the masses.

A second use for VoIP is shown in Figure 30-2, which uses a PC to telephone call. This is slightly less desirable because the PC uses an integrated
microphone and speakers for the most part. Because they are not designed
specifically for this application, they pick up all the noise in the background.
Further, to prepare the real-time voice through the PC, the compression
and conversion process occurs inside the PC, causing some clipping and

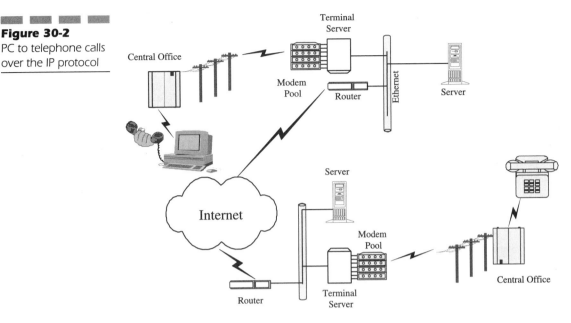

Figure 30-2

PC to telephone calls over the IP protocol

chipping, echo, and other forms or latency and distortion. It is not the best solution in the industry, but it does work.

The third way to handle this is to use a PC to PC connection, which is the worst of the three scenarios. This one introduces the use of the PC substandard technology on both ends of the connection, compounding the problems of echo, delay, clipping, and chipping. This is not the best way to experiment with the technology. Moreover, I have seen people using PC to PC communications, whereby the caller places a telephone call over the PSTN to the party desired and tells the recipient to turn on their PC so they can be called. This is ludicrous: If the IP system does not know whether the called party is active and powered up, then you have to place a telephone call to tell the person to turn on their PC. Why not just have the conversation while you have the called party on the line? This is just an aberration of the technology, but it highlights the fact that VoIP is not yet ready for prime time. There are several tools that IP telephony can bring to the table allowing the use of VoIP to handle

- Voice to voice calls
- Fax to fax
- Voice to call centers
- Voice to conference calling arrangements and PC to PC

Figure 30-3
Various ways of
using VoIP

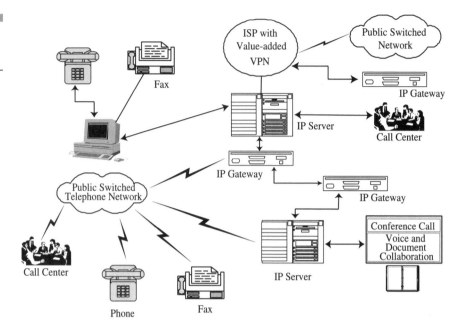

These various paths for handling VoIP are shown in Figure 30-3. The main point here is that there are alternatives to VoIP being used as more than just a voice call. The benefits are the enabling technologies to handle myriad options for now and in the future.

VoIP equipment must be equipped with the capabilities to handle the various options designed around the PSTN and cater to a wide range of configurations. Some of the applications are summarized in Table 30-3 as a means of placing some semblance of order on the VoIP equipment market.

H.323 Protocol Suites

The IP telephony architecture must emulate the PSTN to accommodate the various ways of interconnecting. In Figure 30-4, the ITU has set the H.323 specification to accommodate the various ways the IP telephony industry must work. Defining gateways and gatekeepers, H.323 is a mechanism for the manufacturers to produce the necessary devices and protocol compatibility for interoperability in the PSTN and the Internet convergence.

Figure 30-5 is a different view of the H.323 specifications for the VoIP standards showing the various other protocol suites needed to support the

Table 30-3

Summary of areas where VoIP equipment must work

Applications	Discussion
Internet Telephones	Enhanced telephone sets, both wired and wireless, provide basic telephony services across the Internet instead of the PSTN. These may use a display that will display CLID and possibly some text messaging capabilities.
Remote Access Services	Branch office or SOHO users can gain necessary access to the corporate voice, data, video, and multimedia applications using an Intranet. The application here is the home-based worker or telemarketer.
PC-based calls from a travelling person	Using the PC with an Internet connection (dial up from a mobile PC or connected through a CATV modem in a hotel), a travelling user can access all of the corporate services. Excellent for the road warriors in today's generation.
Web-enabled Call Center Services	Call centers can be equipped with access to the Internet. As a user surfs home pages and wants to place an order for a product, or if a question arises, the user clicks a button on the PC screen to reach a real-time operator. Customer service applications like this will enhance the overall e-commerce industry and the telephony applications.
PSTN gateways	As shown in Figure 30-3, the gateway function allows the interconnection from the PSTN to the Internet over a gateway. A PC-based telephone can access the PSTN. Another application is for a USB port telephone connected to a PC that will have unlimited access to customers on the PSTN.

standard. In this case, the terminal devices running on various telephony protocols such as ISDN terminals, H.320 terminals, and video conferencing devices are all accounted for in the H.323 specification. The standard also addresses the various protocols for the gateways and gatekeepers on a LAN, as well as any compression techniques to manifest the VoIP capabilities in various points along the network.

What also has to occur in the network is to use a layer two protocol to carry the IP traffic. Traditionally, an HDLC frame or a PPP protocol frame is used. However, to assure the QoS, the standards are recommending the use of ATM at the layer two switching architecture. In most cases, ATM introduces added overhead by producing the cells to be mapped onto a SONET frame at the layer one. However, what ATM does assure is the guaranteed bandwidth to deliver the necessary QoS. Only recently has a guaranteed QoS in Frame Relay been introduced. Figure 30-6 is a summary of the bandwidth needs and the QoS issues using the ATM or FR networks

Figure 30-4

ITU H.323 specifications define the various roles in IP telephony

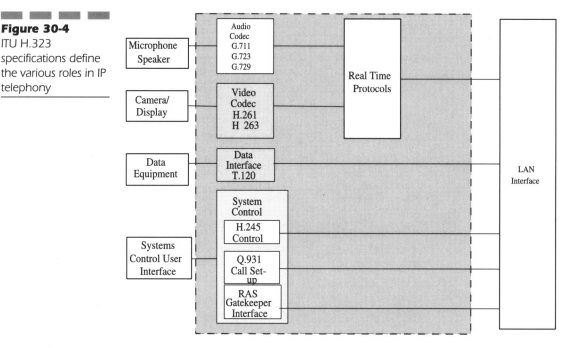

Figure 30-5

H.323 in action

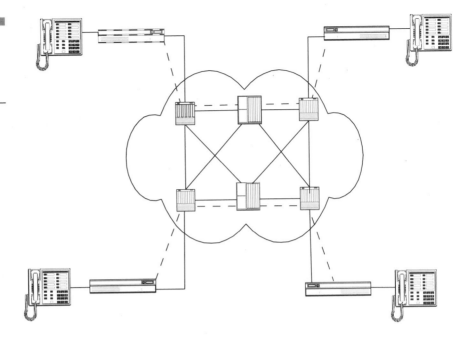

Figure 30-6
The use of ATM or Frame Relay guarantees the QoS today

to carry the VoIP traffic. It is through this combined use of Layer 2 switching and Layer 3 network addressing that the systems are able to map and multiplex the traffic on the data networks with any reliability. By the time this book is published, there will probably be three or four new approaches that will emerge, each claiming to be the better way to handle the traffic for VoIP. In most cases, they will be efficient and better than what has been available previously. That doesn't mean a thing, because each of the techniques that exist today will be implemented and upgraded or improved over time. No one technology or technique stands to be the winning or "killer application" today.

Delay and Jitter on VoIP Networks

Two of the major concerns today are the problems experienced with delay and jitter. The delay factors can be the most disconcerting. In the past, the PSTN has been delivering real-time voice across the network with an average delay of 30–50 msec. When we experimented with satellite communications in the early 1980s, the end user community was disgruntled with the potential delays for real-time voice applications. The average delay on a satellite-based network in geosynchronous orbit approximated 250 msec, which was intolerable for the end user. In the VoIP delivery today, the average response and delay factors must be dealt with. On average, if a call is placed across the Internet, the delays are from 800 msec to 2 seconds domestically. This is totally unacceptable for the user community. It certainly does not lend itself to the conducting of business conversations. When placing a call across the Internet overseas, the delay can be as much as 5 seconds! Again, this will never do.

The delays have been the result of all the processing required throughout the network. One problem that results from high end-to-end communications delay is the increase in echo whenever the round trip delay exceeds 50 msec. Because echo is viewed as a quality problem, VoIP systems must accommodate echo control in order to be accepted. Talkers will overlap themselves when the delay increases if one-way transmission exceeds the 200–250 msec range. This was a major problem with satellite transmission. Too often talkers were over-talking each other, causing total chaos during the conversations. The budgeted delays for VoIP must be within the 200–250 msec range, and that is strictly the worst case. The VoIP systems have to strive for better response and delivery.

Jitter (variable delay) is a variation of the interpacket delivery introduced by the processing of each packet across the network, coupled with transmission delay across the medium. To solve the problem for real-time applications, packets must be collected into a buffer and played out as needed. By holding the packets just long enough to accommodate the longest delay in packet delivery, the added delay can be devastating. This also requires that the VoIP equipment have sufficient buffers to accommodate these interpacket delays on an ongoing basis. The bigger the buffers, the longer the potential delays.

Another major problem that one must contend with is the acceptable levels of packet loss. Several studies have been conducted indicating that acceptable packet loss is less than 10 percent. Anything greater than 10 percent packet loss will be intolerable. The network can drop packets when peak loads are being experienced, causing some degree of loss. Moreover, packets can be discarded in case of buffer overrun at the routers and switches across the network. These combined problems may move the packet loss levels to an unacceptable level.

If we combine the various levels of processing delay and the risk of packet loss, we can see where the networks must accommodate the delay, along with the need for VoIP equipment to address the issue. Figure 30-7 is a representation of the delays that can be cumulatively added to a packet delivery across the network.

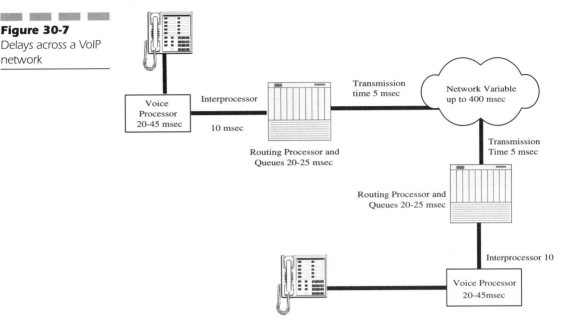

Figure 30-7
Delays across a VoIP
network

Protocol Stack

The VoIP protocol stack uses the benefits of a TCP/IP protocol suite in which to run over the network. Figure 30-8 is a representation of the protocols as they stack on top of each other. The primary areas are as follows:

- In Layer 1, we typically use SONET and DWDM for the physical architecture. Although other variations can be used, this is the more common today.

- At Layer 2, we typically will use ATM or Frame Relay, but the PPP protocol can also be used if some other form of link architecture is employed

- Layer 3 is where we will find the IP layer (Ipv4 or Ipv6), using the network layer to handle the datagram protocols

- At Layer 4, we use UDP instead of TCP for the real-time applications. Let's face it, UDP serves as a better set of protocols to use when we can ignore packet loss, or the reliability issues are addressed at a lower level.

- At the upper layers (5–7), we see the use of the *Real-Time Protocol* (RTP), the *Real-Time Control Protocol* (RTCP), and the *Resource*

Reservation Protocols (RSVP). Others will include the *Network Time Stamp Protocol* (NTP) and the *Session Initiation Protocol* (SIP) as others.

■ At the application, we see the H.323 protocols sitting on top of the heap.

Figure 30-8
The VoIP protocols stacked up

H.323	Layer 7 Application
RTP/ RTCP/ RSVP/ SIP/ NTP	Layer 5-7
UDP	Layer 4
IPv4 or IPv6	Layer 3
AAL	Layer 2
ATM or Frame Relay	
SONET	Layer 1
WDM or DWDM	

Intranets and
Extranets

Introduction

We begin with an attempt at a definition. Most of the buzzwords around the Internet are easy to define until we get to made-up words like Intranet and Extranet. These words were invented by journalists to describe Internet technology employed within an organization rather than outside the organization.

"Intranet" is essentially an internet with a small *i*. In particular, it implies the Web implementation on our internal internet. Now, what do we do about telecommuters of all descriptions? An answer is to give our telecommuters access to our Intranet via the Internet, again using Web technology.

Extranets are simply our Intranet extended to permit access to our organization by our business partners. That is, a Web or a Net interface designed specifically to support our business partners and field offices. This also could be a VPN if that level of security were required. You will note here the mixing of field offices and customers. If we were consistent, the field offices would be covered under Intranet, while customers were covered under Extranet.

Here again, when journalists make up words without clearly defining them, you get these interesting blurring of definitions.

By our definition, if the external employee (telecommuter, and so on) has access to our internal network applications, then we are providing an Intranet. That is, access isn't limited to our external Web site, but extends to our internal Web site and perhaps to FTP, telnet, and other applications. If our external business partner, or field office, only has access to our external Web site or to an external FTP server, then we are providing an Extranet.

Please note that these definitions are not universal. What you call it isn't near as important as what it does. The goal of our Intranet is to facilitate our business—that is, to help it work better internally and with our correspondent employees.

Whether your organization is geographically diverse or simply growing rapidly, maintaining uniform communications across the organization is the challenge. Distributing information—for example, policy and procedures manuals in paper format—proves to be very ineffective. First, they aren't read because people have work to do that they consider more important. Second, by the time they need the information and read the manual, it is long since out of date. Large companies can keep a full time employee busy running around putting change pages in all the policy and procedures manuals. Therefore, it is no surprise that the policy and procedures manual is one of the first to be implemented on our Intranet. Putting it on Web pages with a good index and many cross-references makes it easy to use and find related topics. Keeping it current is also as easy as updating the necessary pages.

Notification of the changes to the manual with a hot link to the affected page may be sent by e-mail to make sure that the latest procedure is followed. This is a lot better than implementing the old change page mentality by using e-mail. That is, instead of sending our change packaged, the changes are sent as e-mail attachments. While this may seem easy to the sender, it has the effect of killing the e-mail system. These attachments (typically huge files compared to normal e-mail) use up the disk space quickly. If the system administrators are on top of things, they will get more disk space. Most mail systems (actually any system) don't deal well with very full disks so if they let it slide, the mail system will continue to crash. This reduces the perceived reliability of the whole computing environment.

Managing the Intranet

The group that is selected initially to develop the Intranet is the network group. After all, who knows more about networks? And isn't this an Intranet? Unfortunately, these folks usually have their plate completely full of moves, adds, changes, and upgrades of all kinds. Their orientation is to get it working and get the heat off.

A better suggestion is to create or give the task to an infrastructure group or tools group whose position in the organization is such that they have an overview of the organization's overall business objectives and direction. To be successful, this group absolutely must have management support. This group will then "design" the Intranet. The group will specify what design applications will be served and from which locations each service or capability will be accessible. If this group lacks the technical know-how to do this, it must set forth the overall objectives and hire a consulting firm to figure out a plan and help make cost/performance tradeoffs.

There is also a tendency to make this group, which soon becomes dubbed the Webmasters, the bottleneck for all changes to any of the Web pages. Their job is to set standards and formats so that the Web pages from the various groups have a consistent look and feel. Their job should not be to make and approve all changes to every page. With current Web tools, it is possible to design and specify stylesheets, forms, and templates to be used by the individual departmental Webmasters. To maintain consistency in our Web appearance, we do need to have a coordinated set of Webmasters. The department level tends to make the most sense. (Here "department" could also mean "project" in a project-oriented business.) Even in this case, where departments are large, the Webmaster tends to become a bottleneck for changes to the pages. Remember that the more dynamic the page (meaning it is frequently updated with new information), the more it will be used.

Another idea is to (within the design guidelines) automate Web page creation so that updating the Web pages is almost as easy as sending an e-mail. Time spent in getting the proper tools to make Web publishing easy will pay off because more people will use this powerful tool. Web publishing should be as easy as drag and drop to encourage folks to communicate via the Web pages.

The secondary benefit is that as more information is contained on our Web pages, the less need there is to send e-mail attachments.

Web Page Organization

A common complaint is: "I can't find anything!"

There are several ways to attack this problem. The first is to calmly sit down and design the organizational structure of the information on the Web. A second, less organized way is to cross-link information everywhere. This cross-linking of pages will cause problems in the future as we outgrow one Web server and want to distribute the load across two or more. The problem is particularly acute if each department has not only its own Web pages but also runs its own Web server. True, the Web was designed to permit access of any machine from anywhere, but when we start to put boundaries around our Web in the form of firewalls and proxy servers, each machine is essentially treated as a separate entity with separate permissions.

The point for our discussion topic here is that moving pages around can break links to other pages. This just causes a lot of hatred and discontent (not to mention the drudgery of finding and repairing all the broken links), which could easily have been avoided by applying a little design thought ahead of time. Here again, consultants can help. Cross-linking Web pages is not the ideal solution, but if properly indexed, it can serve nicely for a relatively small Intranet.

The better solution, in general, is to purchase one of the many document management systems. Some of these are also marketed as knowledge management systems. These systems typically provide a nice Web-like GUI on top of a database engine. There are many benefits to using a relational database as the actual storage vehicle for the documentation. It permits referencing that same document from multiple different apparent locations. The updated document is automatically seen from all of the different locations because there is really only one document in the database. These document management systems permit the storing of all different kinds of documents and drawings in all different formats.

It is often shortsightedly dictated that we will standardize on a given vendor's office automation suite and put everything in that format. What then happens is that we acquire a nearly, equal-sized organization that is using a different operating system and different applications. We now have the cross-platform, cross-application problem.

We have actually experienced the following types of problems with e-mail attachments. The Mac people send an attachment to everybody. The PC people who are running the same office automation software can't open the document because it doesn't have the proper file extension, for example, .doc. The Mac people never heard of file extensions so they never remember to put them on in the first place. When the Unix users get the document, they just delete it and never try to read it because it just comes up as garbage. The Unix people have tools that they can use to decode and read the document, but it isn't as easy as point and click. The Unix users have to save the document as a file, launch a compatible program on another server, and then transfer the file before they can read it. It is much easier to throw it away! After all, it is only administrative nonsense!

The document management system does not automatically alleviate the above problem. Proper design and standards can help this cross platform problem. Specify the standard for the dm system as some universally readable format, such as html or Adobe's Acrobat. All platforms can read these document formats directly from the browser. (True, the browser must launch the Acrobat reader, but it can easily do that on all platforms and operating systems.)

Again, the document management system makes keeping and organizing different types of documents easier, but it does not do the organizational thinking for you! You must still design the system. There is no substitute for thinking about the organizational structure beforehand. Plan what kind of data are to be put in each section and who needs to insert, read, and modify each piece of data.

Document management will also permit the storing of any type of document. If desired, each document can be stored in its native word processing format, as well as an Acrobat or html page.

It is assumed here (not always a valid assumption) that everyone in a given department has the same desktop environment so that those working on a document together have also the tools to modify it. Snapshots of the document can be published in Acrobat or html format for others to review.

To help with the organization of data with the Intranet, several companies have found librarians to be very helpful. Librarians are skilled in the compilation and organization of vast amounts of information, not all of which comes in the forms of books and periodicals. They are also skilled at designing retrieval systems.

Document Security

In this context, security only means access security, for example, who has permission to read and write to each document or folder. Document management systems let you set security levels on each folder and each document in each folder. Access can be defined for groups and individuals. (The Unix operating system file structure enables you to set up all these same permission structures, but it is a manual process best left to your local Unix whiz. This does mean that you can structure permission on Web pages with as much flexibility as with a document management system.)

Collaboration

It is also wise to set up a collaboration workspace within your document management system. The intent here is that this is where work-in-progress documents are kept. Typically, these would be accessible by the folks within the department working on the project. Others within the department could read, but not modify the documents. A separate sub-area is set up for leaving comments about each document. Others outside the department perhaps aren't even allowed to see the work-in-progress files. Here again, having this collaborative workspace eliminates the need for e-mail attachments.

Maintaining Interest

We now leave the department workspace-oriented Intranet and consider it as a mechanism for disseminating information throughout the organization. Each group (typically department) that is publishing information on the Web needs a Webmaster to maintain a consistent image within the department. To the extent we want our Web page to be the first resort for folks seeking information about what we are doing, we must maintain an interesting and up-to-date Web page. To that end, our Webmaster or our page editor must continually provide new articles about what is going on in the department. (A history of these articles serves as a good orientation tool for new employees!)

Jokes

Love them or hate them, but they are a reality! Here again, one can offload a burdened e-mail system by creating a page exclusively for jokes to which anyone can submit a joke. Jokes are then "told" or sent by simply including the hyperlink to the joke on the Web page.

(You can palpably feel the e-mail system administrator breathing easier.)

There is, of course, the problem of "taste" in jokes vs. company policy. This can be spelled out on the jokes title or index page with a hyperlink to the policy and procedure manual that outlines company policy as to joke content.

Forms

The bane of everyone's existence is having to fill out forms to order everything from pencils to software. Our Intranet is a perfect mechanism to put all of the forms online with required and optional fields. When the form is properly filled out, it is then submitted. Submission can take several forms. In a Web-oriented system (as opposed to a document management-based system), the form is typically e-mailed to the required department. In a document management-based system, the completed form is simply saved in the proper work-in-progress area. The structure of html forms is such that it is easy to have required and optional fields.

These forms can extend to travel request forms, making reservation preferences (if the corporation has an internal travel office), and filling out expense forms.

Transition Intranet Solutions

These applications could be placed in either the Intranet or Extranet category. Here, we are talking about tightly integrated business partners or field offices who resell our product or services and who need real-time access to the status of their order.

Portal Products or Customized Web Pages

Companies that market their products through competing resellers need to support their customers in an evenhanded manner. The reseller will frequently sign onto our Web page in the presence of a customer to show that we are an integrated partnership. Typically, this is done by showing our Web page with the partner's business logo prominently displayed. Clearly, the provider doesn't want the customer to know that this same product is available from a cross-town competitor. We can do this with customized Web pages.

When reseller "A" signs onto our Web site, we show our Web page with our and his/her logo along with the product index of our products that he/she is authorized to resell. When reseller "B" signs onto our Web page, we show our Web page with a different logo and products.

This can be accomplished by using specialized software (called portal products) designed to present a different face to each customer, your document management system or your Web server software. Each provides the ability to require a login and password to view certain pages.

Along these lines, other passwords may be required of dealers to see their individualized price lists. These can be structured as wholesale and retail sheets so that the sales office can show the customer the retail or wholesale price sheet as required.

These portal products, as summarized previously, don't really do anything you can't do by using your document management system or your enterprise server software. So before you invest, you need to ask the question: "Are they easier to learn to use than using the tools you already have and that your staff is accustomed to using?"

The key to the question lies in the fact that the pages supporting our sales organizations need to have up-to-date information. Old information quickly becomes unused and ignored. When building these Web pages, there is no substitute for understanding how information is used within your company.

Building a Community

In large organizations with widely varying territories, we still need a sense of community and the feeling that we are all part of the same team. Sales representatives in different markets need different information because the competitive environment is different in different sections of the country or world. Yet, they are all selling the same products. Our Web site, therefore, needs to provide the specific market-related information, and it needs to be up-to-date. At the same time, we need to build a sense of community and share, for example, sales strategies that work well.

It is also possible that our direct marketing team and our resellers may be addressing the same customers. (This scenario tends to occur in rapidly growing companies before their marketing plans and structure are fully mature. This condition also occurs with large chain stores that do both wholesale and retail business. The stores may go after the same customer as a corporate account in a common geographic area.)

An Intranet can help tremendously with information, if it is used. Therefore, before deploying the Intranet, it is wise to contact all departments, agencies, and branches involved to get input about what they would like to see from the Intranet. This gives them an opportunity to feel that they are a part of the development, and may yield some ideas that have not been thought of by the home office.

Warning: Look out for corporate office arrogance! The head-shed people who were given this project are full of themselves and their new high-visibility project. The reality is that no one knows better what he/she needs to make sales than the workers in the trenches.

Warning number two: Look out for "too-busy-itus." The field offices have a full plate and when presented with a request for input, they frequently do the following:

A) Don't really understand what corporate is up to

B) Don't realize how important their input is

C) Won't take the time to understand the issues and create thoughtful input.

The predictable result is that when the system is made operational, there comes a hue and cry from these folks about what is wrong and how it ought to be changed. Because the system is really for their benefit, as well as the corporate benefit, their involvement is essential from the beginning. Therefore, we recommend forming a working group (ideally, with a working area on our Web server or document management system) to work out the design goals and structure of our community-building Intranet.

When introducing the Intranet to a non-computer-oriented business, it may require hands-on training. This may have to extend down to how to start up the computer, launch the browser, get online, find data, save or use data, and then sign off. Do not underestimate the amount of training involved! This even goes for high-tech companies. The problem is that current software products are so complex and will do so many things, that the operation of the software can't be totally intuitive. Because of the flexibility offered, even setting the initial options requires a systems administrator.

If the system is used for order entry, entering those orders ought to be as easy and as similar to the existing order entry process as possible to minimize training. Building html forms from the paper forms minimizes retraining. (If the data needs to be in a different format for the back-end system, let the computer system reformat the data.) Users feel comfortable seeing their old familiar form with a new electronic face. They are not so intimidated, and you will not get as many service or help-desk calls.

For non-computer-oriented business, outsourcing development of the Intranet is the best idea. It will be faster with fewer hassles than developing the skills and technology in house. In any case, it must be treated as a full-blown project with a formal written specification for what the vendor is to do and what constitutes success. The other alternative is a time and materials relationship with your consultants to keep working on the problem until you are satisfied. Understand then that this will take a while and cost more than a predefined contract.

Bulletin Board Service

Although we are touting the benefits of an Intranet here, some companies can benefit from a simple bulletin board service where information is posted on the Web to inform people. This is not a document management system, but a simple posting of training seminars—when and where they are held, and what the prerequisites are (if any). It is almost guaranteed that once the Intranet is built and information is available, the implementation group will be deluged with suggestions on how it can be further used to improve business operations.

Customer Service

Although we discuss customer service to external customers in the next section, the same issues and attitudes need to be applied to our Intranet. The suggestions will come rolling in. To maintain the involvement of everyone using the system, it is imperative that you have a system set up to address and accept these suggestions.

Providing a prompt e-mail response is a good way to acknowledge the receipt of a question, complaint, or suggestion. Having a formal system for posting the arrival of a suggestion, and posting the disposition, keeps everyone involved. Simple suggestions may be handled quickly. The implementation of the change can be posted with a "thank you" for the suggestion note to the originator. Some suggestions may require the formulation of a project team to implement them. That fact and a proposed schedule should also be posted on the system designer's (tools or infrastructure groups) Web page.

Thin Clients

Depending on your environment, a thin client Intranet is definitely worth considering. Thin client means that each of the client machines (typically a PC) has (in the extreme) no local hard drive and no floppy drive. It is a network-connected device only. When it is turned on in the morning, it receives its download from the server. The server provides it with all the operating software it needs. There are actually two implementations of thin client. One is like a Unix terminal where the terminal device only knows how to run the display. The entire application runs on the server (application server). The second implementation is where the desktop device is a PC that is intelligent enough to run the application (for example, a browser), but has no local storage and receives the browser and operating system software as a download from the file server.

The big advantage is that our Intranet is now virus free (more on viruses later). The client systems can't introduce viruses from floppy disks because there aren't any. The main server is managed by the computer services department and backs it up every night. Desktop hard drive crashes are a thing of the past. The downside is (and this is very much a function of your operating environment) that when everyone arrives at 8:01 and turns on the PC, they all request a download at once. This taxes the network and the file server. The bigger the environment (number of PCs), the greater the load on the system is. A compromise position is to provide the local PC with a hard drive to hold the operating system and the local applications.

Using the Unix dumb terminal with an application server can (again depending on your environment) save the expense of a fulltime employee who would otherwise be needed to perform system administration on all the desktops.

Extranets

We now take the advantages of Intranets and extend them to our business partners and clients. They, like our Intranet clients, will access our network over the Internet. Here, we describe the applications and don't worry about the implementation. Later, we will cover design of an Extranet interface and security measures you may want to implement. Our Extranet must project our image to the business community with which we are dealing. It should inspire confidence and project an image of trustworthiness. We should indicate what security steps are taken when handling credit card transactions, for example.

Inventory Management

Wholesale

The first example of an ;wholesale management ExtranetsExtranet is a wholesaler who must maintain inventory to handle transient demand, as well as his contract quantities. The normal cost of distribution is about 18 percent of the cost of goods. This can be reduced to about 5 percent with the application of an Extranet and electronic commerce. The wholesalers can now do just-in-time inventory control. Depending on the goods involved, the wholesaler may be able to get overnight shipment from providers, thus reducing the need for large inventories. The goods can be drop-shipped directly to the client's location. With a good e-commerce Extranet, the wholesaler also has visibility into the source's inventories to immediately determine what promises for delivery to make to clients.

Secondary Markets

The next revolution in e-commerce is in secondary markets. The source (manufacturer or grower) made a product, satisfied major contracts and has remnants left over from the production run or parts that don't quite meet the original specifications. These could be textiles with flaws or paint off color for the major buyer.

Integrated circuits (chips) don't perform at 600 MHz but do perform at 300 MHz. Historically, these circuits were either thrown out (which effectively raised the cost of production on the remaining "good" parts), or a wholesaler (speculator) would buy the stuff at deeply discounted prices in hopes that they could be sold.

The Extranet offers the ability to put the inventory online and accept bids. This can be done by the *Original Equipment Manufacturer* (OEM) or by a wholesale dealer who markets the parts on the Web without having to transport them from the OEM location. When the product is sold, it is drop shipped by the OEM.

Privacy Issues

One could say that privacy is at odds with security. The better we can verify who the calling party is, the more secure our Web (Extranet). The trick is to verify identity without compromising or appearing to compromise the privacy of our customers. This is a double-edged sword. You need the tools to maintain info about your visitors to verify their identity, but these tools can also be abused.

In an attempt to improve security on the Web and to facilitate tracking down hackers, Intel put a serial number in each chip that could be read by browsers (any software really). Thus, we could track the users of the machine as they visited various Web sites if the browser put that number in each transaction. Although much excitement was created over tracking individuals, remember some machines have multiple users. Therefore, we really only know which computer visited the site. That is okay for tracking hackers (how many times does a burglar sneak into the house to use the Internet connection?), but it is insufficient for tracking purchasing profiles.

The privacy advocates Primary Information Center and Junk Busters lead the attack against this invasion of privacy. The *World Wide Web Consortium* (W3C) has developed a platform for *Primary Preferences Project* (P3P). The intent is to build an industry-wide mechanism for controlling privacy information. The goal is that the Web site publicly publishes its privacy control mechanisms to every browser that visits the site.

The site's privacy practices profile might define what data they collect and what their rules are for passing (selling?) that data to other marketing entities. (The obvious question is what is the enforcement mechanism for this policy? What assurances do we, the customer, have that it is being followed?) Here is where we as a site provider must engender confidence and trust.

When configuring your browser, you specify or set up the preferences for privacy practices that you expect from the sites you visit. You also fill in a user profile that stores some or all of your personal data. These data are then sent to Web sites whose privacy practices match your privacy practices profile that you set up. These data could include e-mail addresses, name, billing information, and personal preferences in clothing, music, food, hobbies, and so on. Our privacy profile might include the conditions under which we permit our data to be shared, sold, or used. We could specify the other types of sites with whom we would permit our data to be shared.

P3P is really a plan to create a common protocol between the browser and the Web site (server) to collect and exchange privacy data. It defines the structure and words that define privacy practices. It also defines a protocol used for exchanging these data between browser and server. Thus, the privacy data is only transferred if the privacy profile set by the user matches the practices published by the site. Both Netscape and Microsoft intend to support the P3P standard.

This is a sensitive subject and therefore getting agreement between those who want the information and want to traffic in it and those who want to protect the information totally will be in perpetual conflict. Our observation is that, the probability that the browser profile and the site profile will indeed match up goes down geometrically with the number of elements or rules in the profile.

(Just for a fun example: If the site has four rules A, B, C, and D each of which can be true (1) or false (0) and the browser has the matching four permissions, then the probability of all four matching up is $0.5 \times 0.5 \times 0.5 \times 0.5 = 0.0625$ or about 6%. If we add one more rule E, then it halves the probability to 3%).

Our prognostication is that acceptance of P3P will be slow because of these conflicting goals. More on this can be found at www.w3c.org.

Perishable Goods Application

Most producers of perishable goods (for example, fruit, vegetables, and flowers) have long-<$perishable goods Extranets>term contracts with transportation and distribution companies who market their goods and provide a guaranteed income for a certain quantity of goods. What becomes of surplus production? It typically goes on the spot market. That is, the goods are transported to a physical intermediate market/warehouse where they are then purchased by wholesalers.

This can mean that the goods are sold at deeply discounted prices, or that they are put on a consignment spot market. The producer has no idea if the product will sell or not and absolutely has no price guarantee. The consignee may sell the entire product and tell the producer that only half of it was sold. The ultimate buyer has no idea of how old (fresh) the product really is, and has no real recourse if the product turns out to be substandard.

Several companies are using Extranets to provide a direct link between the source producer and wholesalers. The trick is to contact a reliable transport company that provides timely reliable delivery. This is another case of seeking to eliminate the physical transportation to an intermediate staging point. This takes time and diminishes the freshness of the product.

Offering the goods for sale on our Extranet lets the wholesalers buy directly, thereby improving the product freshness at the retailer by days to a whole week. Working financing into the deal allows the producers to get paid upon shipment, while giving the wholesaler 30 to 60 days to pay. The product is fresher and therefore worth enough extra to cover the financing costs.

Here, we have a product-clearing house that greatly cuts down on time to market and at the same time matches up buyer and seller. When this business is backed by a financial clearing house, it instills confidence in the producer and buyer that the transactions will be handled properly in a timely manner and everyone will get paid.

Purchasing Cooperatives

The ability to form purchasing cooperatives that can negotiate volume pricing with suppliers is another growing Extranet application. Smaller businesses from parts stores to building contractors can pool their requirements and approach the supplier with volume deals.

The second major advantage is getting the orders correct. This almost can't be emphasized enough. It can take a fulltime employee to verify orders, process the paperwork, send back the incorrect stuff, and get credit for it while reordering the proper parts.

Finding out immediately if the product is in stock and when the delivery time will be cuts down on local inventory requirements. (Many parts and even drugstores practice just-in-time inventory where they will order the part for next day delivery.) The ability to see the supplier's inventory online permits shopping at alternate sources should your normal supplier be out of stock.

This system can work by using the Web, or it can work in the case of banks and brokerages houses the Extranet uses special software the client of which runs on the client's system and the server portion runs on the server.

Brokerage houses, for example, supply their good clients with desktop software. In this manner, they standardize the software and guarantee interoperability. The same model works for construction projects. If the customer (for example, a construction company) uses the standardized project management software, the supplier can then read the bill of material required and bid on supplying the components. The suppliers then become business partners with their customers. This also helps the suppliers plan their production schedules and inventory levels. (Ford has had such a system in place for over ten years. The suppliers see the production schedule and by contract supply the parts just in time. The reduction of on-site inventory has more than paid for the implementation of the system. The original system used the X.25 network and custom software.) The objective is eventually to be able to run paperless projects.

Boeing's 777 was the first paperless airplane. They were able to maintain schedule and, more importantly, the parts fit better, minimizing production line rework.

Outsourcing

Whether the problem is to manage a large project with just-in-time delivery of component parts or subcontracted services or a just-in-time maintenance service, all can be done over the Net.

The requirements are the following:

- A common database
- A common mechanism (interface) for accessing that database
- The ability to set up different views of the data

Each contributor to the project needs only to see that set or subset of data relative to his portion of the project.

A good example of this might be an airline that outsources some or all of its services. For example, it has a schedule for the following:

Flight crew

Cabin crew

Airframe maintenance

Engine maintenance

Reservation agents

Gate agents

Ticket counter agents

Flights themselves

Seat reservations

Meals (catering)

Wheelchairs

Cabin cleaning

Refueling

The list is almost endless. Most airlines outsource catering and fueling services, but they could outsource everything. Although the crews typically work for the airline, they are treated almost as outsource subcontractors. The flight and cabin crew members individually bid on each flight. Their bid is accepted or rejected, not on price, but on seniority. Maintenance could be handled the same way and for small startup airlines typically is. (Small startup airlines, to keep initial investment costs down, outsource many more services than the major carriers.)

The flight hours of the plane are published on the Extranet page. Then the maintenance company prepares a bid and crew to handle the scheduled maintenance stop. Note that the maintenance company doesn't need to see the catering schedule, nor almost any of the other schedules. Each contributor to the overall project has its own view of the common project database that contains all of the schedules. The ability to provide these different views, as discussed, can be implemented by using the inherent system file permissions, a "portal" software product, or a data management system.

Watch for these outsourcing trends using Extranets, not only for airlines, but also for design and construction projects both large and small.

Computer Hardware Vendor

Typically, the delivery of a complex hardware server system (especially complex customized configurations, containing multiple product lines) requires months of delivery time. When the customer is dealing with a local representative who may be halfway round the world from the productions source, getting the order right and delivered on time is nearly impossible.

The field office creates the order for the system, doing the best job it can in specifying the configuration. Unfortunately, they typically have no idea of what the production schedules or inventory levels are. This can result in long delays in the fulfillment of the order.

Here is a typical example: The customer orders a system with two, 9-gigabyte hard drives. What the configurator in the field office couldn't know is that the factory has stopped ordering 9 GB drives and has gone to 18 GB drives as the standard drive. The order waits on the shelf until some 9 GB drives can be ordered and delivered. Note that the order processing folks can't take it upon themselves to substitute an 18 GB drive. Two 9 GB drives were ordered, and that is what must be delivered!

Now let's examine the same delivery problem with an Extranet, where the production schedules and inventory levels are available to the field offices by signing onto our home office Web site.

The sales representative, or technical support representative, can log into the site with the customer right there. The parts and price list are right there.

They can use the online system configuration software to select hardware components, including redundant power supplies and processors, if required. When they get to specifying the two 9 GB drives, they find that inventory is nonexistent, and there is a 60-day wait if they are backordered. They find that the 18 GB drives are in stock. The customer, on the spot, opts for the 18 GB drives because it means a difference in delivery schedule from two weeks to two months. (The customer planned on mirroring the data on the drives, and therefore needed two physical drives instead of the single 18 GB drive.)

The customer is ecstatic with our ability to be responsive to his/her needs and to deliver the equipment so quickly.

Automating Customer Service

The two issues to be addressed here are the following:

- Improving the responsiveness of our customer service organization or function
- Reducing the cost of providing the service

(Some companies have found the ideal solution. They charge their customers to provide service that a well-designed product wouldn't require.)

The basic problem is that having technical folks respond to e-mail customer service queries is boring for the individual. They soon develop a bad attitude and quit. Having nontechnical personnel respond has the danger of responses not being technically correct (or even close in some cases). Therefore, the need to create a low-cost, accurate automated e-mail response system is absolutely necessary.

The answers to simple and standard e-mail queries can be handled by a completely automated system. An example of this is that a customer can order parts by e-mail. If the customer is willing to include the part number and catalog number, it greatly minimizes the probability of error. Although this is an idealized example, of a query/response from an automated system point of view, it illustrates one end of the spectrum of capabilities.

In real life, customers will mix multiple unrelated topics in an e-mail. (We know we do it all the time.) The trick then for the automated system is to accurately parse the content of the e-mail for subjects, verbs, and objects. The analysis of the query then eliminates the chaff and identifies the important content. The success or failure of the system depends upon the complexity of the topics and issues. Take a test drive before you buy!

The two goals of improving service while lowering costs are often in conflict. Our opinion is that good responsive customer service should outweigh the cost consideration. Who is the company to the customers? Well, at first it is the salesperson who sold the product (unless it is bought off the shelf in a store) and after that it is the customer service representative/system. Future sales (via word of mouth) rest upon the responsiveness of our customer service.

Ever more, our corporate image is our Web site backed up by the customer service system. As technology improves, more and more automation can be applied to the customer service problem. Our preferred design approach for an automated system is to automatically scan and parse the e-mail looking for issues. It's important to fabricate an immediate, automatic response outlining the issues detected. This response always includes a hyperlink to the specific Web page that addresses those issues. This would include a pointer to frequently asked questions. For those subjects not recognized or identified, that fact is relayed to the customer. "The following issues have been sent to our research department."

If more information is known to be required, then the automated response can make that query. The goal is to keep the customer in the automated system while still being responsive. For some number of queries (ideally, a small number), a human will have to respond. This implies a hierarchy of technical ability, each of which in turn analyzes the query, asks for more information if needed, and passes the request to the next level if it can't resolve the problem at that level.

Within a short while, you will learn what standard customer queries are and be able to automate the responses to these. The trick is to detect the "out of the normal" queries and quickly pass them to the human hierarchy of knowledge response team.

This system structure provides the customer with an immediate response that is correct (or nearly so), or a response indicating that this problem requires some research. What is most important is that the customer feels that he has been heard and his problems are being handled. Ideally, the customer is also told approximately how long it will take for the answer to his/her problem to be researched and formulated. Our experience is that customers are willing to wait a couple of days if they know that the response they will receive will be accurate and to the point.

Our interaction with vendors who have a poorly designed human interface on their product is that they have many more requests for help than if the product were more intuitive. Their product is fairly technical, their customer service is overloaded, and their business is handled by humans in a hierarchical system. The problem is that because there is no continuity to a given customer's request, a sequence of queries about the same subject receives answers that are mostly tangential to the query at hand. This does not instill confidence in the customer base. Needless to say, this customer is losing market share.

Implementing Extranets

In order to provide some sort of consistency, some definitions are required. These are not universal (and will be violated frequently in the trade press), but will help with the following discussions.

Intranet

An internal network is comprised of one or more LANs, interconnected with one or more routers running the IP protocol. (So far, this is only an internet with a small *i*.)

The bulk of our traffic on this Net is Web-based. That is we have one or more Web servers, and all of the clients are running a browser. Our primary means of communicating is using the Web. (Technically, mail is a different set of application protocols, but most browsers also provide a mail capability.) Perhaps we should say that the primary interface that our users use for their data is via their browser. This network may be geographically far flung where dedicated lines (or Frame Relay) are used to connect our routers. There is no connection to the Internet. Now, here is where things get touchy. If we were to use VPN technology between our corporate locations, one could argue that it is still an Intranet.

Extranet

We are connected to the Internet and specifically permit access to our network (or portions of our network) by our customers.

We are now at risk from all sorts of hackers, crackers, and freaks. The goal then is to keep them out (or at least slow them down and make them work for it), while giving our friends and customers "free" that is, unfettered access.

The first step then is to install a firewall. A firewall is simply a machine whose job it is to examine each packet arriving and verify whether it is permitted, based on a set of rules. This set of rules is referred to as the rule base. Firewalls essentially operate at the IP and TCP levels of the protocol stack. This means that the firewall examines each packet to see if it is coming from and going to the proper IP address, and that it contains the proper or permitted TCP port number. You can set up firewalls to be a one-way valve. If you do this, your employees can surf the net at will and send out any kind of packet, but your firewall will not permit any unsolicited packets. (This means that mail won't work, for example, the firewall won't let in any packets including mail.) The other problem with this approach is that it will permit your employees to hack on the Net. If they are caught, the corporation may be liable, so think about this issue. This is not to say that firewalls are not useful, but to point out that simplistic solutions will not achieve your objectives.

Performing basic packet filtering on source and destination IP address is a useful feature when combined with additional firewall capabilities. It was initially believed that a firewall was the solution to all our Internet security problems. The limitations of a firewall-only implementation are now apparent. If we only need to communicate between specific networks or specific devices with fixed IP addresses, a firewall is a fine solution.

TCP filtering

The next level of checking that the firewall provides is to check inside the TCP packet for a port number. For example, the standard TCP port for HTTP is port 80. (Any port can be could be used in this example. Port 8080 is another port that is frequently used by proxies.)

The firewall now checks for IP addresses (which might say any IP address) and the port number in the TCP packet. This means that unless the port is open through the firewall, you can't get that service (for example, HTTP or FTP). This provides for a more general audience. Now instead of filtering an IP address, one can say, let anybody from any IP address come through the firewall, but only if he is doing HTTP on port 80 to a specific server. This means that we could have other Web pages on other servers on port 80 internally, but they would not be available because the firewall limits access to a specific server. Clearly, we can allow access to as many servers as desired.

Our presentation is a little simplistic since it makes the assumption that the operating system on which the firewall software is running is itself secure and unhackable. Not only is this not generally true, but it is impossible to secure the OS perfectly. (For example, there are known attacks to a UNIX based system that is running send mail. Therefore, the send mail application should not be present on your firewall machine. A properly maintained OS will have this door closed—see explanation below.) So, it isn't just enough to run a firewall; that firewall must be run on a "locked down" operating system.

(An aside on attacks: There are two general kinds of attacks against systems. One is called the *denial of service* attack. In this case the hacker sends one or thousands of packets to the server, causing it to spend all its processing power processing these garbage packets so that it has no time left over to do its real job, for example, serve up Web pages. The second kind of attack is more invasive. In this case, the attacker tries to gain access to the machine, preferably as the super user. If the attacker achieves this, the system is literally wide open. As a super user, the attacker can set up accounts on the machine, or read any file, or wipe out any file, etc. Very clever hackers will therefore edit the log files so that they show no trace of the hackers' presence. There are a couple of ways to keep log files. One is to frequently copy the log file to an obscure directory, and change its name so that its name does not contain "log" (for example, call it bob). The second is to encrypt the file. The latter is only convenient if the firewall application provides this option. The former only works if the log file is copied often enough to capture the hacker's presence before the log file gets cleaned.)

Constant vigilance *must* be maintained to ensure that the OS has the latest patches that close off discovered holes. Because the UNIX community is open, as each attack is discovered, patches are written and posted to newsgroups where your system administrator can download and install them.

Despite the fact that UNIX isn't perfect, the rapid availability of fixes for bugs and holes nearly as quickly as they are discovered means those diligent system administrators can maintain a very secure OS. Our security experts generally agree that the preferred OS is UNIX for firewalls and proxy servers. (There are frequently reports of break-ins to Web servers, and the more noteworthy are those belonging to the Department of Defense —because they should know better. In practically every case, the hole (or weakness) exploited by the hacker is a known hole, the fix for which has been available from the newsgroup for six to nine months before the attack took place. The fault lies then with sloppy or nonexistent system administration.) An equal number of holes exist in proprietary operating systems; however, one must wait for the vendor to release a patch. This isn't always a speedy process.

Stand-alone System

It is highly recommended that the firewall run on a dedicated machine. Figure 31-1 is a classic firewall drawing with the external network to the left and our protected network to the right. To save on hardware costs, it is tempting to run the Web server as another task on the same machine. This is not wise because, if the hacker manages to crack the OS, they have access to the whole machine and any applications running on it. As a stand-alone machine, they can crack into the firewall, but can't go any farther. This assumes that the other machines on your network won't allow access (for example, they don't allow telnet) from the now defunct firewall.

To further protect against this problem, we introduce the DMZ (see Figure 31-2). It is highly recommended that you not implement the network in Figure 31-1. The DMZ, as we shall see, is a protected area into which all

Figure 31-1
A classic firewall implementation

Figure 31-2
The DMZ provides a protected area where all traffic flows

external traffic goes. That means, ideally, that there is no external traffic flows through the firewall to our internal network. It all goes to servers in the DMZ.

Figure 31-3 shows the relationship of the DMZ to the logical firewall. Although we think and talk about firewalls as a single entity, when we are using a DMZ, it is logically two firewalls.

Note that the DMZ is another communications port on the firewall (typically, an Ethernet port with its own subnet address). The firewall may now have up to six sets of rules for traffic from the following:

1. Outside to inside

2. Inside to outside

3. Outside to DMZ

4. DMZ to outside

5. DMZ to inside

6. Inside to DMZ

In general then, we will put in the DMZ all services accessible by external folks and allow none onto our internal network. (We all recognize that there will be exceptions to this rule. In all cases, make sure that you design the firewall rules carefully so as not to get any unintended consequences.) Figure 31-3 shows the FTP, mail, and Web (HTTP) servers in the DMZ.

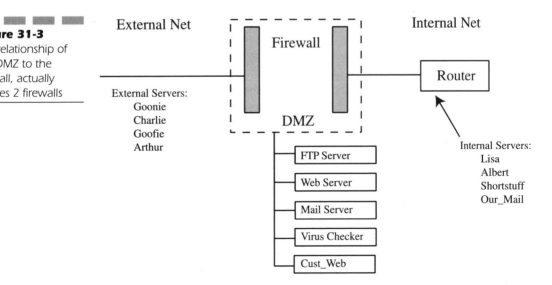

Figure 31-3
The relationship of the DMZ to the firewall, actually creates 2 firewalls

Virus Checking

With the constant threat of viruses arriving in mail attachments and via other mechanisms, some firewall vendors also offer virus checking. Because the processing load on the firewall can be enough for one machine, the virus checking function is offloaded onto another machine that also resides in the DMZ. The process works like this. The firewall recognizes a mail attachment on the mail message. Before forwarding the mail message to the mail server, it first sends it to the virus checker that opens the attachment and scans it for known viruses. It would do the same for a file forwarded to the FTP server. If there is no problem, the virus checker tells the firewall that all is OK, and the message is delivered. If a virus is detected, there are several options available, depending on the vendor's implementation:

1. Discard the message
2. Notify the system administrator by e-mail or pager
3. Clean the file by removing the virus and pass the message on
4. Return the message to the sender
5. Send e-mail to the sender
6. Send e-mail to the recipient

In some cases, these options can be combined. Note also that not all vendors support all these options.

The virus checking function is no better than the latest information it has about viruses. Therefore, if you choose to implement virus checking, it places another task on the system administrators to periodically log into the virus vendor's Web site and download the latest virus profiles. Although this sounds like an easier solution to manage than desktop virus checking, according to the vendors themselves, it is not a substitute for desktop virus checking. Again, the desktop virus checking is only effective if the virus profiles are kept up to date. If you are getting the feeling that maintaining network security is a perpetual struggle and (depending on the size of your shop) requires a dedicated system administrator, you are getting the right feeling.

Firewall Rules Bases

Most firewalls operate (or should operate) from the position that nothing is permitted except that which is specifically allowed by a rule in the rule base. Some early firewall products required that the administrator learn the syntax of the rule base. Most firewalls available today offer an easy-to-use GUI interface. Table 31-1 shows a typical rule base as one might see it from a GUI.

The basic rule here is to put the most frequently used rule first. You don't want the firewall searching through and applying a whole bunch of rules

Table 31-1

A typical rule base viewed through a GUI

source	destination	protocol
any	mail server	http
any	mail server	smtp
client1	cust_Web	http
client2	cust_web	http
goofie	ftp_server	ftp
charlie	ftp_server	ftp
goonie	ftp_server	ftp
mail_server	our_mail	smtp
goofie	shortstuff	telnet
lisa	ftp_server	ftp
albert	ftp_server	ftp
shortstuff	ftp_server	ftp

that rarely apply before it gets to the one that is needed. All this searching takes precious processing power. Therefore, the permission of any source, external or internal, can contact the Web_server machine, but it may only use HTTP protocol. (Since the port isn't specified, it is assumed to be the default port 80.) Note that any source can send mail to mail_server. We have also installed another Web server (cust_Web) especially for our customers. This one is only accessible by devices attached to specific network addresses belonging to our customers (for example, custnet1 and custnet2). There is a list of network addresses in the firewall's rule base—one for each customer. There are normally other rules in the table that tend to be vendor-specific. For example, the tracking column lets you specify if you want a short or long log entry every time this rule is invoked.

Next, the FTP users are at machines called goonie, charlie, and goofie (these are really aliases for specific IP addresses). They are the only devices allowed to FTP to FTP_server. In the last rule of our example, we are also restricting FTP to internal machines called lisa, albert, and shortstuff.

The rule allowing mail_server to talk to our_mail server indicates that mail_server is only a mail forwarder. It isn't the real mail server. Only it is allowed to talk to our real mail server with our_mail. If somehow, mail_server is hacked, it won't stop our internal mail server. In addition, you don't want to put the real mail server or your internal Web server in the DMZ because then all your internal traffic has to go through the firewall increasing its burden (see firewall performance).

The next to last rule allows goofie to telnet to shortstuff. While this seems innocent enough, the question is: Where can the user go from here? This gets into password administration, which is beyond our scope. The issue is that normally a single username/password file is used system-wide so that we only have to administer access permissions in one place. If we allow telnet from goofie, then the user at goofie has a login account and could telnet to other servers in the internal network. The solution is to have (and maintain) separate accounts on the server lisa.

Note that there are no rules permitting anyone to talk to virus_checker. There are also no rules permitting icmp-echo (ping), so that "would be" hackers cannot use this tool to discover servers in the DMZ or on our network.

Firewall Performance (Again)

Although we mentioned it previously, firewall performance can be a bottleneck. The more traffic we run through the firewall, the greater potential bottleneck it becomes. Again, it is essential to have an understanding of the traffic flow. How many packets/second must the firewall handle? In our example, we would add up mail, Web, FTP, and telenet traffic. Remember that mail is handled twice by the firewall and three times if the virus checker is employed. After you have determined the traffic volumes, check with the hardware and software vendors of your firewall package to determine if you have sized the system correctly.

Proxies

Simplistically, you may think of a proxy as an application level firewall. In the case of Web proxies, it can serve to direct, redirect, or limit access through the proxy to servers and specific URLs.

Although the proxy may be placed in parallel with the firewall, our preference is to put it behind the firewall. Remember, we may use several firewalls in parallel to handle the traffic load. Extreme care must be taken to make sure that you have a consistent rule set across all firewalls.

Forward Proxy

There are two kinds of proxies. Your company uses a forward proxy to limit where your employees are permitted to surf. ISPs use proxies to limit the options that their members have. They don't limit destination Web sites (although they could exclude known porn sites), but they limit their members to HTTP and FTP. This limits some of their hacking options. Also, the queries to the remote Web sites appear to be coming from the proxy instead of from the individuals user's terminal because the proxy masks the user address by replacing it with its own.

Reverse Proxy

A reverse proxy is used as a front-end for a Web server. (A large server farm would implement an array of proxies.) Requests to www.anywhere.com are routed to our proxy which forwards the request to server content_server (keeper of the real Web page). The surfer thinks it is talking to server www for all requests. This makes it nice for the Webmaster, who is in the process of replacing content_server with bigger_server. All of our user's links remain the same; we just change where the proxy points. As above, we can limit which URLs the proxy permits. So we could have two Web servers, but only the content of one is visible to the outside world.

The proxy (both forward and reverse) also helps reduce network traffic and content_server load. The proxy caches every requested Web page. The next request to that same server, whether it is from the same or a different source, is satisfied out of cache without having to go back to the content_server. If the ISP has a proxy array, it will eventually contain all the popular Web pages, thus minimizing the number of requests sent out across the Internet. The reverse proxies at the content provider's location also cache popular pages. The rules as to how much to cache and when to go back to the content_server for the latest version of the page may be set up in the proxy server. In fact, at a large site with a large, the popular pages are downloaded into the proxies at low traffic time, and they serve the Web requesters. This improves site reliability and flexibility.

Figure 31-4 shows a typical proxy array behind a firewall. Each of the proxies directly connected to the firewall provides an interface to one or more servers, usually handling a given subject. The proxy in front of the other proxies has servers of related subjects, and the front-end proxy steers the request to the appropriate proxy that is caching that subject. This array then acts to share the load and provides for backup in case of a proxy failure.

Figure 31-4

A typical array of proxy devices sitting behind the firewall

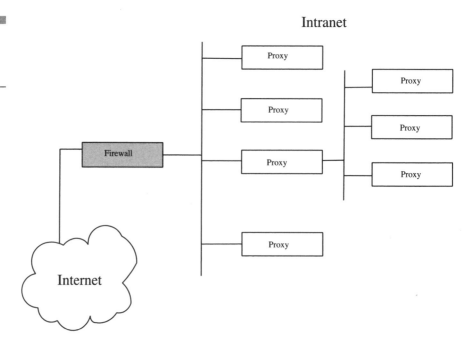

Proxy Security

Figure 31-5 shows a hypothetical configuration. Note that there are two proxy functions. One is a proxy for the internal Web server bigger_server, and the other is a proxy for cust_Web. At this point, our options begin to expand (read "get complex"). Whereas, the proxy could simply map request from the surfer to the server, it could also require that the surfer has an account on the proxy and log in to see a particular server. We can take this to whatever extreme satisfies our paranoia. For example, we could issue account IDs and passwords to our customers. First, they must log into the proxy. Then they must log into cust_Web to see their specific information. (If we were paranoid, we could make them log into the firewall first.) We would only let our remote offices see the Web pages on bigger_server and they must log into the proxy first. This is another safeguard against hackers mucking with our Web page.

The LDAP server is the keeper of the user database of username and password. These users can be segregated into groups such as sales, customer1, customer2, marketing, engineering, and customer support. The proxies then could restrict access, based on group membership. Note that the LDAP function may be a single point of failure. Therefore, consideration should be given to providing access to a backup LDAP that

Figure 31-5

A hypothetical flow through the firewall to the proxy then on to the applicable server in the DMZ

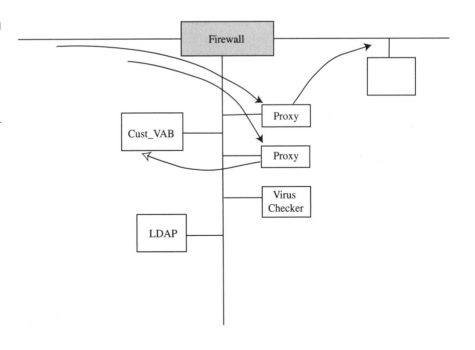

has been synchronized to the main one. Setting up (designing the database structure) and maintaining the LDAP server is a significant effort depending on the experience of your staff. Maintenance of the LDAP database will also take administrative and support staff time.

Although the servers in the DMZ here have been drawn as separate boxes for clarity, some of the functions could be combined into one machine. This appears to be cost effective, but remember, if it is that machine that is being hacked, you could lose all services it provides.

Proxies can also do redirection. When a server is taken down and another takes its place at a completely different address, the proxy can simply redirect the user to the new URL.

We discussed securing the operating system on the firewall. The same should be done for the proxy. It too should be secured with all of the latest operating system patches installed. Applications such as send mail that are not needed should not be running.

Administration

For both firewalls and proxies, the safest way to administer them is to require the administrator to log into the system console. That said, what about remote administration?

Your local proxy/firewall can be administered from inside your secured network. The safest solution here is for the firewall to limit administration access to a few specific machines. Here we mean desktop machines, not multi-user hosts. Few of us work in an organization where everyone is so occupied contributing to the bottom line that he or she has no time or inclination to "explore" (hack) systems on the network.

For remote administration, you are going to have to open the administration port to the world. The difference between the proxy and the firewall function from our discussion is not crystal clear. In addition, in products today, their functionality is melding.

Firewall

In our discussions, we have limited firewall functionality to IP and TCP layers. It is the gatekeeper for our network or DMZ. Its functionality and configuration should be considered *company private* information. The extent that a hacker knows what kind of firewall we are running and its configuration, gives the hacker a better clue as to where start when attacking our network.

Proxy

We have used the proxy as an application level filter that is able to filter (or force) access to specific servers. We use it to hide the true identity of our servers and limit the pages that our visitors are allowed to see and the functions (http, ftp) that they are allowed to perform at each server. When designing a Web site, all references within that Web site should be relative references (for example, relative to the existing server). Web pages containing absolute references will be passed directly back to the user who will now have the name (or IP address) of the server. The proxy won't help because, in this case, only diligence on the part of the Webmaster will fix this problem. It is also a good idea to avoid using IP addresses when identifying servers for two reasons. First, you don't want the external world to know your IP addresses. Second, if you need to upgrade a server from a smaller machine to a bigger one, you have to ferret out every reference to the old IP address and change it—a maintenance nightmare. The solution is to use aliases via and the *Domain Name System* (DNS).

DNS

The Domain Name System is simply a listing of the device IP and the human readable name (alias). A site or company usually runs two DNS systems—one for the internal (protected) network and one for the external network. When a query for www.company.com arrives, the DNS returns the IP of the server (or proxy) that is designated as www. The external DNS then will only point to proxies, while the internal DNS will point to the actual servers.

Internal Web pages that are available or viewable by external users should only reference other Web pages by their external DNS alias. You can have as many aliases as you want for an IP address. If you fail to follow this rule, here is what happens. A page is delivered to a user that references another page in the internal system by its IP address. When the user selects that link, the firewall says permission denied.

If you were to use the internal alias for the server in the delivered Web page, when the user selects that reference, he gets a *no DNS entry* error. Therefore, the rule is when you deliver a page to an external user, that user must have a mechanism to resolve the actual address *and* have permission via the proxy/firewall to access the designated server.

If the pages are viewable by both internal and external folks, then the internal and external DNS entries have to be the same so that all of the links work properly.

Fungible Services

Clearly, the firewall can take on proxy services and the proxy can take on firewall services. We recommend using two separate boxes to improve security. Investigate your firewall vendor thoroughly. Research as much information as you can before making a decision. A beautiful, easy-to-use GUI is wonderful, but what you really want is the best security on the market. Remember, setting up the firewall, proxy, and Web servers is not a "spare time" operation. You will have to dedicate individuals to these tasks. The maintenance of these servers must also be budgeted.

We also recommend designing an internal policy with a review board to review changes to the firewall and proxy configurations. The goal here is to put a damper on: "Oh we'll just open up this port for that function." Indeed, that may be a good idea, but it may also have other security implications.

Since the firewall is the keeper of the keys to the kingdom, its configuration should be treated as company private. The more a hacker knows about your firewall, the easier his job is. The proxy configuration on the other hand is only going to permit access to selected protocols on selected servers so it is easier to manage and maintain. Unfortunately, as new servers and proxies for them are added, firewall adjustments must also be made to permit those connections.

Network Management
SNMP

Introduction

Simple network management protocol is the Internet standard for monitoring and managing devices connected to the Internet or your own Intranet. It defines a data set structure of information that each device may provide called a *Management Information Data Base* (MIB). It also makes provision for individual vendor-based, customized MIBs. A vendor then can provide the standard set of parameters and also extend that set with "custom," for example "vendor-specific," information. These data are collected, kept, and reported by an agent that runs on the managed device. We call this the managed function.

The Management Function resides on a workstation or host, and interacts with the managed function or agent. The management function must have a copy of the MIB profile loaded for each managed function, plus any vendor-specific profile.

In this manner, a standard management function can interact with the standard MIB, while a vendor-specific management function may be used to glean more specific information. These custom MIBs are also registered with the *Internet Advisory Board* (IAB).

Thus, you find standards-based systems, such as HP's OpenView, and vendor-specific systems, such as Fore's ForeView and Cisco's CiscoView, that are compatible and work with OpenView.

Network Management Goals

The ultimate goal of network management is to improve system uptime. The CEO of an organization may see it as another overhead expense. The CIO should see it as a tool for maintaining a near 100 percent uptime.

The higher the system availability, the less equipment is needed, and the lower the cost of running the network and the more efficient the organization will be that utilizes its resources.

Everyone in the corporate management structure needs a clear understanding of the goals of the network management, and the amount of time and effort required to reach those goals. Network management is a great tool for proactively managing the network and collecting alarms and alerts at threshold levels that are below the "Houston, we've got a problem" level. It has often been said that if you do have a failure, then the network management didn't work! Unfortunately, many network management systems are viewed as disaster recovery mechanisms rather than disaster avoidance and prevention mechanisms.

One must also consider the point of diminishing returns. Whereas you may be able to achieve 99.9999 percent uptime, the cost in additional redundant equipment and failover systems may not be justified for your enterprise. Here is a good point to insert a plug for disaster recovery plans. Every enterprise must have a disaster recovery plan. In the context of network management systems, what is the plan of action if multiple (unlikely) failures occur? The likely events should already be handled by the network design. (In this context, the building burning down or an earthquake rendering it uninhabitable is an unlikely event.) At one extreme, the plan may state "We close up shop and go out of business," or "We wait until the flood waters recede, clean up, buy new equipment, and start over". While these aren't recommended plans, they are plans and indicate that management has done some thinking about the problem. At the other extreme is a company, for example, a stock exchange or a hospital, which must keep going at all costs. In this case, alternate facilities are identified which contain computing equipment that can take over if their own equipment fails. A set of network connections is in place to provide connectivity to all of the customers/branch offices including a backup network management. The downtime is minimal. Namely, the downtime would be the time it takes to load the backup configuration on backup machines, plus the time it takes to switch over the network connections and — we're up!

History

In the days of centralized systems, the host or mainframe was about the only smart device in the system. The rest of the devices in the network were relatively dumb and could be reset when necessary by the host.

Figure 32-1 shows a typical centralized network configuration. Most of the remote devices in the network were connected to the host by dedicated lines. At the larger remote locations, a remote concentrator, or controller, is used to multiplex multiple terminals on the single data line. At smaller locations, a single terminal may exist. Dial-up lines may serve some low-traffic locations. But, in any case, there were not multiple paths in the network. It was strictly hub and spoke. Network management consisted of monitoring the status of all phone lines and modems. Dial backup for the dedicated lines was often used when high availability was required. This technique is still used today to back up our Frame Relay connection. Today, an ISDN line can provide 128 kbps dial backup.

Figure 32-1
Typical centralized
network
configuration

Figure 32-2
Typical distributed
processing system

As we have migrated from the centralized model to the distributed computing model, every element in the network is now intelligent.

Figure 32-2 depicts a typical distributed processing system. The WAN technology isn't important per se in that it may be leased as a service from a network provider who will generally not allow our management system visibility into his network components. What is important is that the WAN connection must be available in all of our locations, and that it provides an inexpensive high performance interconnection.

If the WAN is of our own construction, then we consider it to be part of our overall network. That is, we lease lines between locations and provide our own networking equipment, whether it is the Frame Relay switches or Routers.

In either case, the distributed system has many more elements to manage and most importantly that can be controlled managed. The failure of any one of these elements could cause a significant number of our users to be cut off.

In Figure 32-2, the "Router" may be taken to represent a router specifically, but in a general diagram like this it could be representative of other network elements such as a *Frame Relay Access Device* (FRAD), an ATM switch, a LAN bridge, or even a lowly hub.

Network Management Function Interaction

Philosophically, there is a manager function and a managed function. Their physical location in the network is not important. Managed elements may be physically collocated with the management function, or may be at the far edge of the network. The management function is typically located at the headquarters' location. However, it does not have to be that way. As a case in point, the major network providers, such as AT&T, Sprint, and World Com, have distributed *Network Operations Centers* (NOC).

The manager function is the user's interface to the network management system, and it is typically a GUI, which provides status, reports, and statistics gathered from the managed function.

Figure 32-3 depicts the logical relationship between the managed and manager functions. The managed element, or object, is the actual device, whether it is a hub, bridge, or router. The agent function is the program, or process, that runs on the managed system and provides the interface to the manager function. Associated with the managed element or device is a set of attributes. These attributes may include memory size and utilization, interface speed, traffic load, and so on, depending on the type of device.

As one might suspect, there is a protocol defined between the management function and the agent that defines the format and content of each of the commands, responses, and traps. RFC 1905 governs these protocols.

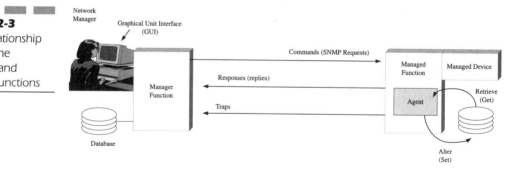

RFC 1907, in turn, governs the content of the SNMP messages. SNMPv2, also known as MIB-II or mib-2, is the dominant definition set today and has superseded the SNMPv1 mib-1. Version 2 greatly expands the detail available from the managed agent.

The agent is the interface to the manager function, and may accept commands to set parameters (such as timeouts, thresholds, and, in some cases, system operational parameters, such as group membership addresses or routing tables), as well as provide status upon request. The agent can also send unsolicited messages (called traps), indicating alarm conditions. (The name "trap" comes from the capability of the manager function to set a threshold (trap), and should this condition be met, the agent sends off an unsolicited message. All other messages sent by the managed function agent are a result of GetRequest or GetNextRequest queries.)

It should be clear that the data reported by a device like a hub is a lot less complex than that from something like a router. The hub needs to report Ethernet collisions on the LAN, the number of frames handled, the number of errored frames, and so on. The router, on the other hand, has routing tables, queue sizes, memory utilization, CPU utilization, and so on.

To keep this all straight, the management function must know about the content of each MIB on each device. In other words, it must have a complete database identical to the one on the managed device so that when it receives the information, it can place it in the appropriate location. When a new device is installed in the network, its MIB content and format must be loaded into the machine (typically, a host or a workstation), providing the management function. Likewise, the agent software must be installed in the managed device.

This emphasizes an important point. Setting up and running a network management system is more than a casual undertaking. It requires a commitment on the part of management to provide the resources and purchase the appropriate tools.

Database Structure

The MIB content is defined by the RFCs listed in Table 32-1.

Table 32-1 is a listing of the relevant standards that apply the network management, in general, and SNMP and RMON, in particular. The complete listing is available from: http://rfc.fh-koeln.de/rfc/html_gz/rfc.html.gz.

It is worth noting that the contents of the MIBs are defined by using ANS.1 *(Abstract Syntax Notation 1)*. This is a language that allows a succinct definition of the content in terms of numeric or alpha characters. It takes a little practice to be able to read the descriptions because ANS.1 is very much like a programming language. If it becomes necessary to delve into the contents of the MIBs, we recommend one of the books in the bibliography.

The content is organized according to a logical structure, called a *Structure of Management Information,* or SMI. The actual SMI structure is defined internationally by the RFC as a tree, using branches of the tree for various organizations. Figure 32-4 is a partially filled-in example of this tree. Each data element is unique because its path from the root through the various branches and twigs to the leaf is unique. Vendors may choose to customize the content of their MIBs under their internationally assigned vendor number.

Those people who are familiar with computer file system structures will recognize this tree. Whether you call the nodes or levels "folders" or "directories," one contained within another allows different leaf notes to have the same name, yet be unique because the path to that data is different. E-mail presents a simple example of this. JohnSmith@BigCompany.com and JohnSmith@BigUniversity.edu are different people with a common name. They are distinguished globally because the path to each is different. (First, you must go to the .edu as opposed to the .com domain and then you go to the BigCompany or the BigUniversity).

Figure 32-4 shows the internationally agreed to structure and the numbers to each level. Like a file system, this sequence is read from the highest level (tree root) to the lowest level leaf, where the actual data element resides. For example, you can find out what kind of a system you are managing by going to 1.3.6.1.2.1.1.1. This is the path: iso.identified-organization.dod.internet.management.mib-2.system.SysDescription.

The SNMP data are contained in 1.3.6.1.2.1.11.

Vendor specific MIBs are under 1.3.6.1.4.1, that is: iso.identified-organization.dod.internet.private.enterprises.

Table 32-1

Relevant standards that apply the network management

RFC	Definition
1155	Structure and identification of management information for TCP/IP-based Internets
1156	Management Information Base for network management of TCP/IP-based Internets
1157	Simple Network management Protocol (SNMP)
1158	Management Information Base for network management of TCP/IP-based Internets: MIB-II
1159	Message Send Protocol
1160	Internet Activities Board
1161	SNMP over OSI
1270	SNMP communications services
1271	Remote network monitoring Management Information Base
1900	Renumbering needs work
1901	Introduction to community-based SNMPv2
1902	Structure of Management Information for Version 2 of the Simple Network Management Protocol (SNMPv2)
1903	Textual Conventions for Version 2 of the Simple Network Management Protocol (SNMPv2)
1904	Conformance Statements for Version 2 of the Simple Network Management Protocol (SNMPv2)
1905	Protocol Operations for Version 2 of the Simple Network Management Protocol (SNMPv2)
1906	Transport Mappings for Version 2 of the Simple Network Management Protocol (SNMPv2)
1907	Management Information Base for Version 2 of the Simple Network Management Protocol (SNMPv2)
1908	Coexistence between Version 1 and Version 2 of the Internet-standard Network Management Framework
1909	An Administrative Infrastructure for SNMPv2

Figure 32-4
SMI structure defined
as a tree

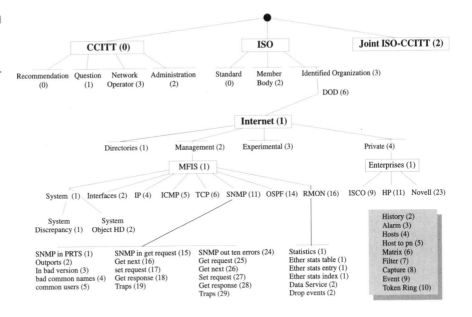

Architecture

Commands are sent from the network management function to the agent, either to set parameters, or to collect status information. When status is requested, the agent replies with a Response message. The agent is also capable of sending an unsolicited status message if some previously defined threshold is exceeded.

Figure 32-5 shows the architecture of the SNMP function and how the functions are related to the ISO's OSI model.

As shown, SNMP runs on top of a standard Internet protocol stack where *User Datagram Protocol* (UDP) is used at the transport. UDP is a connectionless protocol and, like IP, does not set up a connection between communicating entities. In a query/response protocol like SNMP, a connection is not necessary. If the target device fails to respond to a query, simply send the query again. If it were necessary to set up a connection for short queries, it would actually make the system slower.

Figure 32-5

SNMP function and how they are related to the ISO's OSI model

The underlying physical network is irrelevant because it can be Ethernet or a Frame Relay network. From an OSI point of view, SNMP and its interaction with the manager and managed functions encompasses the upper three layers of the protocol stack. A discussion as to which functions belong to which layer usually winds up in a religious argument, so we will consider that SNMP and the management functions are an undivided upper layer set of protocols.

The commands and responses are indicated in Figure 32-5. The actual names (and assigned numbers) for them are delineated in Figure 32-4.

Figure 32-6 shows how the SNMP message is carried nested within the protocol layers. Since our discussion must remain general, we take it that the link layer protocol typically has some leader and header information and some trailer bits. The link layer frame carries the IP packet across the

Figure 32-6

The SNMP message carried within the protocol layers

link layer connection to the next router. The IP datagram header has the routing information that lets the routers direct this packet to its final destination. The UDP header, frankly, doesn't do much. The SNMP message is contained in the UDP datagram. The version field insures that we are conversing with another agent of the same version. The community field is important because it is the security function. Our SNMP function can therefore only collect information from members of our own community. This prevents your competitor from picking the brains of your equipment attached to the Internet. The PDU *(Protocol Data Unit)* type specifies whether there is a GetRequest, GetNextRequest, SetRequest, Response, or a Trap. The error fields are used to identify SNMP errors, for example, tooBig, noAccess, or badValue. The error pointer points to the location of the offending data field. Each *Object Identifier* (OID) then is included, followed by its value (see Figure 32-4).

It is clear that this is a Simple Network Management System. The messages are simple and straightforward, and all of the items in the MIB are defined either by the standards bodies or by the vendors themselves. (If it is so simple, why are the books on the subject two inches thick?)

RMON, or *Remote Monitoring,* although not nearly as widely used as SNMP, is simply another mechanism for keeping a database on the performance of network elements. Figure 32-4 breaks out a small portion of the Ethernet statistics, part of the statistics category under RMON. These tables exist for Token Ring as well.

Network Management System Issues

If one had to ascertain the performance of one's network by evaluating the detailed content of each of these message types, it could drive one to drink. Fortunately, there are network monitoring systems (the network management function) with GUI interfaces. Unfortunately, these products are not fully mature and easy to use. In the following paragraphs, we provide some insight as to what to look for when purchasing one of these systems.

Bundling

It has always been in the interest of the seller to bundle features into feature packages. They can then charge more for the package because it contains so many features. Unfortunately, the package usually contains only a few of the features that we want. It is in the interest of the user then to have things unbundled so that you only need to buy the specific features that you want.

The first question that needs to be asked of the vendor is whether his feature sets are bundled. What is included in the basic package? What features are extra? Is installation included? Is implementation support included? (As indicated above, the installation and configuration of a network management system is not easy for the uninitiated. You can anticipate the need for plenty of vendor help.)

The GUI

The fact that the network management has a Graphical User Interface does not mean that it is intuitive or even easy to use. Each vendor has a different approach. The people who use the tool to watch the network don't necessarily understand the interrelationships of all of the protocols and elements that are being managed. The GUI should permit high level management to be easily accomplished, while providing the capability to see deeper into the system as the operators become more sophisticated.

Network Size

How large a network will the network management system handle? Most network management systems will handle the average to small network with no problem. It should not be assumed that the network management system is infinitely extensible. This issue goes to the core of the network management system's internal database design. (Ideally, the purchaser/user/operator shouldn't have to know or worry about this, but limitations in the database design will affect the number of nodes that can be handled.) We point out here merely that the design of the database affects the efficiency with which data can be extracted. It, therefore, has a direct impact on system performance. Performance can be improved by keeping the entire database in memory. Unfortunately, the size of the memory either limits the size of the database or creates a performance impact when the database can no longer be contained within the system memory. The other alternative is to grow the system memory to match the size of the network. In the worst case, the network management system application may become unstable and crash as the database size is exceeded. (This behavior has been noted in published tests of the leading network management system products.)

The user of the GUI, therefore, should not have to know how the database objects are structured in order to retrieve information. For instance, it is frequently desirable to be able to compare the performance statistics of one device with another similar, but not identical, device. This may be accomplished with varying degrees of success with different products. Although routers and bridges perform different functions, it is frequently useful to be able to compare the total number of frames, or packets, handled by each one as the load on the network varies. Another useful metric is the number of errored frames, or packets, per period of different network loads from these different devices. Unfortunately, not all network management systems permit such comparisons. In some cases, it is impossible; in others, it is merely difficult.

Another important question is: What percent of your network bandwidth is used by the network management system? Ideally, it should be one percent or less. This is significant since most SNMP data are collected by polling (sending GetRequests). Trap data only arrives if the thresholds that have been set in the devices have been exceeded. Therefore, it is important that the network management system have a user programmable polling priority system so that important devices, like routers and firewalls, are sampled more frequently than end devices, such as workstations. This maintains the current status of the important devices and eventually collects information on the end stations, while minimizing the impact on network traffic by the network management system.

Web Enabled GUI

Being able to view reports via a browser eliminated the need to collect, duplicate, and distribute reports. With this feature, anyone who wants the report (and has the proper permission) may collect the report from the network management system. This worthwhile feature is far from universal.

Alarm History

Keeping a history of the exception condition is so useful that we count it as a necessity. It is not only necessary to be able to identify problem areas and equipment, but also to keep a history of similar problems on that and similar equipment. Keeping history alone is not sufficient. The network management system must permit searching the data and displaying historical trends. Trend analysis is the key here. Remember that the goal of a network management system is to maximize uptime. This cannot be achieved by reacting to problems. It can only be accomplished by anticipating problems. Therefore, in addition to keeping a historical record that lets us find out when and where this problem last appeared, the network management system can, with the proper software, do a linear regression on the historical data and predict when (but probably not where) the problem might next appear. (If, for example, the problem were a hardware failure associated with a particular component, the frequency of failure might tell us when it might again appear. If the problem were traffic related, for example, buffer overflow, and the consequence was a system slow down or crash, then the historical traffic data could allow us to identify potential trouble spots that exhibit similar traffic profiles to the failed node. Then we could go back and look at traffic through the failed node from last week/month/year and use that data as a threshold, or trap value, to predict failures of similar nodes under similar traffic conditions.)

Alarm Presentation

Alarms must be programmable and presented in a logical fashion so that they convey the relative importance of the message. Programmable filters are a "must" to permit concentration on the most important alarms first.

Statistics

It should be noted that the more statistics the network management system keeps, the more data it must collect from all components (agents) in the network, and therefore the more traffic will be added to the network by the network management system. Again, the goal is to try to keep the bandwidth requirements of the network management system down to one percent of network traffic. In small networks, this is easy to achieve. In networks of thousands of nodes, a thoughtful use of alarms and a prioritized polling sequence must be employed to achieve this goal.

Free Trials

There is no better way to learn about a network management system than to live with it for a few months before you buy it. This provides the opportunity to find out how user friendly the network management system really is. The vendor's demonstration always looks good. They showcase their strengths and minimize their weaknesses. Their presenter is thoroughly familiar with the product, and can really make it sing. (How long did it take her to become that proficient?)

In reality, it is often difficult to take advantage of a prepurchase trial offer. Although the vendor is willing to let you try the software for a few months, it takes management commitment to dedicate resources within the organization to this trial. A network management system is not something that can be evaluated in someone's (nonexistent) spare time. Remember, it isn't the initial outlay that is important, but the ongoing lifecycle costs.

Network Mapping

Much is made of this "gee whiz" feature. And, to be sure, it is a convenience to have a network management system that will create a network map. (One could argue that we should have been keeping the network map up-to-date as we designed and grew the network. Unfortunately, most networks and especially LANs were never really designed, they just grew.) How the mapping function works and what information is available from the map is more important than the map itself. First, the map should be readable. Second, it should permit zooming in and out. In the "show-me-the-whole-network" mode, the map of a large network will be too dense to read. It should provide the capability to zoom in until the subsystem of interest is displayed.

The mapping function from different vendors works with varying degrees of success. There are two basic ways of implementing the mapping function. The network management system can use SNMP information from each agent, or it can observe the traffic on the network. The SNMP approach yields a map more quickly, but can only display managed devices. Tracking traffic flow can theoretically find every interface or port on the network, but could take months of operation to finally determine a map. (In some cases, it is impossible to create a map.)

Thus, two points must be made. First, accumulating the information to draw the map is not a 10-minute job. Depending on the size of the network, it could take hours to days to months. Second, the accuracy of the map is dependent on how extensively each node on the network supports SNMP. For example, several prominent vendors don't support standard MIBs. As noted, there are standard MIBs and vendor-specific MIBs. In order for your network management system to utilize the vendor specific MIBs, they must be loaded into the manager function. For normal (here read "simple") networks, the mapping function works well and fairly quickly. That means, it works for Ethernets connected with routers and bridges, where each interface on the router is its own subnet. Today, however, we tend to use fully switched Ethernets where each device is on its own switched port from a LAN switch. The advantage is that each port is buffered so that there are no collisions on the network. A second advantage is that one can create virtual LANs (VLAN) where devices on different ports of the switch can be made members of the same broadcast domain or subnet. This permits you to design traffic flows on the various LAN segments in order to minimize response times and keep broadcast traffic confined to those devices that are members of the same logical net. The disadvantage from a network management system point of view is that it is between difficult and impossible for the mapping function to create a map because there are managed devices hiding behind, or within, other managed devices. This is especially true if the LAN switch doesn't support transparent bridging or standard SNMP MIBs. The next fly in the ointment is the development of ELANs *(Emulated LANs)*. These are similar to VLAN *(Virtual LANS)*, where we are emulating LANs by using an ATM switch and edge devices that are LAN switches. Here again, the ATM switch and edge device may not provide the MIB support necessary to build an accurate map of your network.

We tend to expect too much from the mapping function. Consider the difference between the logical and physical network. In the first case (see Figure 32-7), the logical and physical networks are the same. We would expect the mapping function to find or draw this picture.

In Figure 32-8, we have a common configuration for today's networks. All the devices (1–9) are on the same IP subnet. However, for traffic reasons there are only two VLANs. A and B segments comprise one, while the other is made up of segments B and D. For example, devices 1, 2, 3, and 5 are in one broadcast domain, while devices 4, 5, 6, 7, 8, and 9 are in another. This appears logically as shown in Figure 32-9. Different vendors' network management systems will yield different network maps in this case. The best (for example, as close as you are going to get) will be from a network management system that watches network traffic.

Figure 32-7
Logical and physical
networks

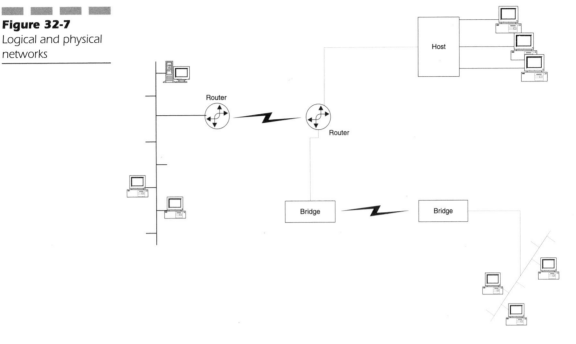

Figure 32-8
The common
configuration for
networks

Figure 32-9
Two VLANs

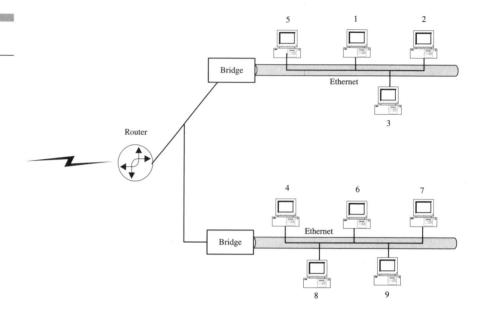

An exercise left to the reader is to make up a traffic scenario and then see what you can learn about this from the other side of the router. If the MIB in the switch has good data, you are in fairly good shape. If it doesn't support the standard MIB, it is hopeless. Although most of the industry has supported SNMPv2 for several years, there has always been contention among the vendors and a significant amount of noncompliance. The noncompliant vendors are opting for their own custom MIB.

SNMPv3

Part of this contention is due to shortcomings of SNMPv2 that require fixing. There are also some additions that are needed to support higher speed networks. SNMPv3 adds a GetBulk command, a better Set command, a unique ID for each SNMP agent, and 64-bit counters to accommodate Gigabit Ethernet. Perhaps the most important new feature is the addition of real security in the SNMP packets. The GetBulk command essentially says "gimme the whole MIB." This is important because it reduces the polling traffic that was created by multiple GetNext command. While this will have a minimal impact on reducing the network management system traffic on a small network, it will be a major benefit to large networks.

Although MIB data can be fetched remotely, there is no way to remotely manage SNMP agents. SNMPv3 adds this feature, along with the ability to describe agents within agents (the problem identified in Figures 32-8 and 32-9). Namely, the problem is that switches and hubs behind other switches and routers are difficult (to impossible) to find. SNMPv3 provides the mechanism to do this *if* (and it is a big "if") SNMPv3 is supported by all the vendors involved. The same problem exists here as with the previous versions of SNMP—standards compliance is not universal. Many vendors have opted for vendor-specific MIBs (shown in Figure 32-4) under 1.3.6.1.4.1. This means that the network management system must have a current complete copy of the MIB in order to be able to manage the device (agent). The *Desktop Management Task Force* (DTMF) is trying to standardize the various data types in a more useful form via the *Common Information Model* (CIM).

Although the forgoing seems critical of SNMPv1 and v2, one must remember the era in which they were born. The functionality and memory in managed devices (for example, hubs) were extremely limited. Today, with cheap memory and processing power in every device, many more capabilities can be added at little or no cost.

Security

The initial design of SNMP included a modicum of security by using the community field (see Figure 32-6) to identify devices belonging to our management community. This was satisfactory before the days of inexpensive and capable network sniffers. With today's technology, it is not difficult to collect SNMP packet header information and either steal the information or substitute the content. Stealing content allows the hacker to build a map of our network for later exploitation. Substituting the content of our Set commands could be disastrous. One hopes that there are no hackers on our internal network, but in today's world, one is never sure. Our paranoia turns to prudence if we are running interconnections of our network over the public Internet without the benefit of a *Virtual Private Network* (VPN). Before becoming too paranoid, one must remember that the hacker must have physical access to the network backbone, a sniffer, the time and will to hack, and the desire to do mischief. The probability of the confluence of all these attributes is relatively low in absolute or real terms.

Given our paranoia, if the hacker is able to collect frames and packets, he can collect the community information and therefore control any of our devices that are preconfigured for remote management.

SNMPv3 solves this problem by encrypting the packet content and by establishing an authentication mechanism. All this is possible today because memory and processing power are readily available in the managed devices.

The new RFCs 2274 and 2275 provide for a *User-based Security Model* (USM) and a *Views-based Access Control Model* (VACM). These RFCs set forth a scheme for today and provide for future expansion to include possibly public key authentication and directory integration.

The User-based Security Model uses a distributed security control mechanism with user name and password disseminated from a centralized system. The user list, or access list, is distributed (securely using encryption) to each managed element. The remote agent does enforcement. Thus, the network management system user must log into each managed agent. Access to each MIB element can be put under password control. The User-based Security Model specifies authentication and encryption functions, while the Views-based Security Model specifies the access control rules. Each managed device can therefore keep a log of accesses and by whom (just as a firewall or proxy server logs each access). While perhaps not important for hubs, it is very important for sensitive devices such as routers, switches, and firewalls. Previous versions of SNMP could not create this important audit trail.

In addition to encrypting the packet contents (for example, using DES, the *Data Encryption Standard*), each packet contains a time stamp to synchronize the network management system with the managed agent. This prevents the man-in-the-middle from recording the queries and commands, analyzing them, substituting his content, and playing them back at a later time. The User-based Security Model also specifies its own MIB so that the passwords and user name can be remotely and securely maintained. Thus, although enforcement of access is done by each MIB agent, the user name and password can be centrally maintained and distributed to each managed agent by using the specified encryption technique. This means that a network authentication server must exist just for SNMP.

In the final analysis, it is your choice whether or not to implement the security features. There are definite benefits and costs associated with installing and maintaining the authentication server and its database.

JAVA

Sun Microsystems is pushing Java as a mechanism for enhancing the flexibility of SNMP. As we have seen, SNMP defines MIB-1, MIB-2, and MIB-3. Even though we can capture data on higher speed elements and control network components securely, if changes are needed to the SNMP agents, software must be loaded (typically, manually) in each managed device.

The Java concept is that now each agent has memory and CPU cycles to spare. Why not have it as a Java virtual machine? The network management system would then download the new functionality to each or all devices.

SNMP was designed to manage elements like routers, switches, and hubs. Managing remote servers, desktop machines, or applications processes was never considered. The introduction of a Java-based system permits the management of higher layer functions, while maintaining compatibility and coexistence with SNMP.

There are several problems to overcome before Java can become a popular network management system tool. The first is performance. As an interpretive language, it is processor intensive and slow. Solutions for this problem involve compiling the Java code on the fly as it is downloaded to the target machine. The code then runs in native mode. Sun's Hot Spot compiler and Microsoft's *Just in Time* (JIT) compiler are designed to do this.

The good part is that the network manager can easily maintain the same revision level of the agents in all devices throughout the network. Unfortunately, this threatens the market for vendor-specific network management system solutions. The best advice is to stay tuned as vendors and standards organizations try to solve the problems of complex networks of today, and the even more complex networks of tomorrow. History has shown that during the early stages of development and deployment, there are multiple solutions from a variety of vendors—all incompatible with one another. As the technology (and markets) mature, the industry hones in on the essential features and functions that are accepted and supported industry-wide. These then become the core functionality on top of which each vendor builds his proprietary extensions.

INDEX

Symbols

10 Mbps Ethernet, 235
100 Mbps Ethernet, 235
1129 protocol, transaction processing, 48
2b1Q, ISDN, 139-140
911 issues, LNP, 71
9CDV, (cell delay variation), ATM, 185

A

A links, (access), SS7, 102
A-Law, E-1, 438
AA, (automated attendant), 19, 80
AAL, (ATM Adaptation Layer), 181–182
AAs, (automated attendants), 19, 80, 120
Access options. *See* connections.
ACDs, (automatic call distributors), 19, 76
ACM, (address complete message),
    ISUP, 108
Active filters, ADSL, 279
ActiveX, (Microsoft), 498
ACTS, (Advanced Communication
    Technology Satellite), 398
Add-drop multiplexing, SONET, 467–468
Address fields
    DXI frames, 207
    FUNI frames, 207
    25 packets, 505
Addresses
    dynamic allocation, ISPs, 517
    IP, 512
        classes, 512–513
        routing, 514
        subnetting, 515–516
        translating, DNS, 497

ADM, (Add–Drop Multiplexer), SONET, 470
Administration, firewalls 572, 571
ADSL, (Asymmetric Digital Subscriber
    Line), 264–265, 272
    CAP data encoding, 280
    DMT data encoding, 277
    filters, 279
    future growth, 248
    splitters, 278
ADSL–lite, 273
Aegis system, (Ericsson), 331
Agents
    managed, 580, 583
    SNMPv3, remote, management, 593
Agricultural uses for satellites, 390
AIN, (Advanced Intelligent Networks),
    5, 9, 44, 53
    applications, real-time, requirements, 51
    combining with CTI services, 46–47
    IPs, 44
        application, processing, 49
        resource, processing, 49
        services, 49
        software, architecture, 50
AIN 0.1 protocol, 64
Air interfaces, mobile IP, 385
Airline outsourcing Extranets, 557
Alarm histories, network management
    systems, 587
Allocation
    ATM bandwidth, 192
    frequencies, cellular systems, 347
AMI, (Alternate Mark Inversion), 270, 443
Analog cellular systems, 335
Analog faxes, ISDN, 142
Analog modems, 267–268
Analog services, DTV, 229

F

H

W

X-Z